SALMON AQUACULTURE

EDITED BY

KNUT HEEN, ROBERT L. MONAHAN,
FRED UTTER

FISHING NEWS BOOKS

A division of Blackwell Scientific Publications Ltd
Editorial Offices:
Osney Mead, Oxford OX2 0EL
25 John Street, London WC1N 2BL
23 Ainslie Place, Edinburgh EH3 6AJ
238 Main Street, Cambridge, MA 02142, USA
54 University Street, Carlton, Victoria 3053, Australia

Other Editorial Offices:
Librairie Arnette SA
2, rue Casimir-Delavigne
75006 Paris
France

Blackwell Wissenschafts-Verlag GmbH
Düsseldorfer Str. 38
D10707 Berlin
Germany

Blackwell MZV
Feldgasse 13
A-1238 Wien
Austria

First published 1993

Set by Best-set Typesetter Ltd., Hong Kong
Printed and bound in Great Britain by Hartnolls Ltd, Bodmin, Cornwall.

DISTRIBUTORS

Marston Book Services Ltd
PO Box 87
Oxford OX2 0DT
(*Orders*: Tel: 0865 240201
Fax: 0865 721205
Telex: 83355 MEDBOK G)

Australia
Blackwell Scientific Publications Pty Ltd
54 University Street
Carlton, Victoria 3053
(*Orders*: Tel: (03) 347-0300)

British Library
Cataloguing in Publication Data
A Catalogue record for this book is available from the British Library

ISBN 0-85238-204-9

Contents

Contributors

Trond Bjørndal is Professor of Fisheries Economics at the Institute of Fisheries Economics, Norwegian School of Economics and Business Administration. He received his PhD in Economics from the University of British Columbia. He specializes in fisheries and aquaculture economics and has published extensively in the field. Dr Bjørndal has taught courses on aquaculture economics in Norway, the United States and Canada. He is the author of the book, *The Economics of Salmon Aquaculture* (Blackwell Scientific Publications 1990).

Bjarne Gjerde, Senior Scientist in Genetics and Selective Breeding, Institute of Aquaculture Research (AKVAFORSK), has degrees from the Agricultural University of Norway. He has been working with quantitative genetics of Atlantic salmon and rainbow trout, and has been visiting scholar at University of Guelph, Canada.

Knut Heen, Associate Professor in Fisheries Economics, Norwegian College of Fisheries Science, University of Tromsø, has been working with fisheries and aquaculture problems for over ten years. He has been with the University of Tromsø since 1973, including two years as Research Director at the Foundation for Applied Research at the University of Tromsø (FORUT). His degree is from the Norwegian School of Economics and Business Administration, and he has been visiting scholar at London School of Economics, Simon Fraser University, British Columbia and visiting professor at Western Washington University.

Kjetil Hindar, a graduate of the University of Oslo, is currently a researcher at the Norwegian Institute for Nature Research. He has held positions at the University of Oslo, a biotechnology company, and has been a visiting researcher to the University of British Columbia and the University of Stockholm. His research interests include ecological polymorphism and genetic differentiation of populations, with emphasis on natural and cultured stocks of salmonid fishes. He is a member of a national editorial board and boards on environmental effects of sea ranching, aquaculture and of releases of transgenic organisms.

Malcolm Jobling is Professor of Aquaculture at the Norwegian College of Fisheries Science, University of Tromsø. He has formal qualifications in marine biology, botany and zoology from the universities of Hull, Wales (UCNW, Bangor) and Glasgow, and has special interests in fish physiology and nutrition. He has published numerous scientific articles related to the feeding, nutrition and growth of fish, has contributed to several books, is involved in work with aquaculture societies, and is on the editorial advisory boards of scientific journals. His research is largely directed towards the solution of problems associated with the farming of salmonids and various species of marine fish.

Ludvig Karlsen is Associate Professor of Fishing Gear and Fish Farm Design, Norwegian Institute of Technology, University of Trondheim. His degree is from the Norwegian Institute of Technology, and he has held positions at the Institute of Fishing Technology Research, Bergen, the Norwegian College of Fisheries Science, University of Tromsø and a land-based fish farm in western Norway. His main research interest has been in fishing gear selectivity and farm cage hydrodynamics.

Jon Lindbergh, private internationally-known aquaculture consultant in Seattle, Washington, has been engaged in commercial salmon culture since 1970. He has worked with salmon projects in the Pacific Northwest, California, New Brunswick, Maine, Argentina, and has been active in Chile since 1977. Jon Lindbergh has an undergraduate degree in biology from Stanford University. He has published several papers on salmon aquaculture and in 1991 became president of the Washington Fish Growers Association.

Conrad Mahnken is Program Manager of the Fish Enhancement Research Program under Northwest Fisheries Science Center in Seattle. His main areas of research are salmonid physiology, genetics, husbandry and pathology. He is past chairman of the US–Japan Natural Resource Panel on aquaculture and is experienced in cooperative aquacultural projects with Japanese research agencies. His degrees in oceanography are from the University of Washington.

Robert L. Monahan, Professor of Geography and Director of the Center for Canadian Studies, Western Washington University, has a longstanding interest in the salmon industry including cannery experience. The focus of his work deals with location of salmon aquaculture. His academic background includes degrees from McGill University, the University of Michigan and the University of Washington. His work includes fisheries investigations in Norway, Finland, British Columbia, and Alaska.

Jan A. Olafsen is a Professor of Marine Biochemistry at the Norwegian College of Fishery Science, University of Tromsø, Norway. He received his degree from the Norwegian Institute of Technology, Trondheim. From 1986–90 he was Research Director at the Norwegian Institute of Fisheries and Aquaculture, and is currently scientific advisor to this institute. In 1991–92 he was visiting professor at the Université des Sciences et Techniques de Lille, France. His primary research interests include comparative immunology, the non-specific defence mechanisms of fish eggs and larvae, and the role of the indigenous microflora of marine fish larvae and invertebrates.

Neil Ridler, Professor of Economics and Chairman of the Division of Social Sciences, University of New Brunswick, Saint John, New Brunswick, Canada, became interested in aquaculture because of its implications for economic development in the poorer areas of Canada. His academic background includes degrees from Oxford and Simon Fraser University, British Columbia.

Ronald J. Roberts is Professor of Aquatic Pathobiology and Director of the Institute of Aquaculture at the University of Stirling, Scotland. He has a degree in veterinary medicine and a doctorate in pathology from the University of Glasgow. He is a fellow of the Royal College of Pathologists and a fellow of the Royal Society of Edinburgh. He is recipient of the Bucklin Medal and the Dalrymple-Champneys Medals for contributions to fish pathology. He has published over 100 scientific papers and is editor of 11 books on fish disease and fish pathology. His work with salmon aquaculture began in Scotland in the early 1960s.

Nils Ryman received his PhD in genetics from the University of Stockholm, Sweden. He is currently with the Division of Population Genetics at the University of Stockholm in the position of a Scientist in Population Genetics at the Swedish Natural Science Research Council.

Richard Schwindt, Associate Professor of Economics and Business Administration, Simon Fraser University, British Columbia, has done research on the industrial organization of fisheries for over a decade. More recently he has extended his work from the Pacific wild-stock salmon fisheries to salmon aquaculture. He holds a doctorate from the University of California (Berkeley) where he specialized in industrial organization and international trade. His research focuses on the organization and regulation of renewable resource industries, primarily forestry and fisheries.

Susan A. Shaw is a Professor of Marketing in the Department of Marketing at the University of Strathclyde, Scotland. She has worked for many years in the area of aquaculture, marketing and economics. Among her publications are the books *Marketing for Fish Farmers* and *Marketing the Products of Aquaculture*. She is a former General Secretary of the Fédération Européene de La Salmoniculture and is currently employed as an adviser on aquaculture to the European Commission.

John Thorpe, Senior Principal Scientific Officer at the Freshwater Fisheries Laboratory, Pitlochry, received his bachelor's degree and doctorate from the University of Cambridge. His research interests over the past 30 years have covered many aspects of salmonid biology world-wide, including genetic conservation, behavioural ecology, physiology, development, and life-history variation in Atlantic salmon, and their application in culture. Until recently he was Vice-President of the Fisheries Society of the British Isles, and is now Editor of its *Journal of Fish Biology*, and on the editorial boards of four other international scientific journals.

Russell S. Uhler is Professor of Economics at the University of British Columbia, Vancouver, Canada. He received his PhD degree (1967) from the Claremont Graduate School, Claremont, California. He specializes in the economics of natural resources and has published articles and books dealing with economics of energy, forestry, and fishery resources.

Fred Utter received his bachelor's degree in chemistry from the University of Puget Sound, his master's degree in fisheries from the University of Washington and his PhD in genetics from the University of California, Davis. He retired in 1988 following a 30-year career with the US National Marine Fisheries Service devoted to applying population genetic information to problems of fishery management. He is presently employed by the School of Fisheries, University of Washington.

Jeff Wallace, Professor of Biology and Technology of Aquatic Resources, University of Algarve, received his degrees from the University of Glasgow. He was on the faculty of the University of Tromsø for many years and was involved with salmonid aquaculture where he developed a special research interest in the aquaculture potential of Arctic charr. He has been visiting professor at Plymouth Polytechnic and at the Universities of Maputo, Oporto and Algarve before moving from Norway to Portugal in 1990.

Acknowledgements

The editors would like to acknowledge financial support from the Norwegian Fisheries Research Council, the Norwegian Marshall Fund, the Fisheries Research Fund at the University of Tromsø and the Norwegian College of Fisheries Science.

A number of institutions, organizations and individuals have supplied the editors with information about different aspects of salmon aquaculture. It would be impossible to mention everybody by name, but we specially thank the fish and salmon farmers association in Norway, Iceland, Ireland, New Zealand and Washington, the former Fish Farmers Sales Organization (FOS) in Norway; Washington Department of Fisheries, Agriculture and Ecology; Ministry of Agriculture and Fisheries, British Columbia; the Royal Ministry of Fisheries, and Fisheries Directorate, Norway; and Western Washington University, Center for Canadian Studies, which assisted and provided office space for Knut Heen during the preparation of this book.

We also acknowledge excellent map preparation by Eugene A. Hoerauf.

Finally, we would like to gratefully acknowledge the constructive comments of a number of anonymous reviewers. The wide range of topics that this book covers made their contributions especially valuable.

Neil Ridler acknowledges financial assistance from the Social Sciences and Humanities Research Council of Canada, Jan A. Olafsen from the Norwegian Research Council for Science and the Humanities and the Norwegian Fisheries Research Council, Kjetil Hindar from the Norwegian Directorate for Nature Management and Nils Ryman from the Swedish Natural Science Research Council. The US National Marine Fisheries Service provided office space and logistical support for Fred Utter.

Preface

The captive rearing of salmon through the entire lifecycle is a concept that emerged dramatically since the late 1970s in temperate ocean waters throughout the world. This development contrasts with ocean ranching where captive rearing is limited to the juvenile stages of life history. Contained salmon aquaculture arose from advantages to the consumer inherent in such controlled rearing (e.g. year-round supplies of a high quality product), and has resulted in a major new activity in many suitable areas where no enterprise of this nature previously existed. This book probes the underlying elements and consequences of this surging industry with the intention of providing the reader with sufficient knowledge and insight to anticipate and perhaps guide its future development in a realistic and responsible manner.

The book emerged as a collaborative effort among three editors with a common interest in salmon aquaculture, but with diverse perspectives on this important topic. Heen has studied the economic aspects of salmon farming, Monahan has worked on the geographical distribution of various facets of the industry and Utter has extensively researched various biological problems over his long career. Authors of substantial international reputation were invited to contribute specific chapters based on their experience and expertise in various areas relating to salmon aquaculture.

The targeted audience for this book is the individual with a strong interest in salmon aquaculture. Undergraduate university and community college students preparing for a career in aquaculture will find the book a broadly-based comprehensive introduction to the challenges of salmon aquaculture. Individuals associated with the fishing industry wishing to obtain a wide-ranging view of the topic will find the book of value in achieving this goal. The technical details and terms have been intentionally minimized to maximize the accessibility of critical information and ideas. However, this focus is not intended to dilute the accuracy of the information given or to oversimplify the presented concepts, and sufficient references are provided to permit verification of omitted technical and conceptual matters. Details relating to artificial spawning and rearing of salmonids are beyond the scope of this book, and the reader is referred to Stickney (1991) for a current and thorough coverage of these areas.

At this point we note that the name 'salmon' is ambiguous, referring to a number of species within the family Salmonidae – the salmonid fishes. Within different chapters of this book are references to species in three salmonid genera. The scientific and common names of these species are given in the table.

SCIENTIFIC NAME Genus Species	COMMON NAME
Oncorhynchus	Pacific salmon
clarki	Cutthroat trout
gorbuscha	Pink (or humpback) salmon
keta	Chum (or dog) salmon
kisutch	Coho salmon
masu	Cherry (or masou) salmon
mykiss	Rainbow trout (reclassified from *Salmo gairdneri*) (searun fish are called steelhead)
nerka	Sockeye (or red) salmon
tshawytscha	Chinook (or king) salmon
Salmo	
salar	Atlantic salmon
trutta	Brown trout
Salvelinus	Charrs
alpinus	Arctic charr
fontinalis	Brook trout

A perspective chapter providing an overview of the current situation in salmon aquaculture serves as a foundation for the remaining chapters. The reader is alerted to historical and present activities including issues related to the farms and to their interaction with the environment and society.

The extent of salmon aquaculture is highlighted through the descriptions in Chapter 2 of the geographic distribution of salmon aquaculture in Europe, North America, Japan and the southern hemisphere. It has been possible to include recent developments and production figures for 1992 for some countries. The main development in 1992 was that Chile became the second largest producer of farmed salmon. There has been a large reduction in the number of farms in the Faroe Island after 1991. This reduction is expected to continue in 1993.

If the rapidly growing salmon aquaculture activity is to mature into a persistent and healthy industry, the growth must be accompanied by assimilation of efficient and innovative technologies in many areas. Approaches to problems of rearing in open and closed marine rearing systems as well as in land-based operations are discussed in Chapter 3, along with presentations of techniques dealing with transportation and feeding. Details of the critical field of nutrition and diets are described in Chapter 4.

The environment must be considered both from the perspective of its effect on the aquacultural activity and vice versa. Chapter 5 deals with physical consequences of environmental factors on aquaculture where successful rearing is dependent upon considering the effects of variables including temperature, pH, dissolved and suspended molecules and particles, and the flow of water through the system. The effects of salmon aquaculture on the environment are also considered in Chapter 5. The volatile issue of physical pollution includes the effects of organic matter added from faeces and uneaten food as well as those from chemicals used in facility maintenance and medicants used in pathogen control. A distinct set of environmental concerns pertains to genetic problems associated with salmon aquaculture, particularly with regard to the threat of eradication of natural populations due to intentional or accidental releases of cultured fish. This topic is outlined in Chapter 6.

The rearing density of salmon culture creates an environment conducive to the spread of pathogens. This problem and its treatment is considered in Chapter 7.

As with any cultured organism, cultured salmon that are specifically adapted to a particular set of rearing conditions will provide optimal performance and thus greater profitability to the grower. Chapter 8 discusses the genetic principles involved in the development and maintenance of such adapted populations of cultured salmon.

The success or failure of any industry is ultimately dependent upon economic factors. The book closes with four chapters that deal with regulations and economic considerations of salmon aquaculture. The inevitable direct and indirect side-effects of these endeavours have led to the need for and the existence of regulations dealing with aquaculture-related activities: the effects of these regulations are discussed in Chapter 9. Detailed discussions are given of three relevant economic issues including cost structure and profitability (Chapter 10), sea farm management (Chapter 11) and marketing (Chapter 12).

The independent considerations of some of the same topics (e.g., environmental considerations in Chapter 5 and Chapter 9) provides the reader with important data and ideas presented from different perspectives. Through such diversity of experience and in some instances, opinion, it is our intention to leave the impression of the present reality of salmon aquaculture; i.e. an established industry with a potentially bright but uncertain future. Although we do not pretend to see this future with any greater clarity than that seen by the intended audience of this book, we hope that this publication will assist in guiding the responsible development of salmon aquaculture to the mutual benefit of all resources and individuals affected by this activity.

Knut Heen
Robert Monahan
Fred Utter

Reference

Stickney, R.R. (ed.) (1991) *Culture of Salmonid Fishes*. CRC Press, Boca Raton, Florida.

Chapter 1
An Overview of Salmon Aquaculture

ROBERT L. MONAHAN

Introduction

Salmon aquaculture or salmon farming refers to the rearing of salmon under controlled and supervised conditions from the egg to the market. Although eggs may be taken from wild fish as well as farm-reared broodstock, the hatching of the eggs is done under prescribed conditions as is the rearing of the fish. The fish farmer attempts to protect the fish from predators and from disease and promotes rapid growth. The farm personnel work to prevent disease through immunization and they treat disease when it appears in their fish stocks, often with the help of pathologists specially trained in fish care. The routine of the fish farm is designed to maintain the well-being of the fish, to promote efficient growth, and to minimize the stress which increases the susceptibility to disease and injury. Some farms are also involved in the slaughter and processing but this aspect depends on the location and the size of the farm.

Ocean ranching, another way of producing salmon for market, differs from farming in several ways. Ranched salmon are reared in a hatchery to smolting size and then released to the ocean to find their own food and to return from the ocean to the 'ranch' as sexually mature fish ready for harvest. The advantages of this system include much lower capital investment plus smaller feed and labour costs because supervision and feeding are done for a relatively short period of time. The disadvantages are a much lower returning number of fish, (0.1–10% with most ranches in the lower part of that range) with less control over the marketing time because the salmon must be harvested soon after their return. Also, ranched salmon are available to all fishers prior to returning to the release site. In exposed areas and in locations which preclude pen operations, ranching is a viable option and often the only possible way of expanding production. Private entrepreneurs have had limited success in this economic activity with many bankruptcies. Governments concerned with producing more salmon for the market and willing to allow harvesting by many individual fishers seem best suited for ranching operations.

History

Aquaculture originated in eastern Asia some four thousand years ago. One of the several members of the carp family is the most likely candidate for the first cultivation of fish which probably started from fishermen storing their surplus catch in baskets or small enclosures in ponds. Someone undoubtedly deduced that if the fish were fed and protected they would grow, and thus aquaculture was born. Records of Chinese aquaculture, the probable source region for the practice, reach back three thousand, five hundred years (Ling, 1977). Fish is a highly regarded food in the Chinese culture and it is speculated that the desire to please royalty may have been a force in the initial development of special fish in Chinese culture. It is likely that the emperor expected carp to be served to him and to his guests on demand. For year-round availability, his ministers learned to break the ice on ponds, pour in hot water, and capture the fish that flocked to the warm water.

Salmon aquaculture, a relatively new branch of an ancient form of animal husbandry, had an explosive rate of growth in the 1980s. The worldwide growth of salmon farming is due to several factors. The expanding recognition of the nutritional advantage of fish as a source of protein in the diet has led to an increasing demand for the product, especially in the United States. In addition, a stronger propensity to dine out has led to an expanded demand for fresh salmon in the restaurant trade which has favoured farmed salmon, the sector of the fishery which can deliver a high quality fresh product on request. The market factors coincided with the availability of a substantial fisheries technology applicable to the rearing of salmon that developed in Norway, which together strongly influenced the rapid expansion of salmon aquaculture throughout the world.

Salmon farming today is the outgrowth of the trout culture which developed in Denmark around 1890 and was introduced into southern Norway in 1912. Aquaculture progress in Europe was slow until the expansion of trout farming in sea water which occurred in the 1960s and 1970s. Norway is not an ideal environment for trout culture due to low water temperatures in winter and some companies speculated that the same technique might be used with Atlantic salmon. In 1965 A/S Mowi started the rearing of salmon in a closed-off part of the sea and salmon farming was underway. In 1969, floating pen culture was started on the Norwegian island of Hitra (Edwards, 1978). The pioneering work with floating cages has set the pattern for the industry today.

Environmental conditions and general location

Salmon are particularly demanding in terms of oxygen content, temperature and are intolerant of pollution. Human modification of the landscape is detrimental to salmon and, as a general rule, the fewer the humans, the better the living conditions for this sensitive fish. As a consequence, salmon farms are usually found in areas of low population density or where pollution abatement is well established and people, especially seaside dwellers, are tolerant of salmon as neighbours.

Salmon require high levels of dissolved oxygen in the water. Flow of water through densely stocked pens must be good to maintain adequate oxygen supplies. Water temperatures above 22°C are usually lethal for salmon as are waters below −2°C. An optimum range is generally regarded to be 4–15°C.

To meet these environmental requirements, salmon farming is limited to locations between the latitudes of 40° and 70° in both the northern and southern hemisphere (see Fig. 1.1). The single exception is on the island of Honshu in Japan where special husbandry practices have been developed (see Chapter 2). In the southern hemisphere, activity is strongly concentrated between the latitudes of 40° and 50° while in the northern hemisphere, it extends from 45° to 70°. The distribution of

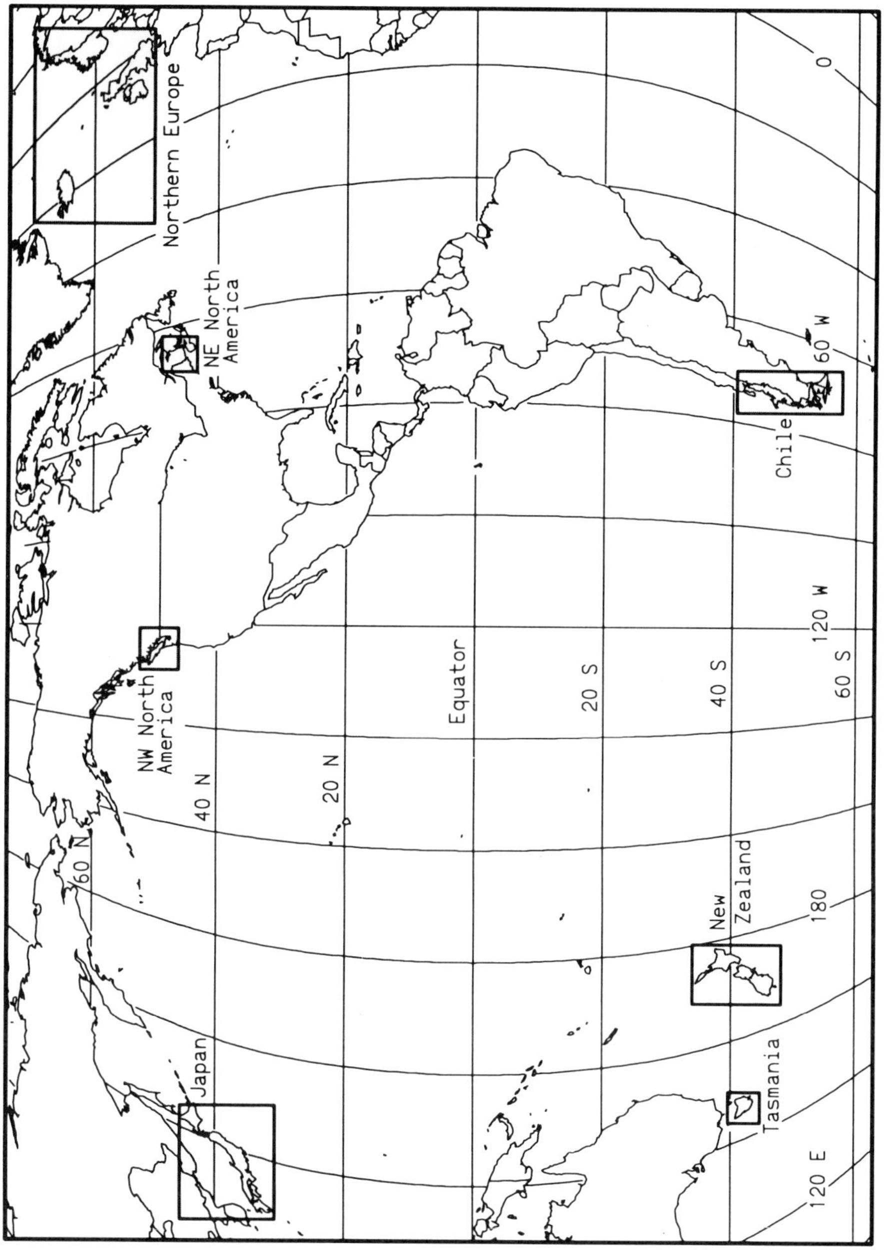

Fig. 1.1 World distribution of salmon farming.

salmon farming as far north as 70° in Europe is due to the presence of the warm Gulf Stream. The absence of salmon farming in the more northerly coastal areas of eastern North America and eastern Asia is due to the cold Labrador and Kamchatka currents. The coast of China, like that of Korea, Russia and Hokkaido, has winter temperatures too low for sea-based salmon aquaculture. Special husbandry techniques may permit expansion into the area at some future date. The salmon farming environment is favourable north of 54°40′N in western North America (the southern boundary of Alaska) but is not permitted under Alaska state regulations.

Plankton blooms represent a threat to penned salmon. Some species of plankton cause gill damage and the dying plankton may reduce the oxygen levels and cause suffocation. Reduced visibility due to high plankton content increases feed loss.

Stages of the production process

The biological aspects of salmon farming are subdivided into separate stages of the production process. In Fig. 1.2, the first four boxes show the biological system. This

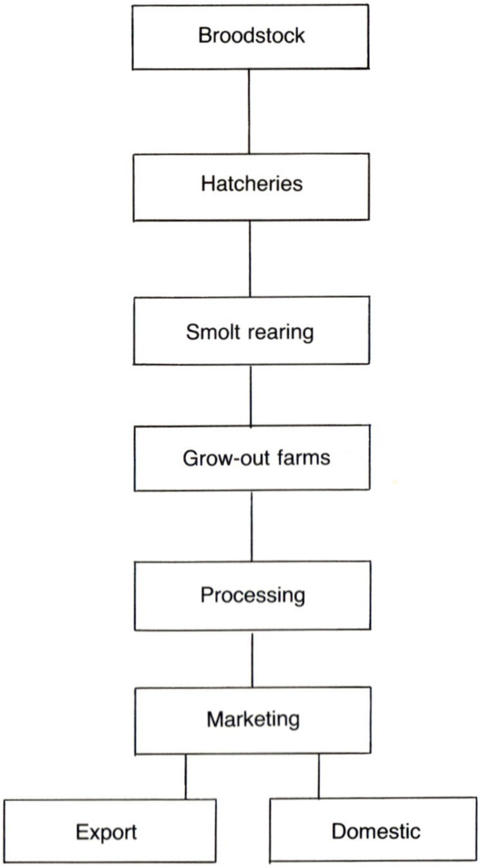

Fig. 1.2 Stages in aquaculture production.

includes the broodstock which is the source of the eggs, incubation and hatching of the eggs, and the rearing through the smolt stage. Smoltification, the physiological change in the salmon moving from fresh to salt water is normally carried out by the hatchery. Careful control of the water environment is particularly important during smoltification and the saline content is closely monitored.

Smolts are transferred to the farms for the grow-out stage. Here fish are carefully monitored for feed uptake, disease and general well-being. Harvest is carried out before sexual maturity.

The last two boxes indicate the post harvest, on-land processing and distribution facets of production. Processing involves the slaughter and preparation of the fish for market. Marketing refers to the distribution of salmon from the farm to the consumer. The two main divisions of the marketing process are domestic and export.

Choice of species

The choice of the kind of fish for farming is a vital question and needs careful attention by the aspiring fish farmer. In the context of commercial production, the price of the fish must be sufficiently high to cover the costs of production which include capital and operating costs. Competition from other farms and the wild fish hunters are considerations. Efficient conversion of feed and rapid growth with an ability to tolerate dense stocking are important characteristics for success. Desirable domesticated fish must be able to reproduce in captivity and have a simple larval development amenable to human care and supervision. (See Shepherd and Bromage, 1988, for an extended discussion.) Of the thousands of fish species only a small number have market prices high enough and production costs low enough to justify farming.

In salmon aquaculture, both Atlantic and Pacific Ocean species are grown. The Atlantic salmon (*Salmo salar*) is native to the North Atlantic Ocean and is farmed in European and eastern North American markets. Two species of Pacific salmon, coho (*Oncorhynchus kisutch*) and chinook (*O. tshawytscha*) are important in western North American, Chilean and Japanese farms.

The market preference in Europe and eastern North America is for Atlantic salmon. This species has been farmed more than 20 years and many strains have developed useful farming characteristics. Atlantic salmon are more docile than their Pacific cousins. As a result, the stocking densities can be higher, reducing the capital costs for pens and associated equipment. They tend to have fewer losses from disease and injuries in the salt water phase of their life cycle. However, Atlantic salmon are more difficult to raise to the smolting stage and are more vulnerable to plankton blooms.

Industrial structure and economics

The industrial structure develops on the basis of the stages in aquaculture production shown in Fig. 1.2. In addition, a number of economic and institutional factors may influence the industrial structure in a country. The industry may develop as vertically integrated, which means that all stages in the operation are controlled by one company. Also it may develop as a horizontally integrated industry which means that one company operates a number of units as the same stage of activity.

There are many variables which influence the profitability of the industry and not all are under the control of the farmer. The price of salmon is influenced by the world supply in which the volume of wild harvest plays an important role. Foreign exchange variations, import restrictions and other non-tariff barriers influencing the movement of fresh salmon also affect the profitability. Changes in the price of feed, labour, smolts, and other services are additional variables influencing profit. Disease losses, storm losses, storm damage to equipment and predation of stocks have an important effect on profitability even when partly covered by insurance.

In farmed salmon, it is possible to guarantee a high quality product, harvested and processed under optimal conditions. It is also possible to guarantee delivery of a specific volume on a specific day and even at a stated price, decided advantages for restauranteurs or retailers. The advantages of farmed salmon are high quality, year-round availability and dependability of supply which generally provides a higher price for farmed salmon than that of the wild catch.

Most farmed salmon is sold fresh to consumers or caterers. The processing always includes slaughtering and gutting the fish for shipment. With expanded production some is smoked, frozen and packaged in special ways.

Salmon farm husbandry

Salmon farming is an increasingly complex economic activity and the requirements for success increase with each passing year. Increasing competition has narrowed the profit margin and successful farmers need to employ more efficient production techniques. Environmental concerns and environmental pressure from other groups of water users have also forced the salmon farmer to have increased levels of skill in managing various aspects of the physical and cultural environment. Expansion of the industry has taken place in more open waters and has required the successful farmer to be very informed and vigilant about anchoring and the construction of the sea pens for rearing salmon. An examination of some of the key factors of farm husbandry follows.

Feeding

In addition to the traditional feeding methods of hand spreading of fish feed and the current use of the automatic feeders which spread feed mechanically, the most recent development in salmon feeding is 'technician controlled feeding'. This is done by a person skilled in observing fish and handling the feed pumping equipment. The feed is brought alongside the pen, either by a support vessel or by a vehicle moved along the walkways. The feed is delivered mechanically while the technician observes the uptake by the fish.

One interesting development is the drop in frequency of feeding, in general, to twice a day, with careful observation of food uptake. It has been found that the growth rate is essentially the same as when more frequent feedings were employed. The lower frequency reduces labour costs and allows a skilled technician to tend an increasing number of pens.

Dead fish collection, assessment, and removal

The removal of dead fish has always been an important issue in salmon farming and the problems have become greater as the pen size and pen depth have grown. Most dead fish sink and the form of the pens is that of a cone to allow the dead fish slide to a central point on the bottom. The mortalities can be removed from the inside of the pen by either lifting the net or diving to collect the expired fish. The preferred way is to remove the mortalities from outside the pen to minimize the disturbance and stress on the living fish. Following the removal, the fish are examined to determine the cause of death and then placed in containers and taken from the site for disposal, usually under the guidelines of government regulations. Most producing areas now have regulations to control the removal and disposal of dead fish from salmon farms.

Growth control and assessment

The key to successful salmon farming is to maximize gain and minimize feed and fish loss. To monitor this it is necessary to count and weigh the fish at selected intervals. Skilled personnel are required to select population samples and accurately weigh and measure fish growth. It is also necessary to keep a close tally on mortalities and to manage the operation to minimize fish losses and maximize weight gain. This process is being simplified by the development of automated equipment which can measure the size in the pens or tanks and reduce the stress on the fish.

Fish handling and treatment

Fairly frequent handling of the fish is a common requirement in salmon farming. It is associated with internal transport and redistribution of the fish populations, size grading, medical treatment, and salmon lice removal. Handling the fish increases the stress which adversely affects the weight gain and often increases mortality. The development of bar graders which can be used in a pen with minimal disturbances and automatic sizing equipment are important developments because they reduce the handling stress for the fish.

Other important husbandry activities include the maintenance of adequate fish space within the cages and a good through-flow of water to maintain an adequate oxygen supply. Anti-infection procedures are important. They include the washing and disinfecting of the nets used to pen the fish. Careful control over cleanliness of work clothing and boots is used to minimize the opportunity of infection and the spread of infection on the fish farm.

Cage net maintenance

A constant enemy of the salmon farmer is the marine fouling of the nets which contain the salmon. It is usually necessary to change the nets every one to two months and the problems involved include minimizing the stress on the fish and the cost to the farmer. Sometimes two net pens are hooked together and the fish moved from the fouled pen into the clean one. The problem is to carry this out without losing fish. Fish loss is a problem because it reduces the profitability of the operation and is regarded as an issue of increasing importance because of the genetic pollution of the wild stocks. A second method is to sink the net and bring it up around the fouled net and then to spill the fish from the old net into the new. Another technique is to use deep net pens and then lift the top three to four metres and let the organisms die in the open air. It is then possible to remove the dead organisms and maintain good water flow through the upper portion of the pen. There are firms which specialize in the cleaning of nets when they are removed. They ordinarily employ a drum washer which removes the organisms from the net surfaces. Some farmers also maintain this equipment for their own operations. Another technique which is employed is high pressure water jets. By lowering a pump with high capacity flow into the pen and blasting the organisms off with a powerful stream of water the net can be cleaned in place.

Cage and farm damage control

The control of pen damage is an issue of increasing importance to salmon farmers. The public concern with escaped fish and the economic loss that this occasions as well as the damage to the capital investment in equipment are reasons for increasing vigilance on the part of the salmon farmer. Careful daily inspection of the pens and frequent inspection of the anchors, mooring lines, nets and plant structure by divers is a growing area of attention by the prudent and successful salmon farmer. Insurance companies, governmental bodies and investors are moving to develop higher technical standards for fish farming equipment. Companies which specialize in farm inspection are becoming more common to provide this specialized service for the salmon farmer. Government regulations are increasing to minimize the escape of farmed fish and insurance companies are insisting on higher standards of pen construction and anchorage to reduce loss claims against them. It is an obvious area of concern for the fish farmer and is receiving increasing public scrutiny.

Farm staff qualifications

From the above discussion it is apparent that the staff on a salmon farm must be skilled in a variety of specialities. Training is commonly available, sometimes at the high school level and more commonly in community colleges or technical school format. Courses extending for two or three years are increasingly available and include practical training in the various areas that have been discussed previously, as well as laboratory training or specialized techniques to increase the efficiency of salmon farming. This is a natural evolution from the 1970s and 1980s when many salmon farmers were self-taught.

Aquaculture and wild fisheries

The aim of aquaculture and wild fish harvesting is to maximize the sustainable yield of organisms from the aquatic environment. Traditionally, the emphasis in fish harvesting has been on the biological yield, but there has been increasing awareness of the economic yield. In wild fish management models, the focus is on the recruitment rate, growth rate and mortality, and the aim is to maximize the yield by managing the fishing effort and thereby fishing mortality. Aquaculture goes beyond commercial fisheries models to achieve increased yield by controlling recruitment and growth as well as mortality. Great advances have been made in fish husbandry through progress in such fields as nutrition, genetic selection, disease control, and farming technology. Collectively, they provide for improved yields in salmon aquaculture.

In management, a critical difference between capture fisheries and aquaculture is the 'common property' nature of the wild fishery. The common property characteristic of wild fish forces the fishers to capture them when they have the opportunity. Delay for a higher market price or extra weight gain is not possible because another fisher may harvest the catch. In aquaculture, the farmer owns and controls the fish and the harvest time will depend on both biological and economic factors including growth, price, costs and discount rate. The farmer will tend to adjust to these factors to secure maximum economic yield.

References and selected bibliography

Beveridge, M. (1987) *Cage Aquaculture*, Fishing News Books, Oxford.

Brown, E.E. (1980) *Textbook of Fish Culture*, AVI Pub. Co, Westport, Connecticut.

Edwards, D.J. (1978) *Salmon and Trout Farming in Norway*, Fishing News Books, Oxford.

Ling, S.-W. (1977) *Aquaculture in Southeast Asia: A Historical Overview*, University of Washington Press, Seattle.

Shepherd, C.J. and Bromage, N.R. (1988) *Intensive Fish Farming*, Blackwell Scientific Publications, Oxford.

Chapter 2
The Distribution of Salmon Aquaculture

KNUT HEEN, JOHN THORPE, NEIL RIDLER,
ROBERT L. MONAHAN, CONRAD MAHNKEN,
and JON LINDBERGH

Editors' introduction

This chapter provides the readers with an overview of world distribution by countries and regions where salmon aquaculture is practised. The growth and development of this new form of aquaculture is indicated through annual production figures and farm numbers. Government regulations influence location, industrial structure and other factors determining the profitability of the industry. These regulations are described. Factors encouraging the expansion and deterrents to growth are also discussed. Economic factors are included in Chapter 10.

Some countries and regions will be given a more comprehensive presentation than others. To understand the present status of the salmon aquaculture industry, it is necessary to describe and to pay particular attention to the development of the industry in Norway since Norway is the pioneer country and world leader in salmon aquaculture with more than 50% of total production in 1991.

Two other regions about which European and North American readers usually have only limited knowledge, Japan and Chile, are also given special attention. Both are important salmon producers with more than 20 000 tons in 1991. Japanese salmon farmers produce solely for the home market. It should be noted that Japan is an important market for salmon in general and as shown in Chapter 12 (Table 12.2), it is the largest importer of salmon. Most of the Japanese imports are currently wild salmon, but farmed salmon may become of increasing importance. The future prospects of the domestic farming industry is therefore of importance for other producing regions. Secondly, the salmon producing regions in Japan are located at a lower latitude from other producing areas. Due to the high summer temperature, Japan has developed a special husbandry practice. Chile, however, relies heavily on export. Chile has the potential to become a dominant producer in the future.

Different criteria could be used in classifying salmon aquaculture by region. The regions in this book are classified by continent in the northern hemisphere with the southern hemisphere forming one region. The regions are:

(1) Europe
(2) North America
(3) Japan
(4) Southern hemisphere

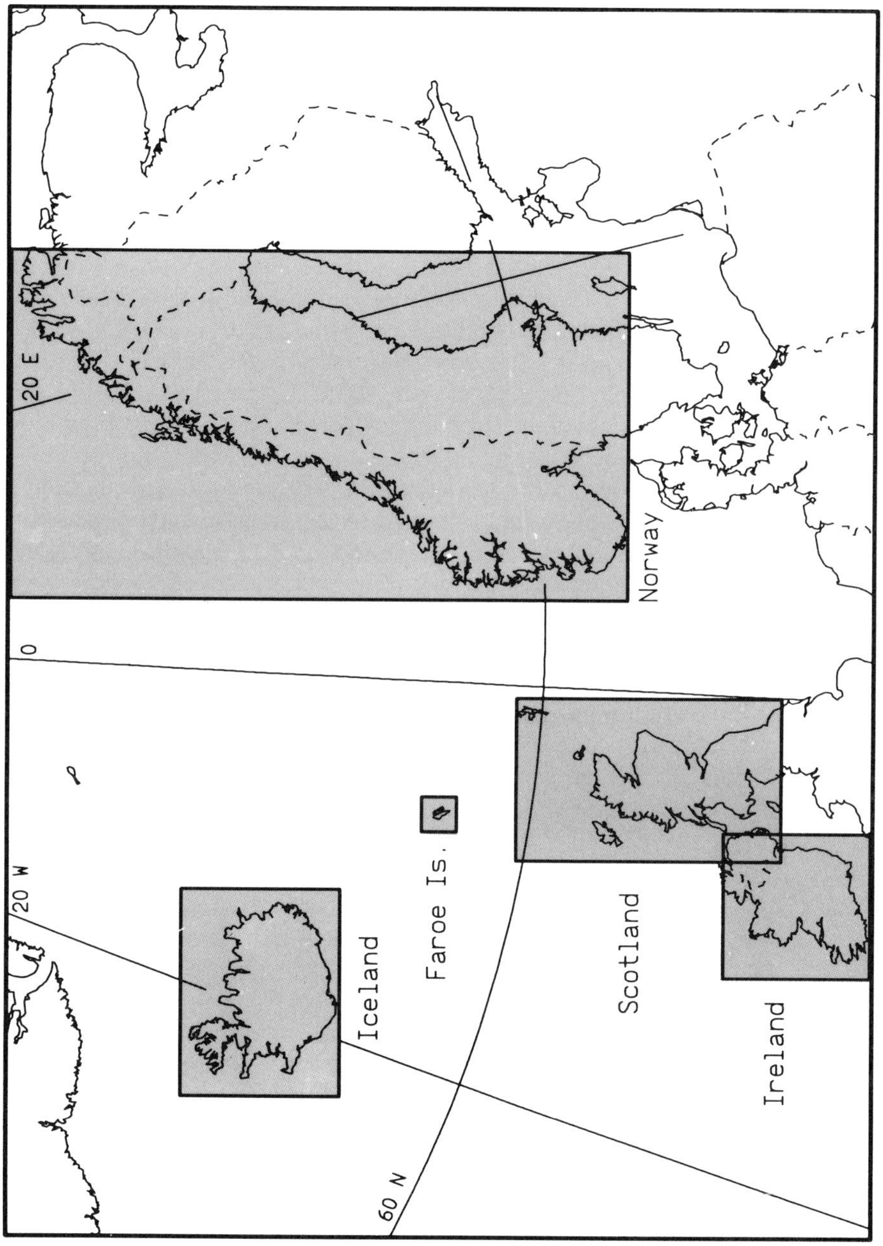

Fig. 2.1 Main salmon farming countries in Europe.

EUROPE

Production in Europe is mainly found in the Northern European countries of Norway, Scotland, Faroe Islands and Iceland. Ireland is the only European country located south of these locations with any significant production. See Fig. 2.1 for location.

Norway by Knut Heen

Introduction

Salmon farming is a new industry in Norway. Even though a number of hatcheries were established in the second part of the last century, farming in cages in salt water started only in the late 1960s. Production increased rapidly and Norway today is by far the most important producer of farmed salmon in the world.

The rapid development of the industry in Norway is the result of excellent environmental conditions with many fjords and islands that protect the cages from storm and wave action. Temperatures are good for salmon farming even though the lower water temperature in the winter in the northern part may lead to slower growth rates. The lower temperature, however, results in less seaweed growth on the facilities and fewer parasites on the fish. Norway also has an abundant skilled labour force, a good transportation system and a well-developed infrastructure. The salmon farming industry operates under a number of government regulations, and the major question is the ability of Norway to meet the growing competition from other countries including Scotland, Canada and Chile. In the presentation of aquaculture in Norway, special emphasis will be given to government regulation. This is necessary in an evaluation of the future prospects of the Norwegian industry.

Table 2.1 Number of farms which have harvested fish, and salmon production in Norway 1972–91.

Year	Farms	Production (tonnes)
1972	5	46
1974	13	600
1976	61	1 400
1978	116	3 500
1980	173	4 300
1982	263	10 700
1984	354	22 000
1985	414	29 500
1986	498	46 500
1987	542	47 600
1988	653	83 300
1989	696	115 000
1990	729	158 000
1991	N.A.	155 000
1992	N.A.	140 000

Production

Table 2.1 gives the number of farms harvesting salmon and production in tonnes by year in Norway.

In 1972 there were only five farms slaughtering salmon. The number increased rapidly in the late 1970s. From 1980 to 1985 the annual increase was about 50 farms. In 1990 the total number of farms which had harvested fish reached its maximum of 729.

In 1974 the production of farmed salmon was 600 tonnes. By 1980 it increased to 4300 tonnes and salmon aquaculture became the fastest growing industry in Norway during the 1980s. The annual growth in the period 1980–90 was approximately 15 000 tonnes, and in 1990 production reached 158 000 tonnes with a slight drop in 1991, 155 000 tonnes. In 1992 the production fell to 140 000 tonnes. The markets for Norwegian salmon are given in Chapter 12.

Location

On the location map in Fig. 2.2, salmon aquaculture in Norway is divided into four regions: North Norway, Central Norway, West Norway and South and East Norway.

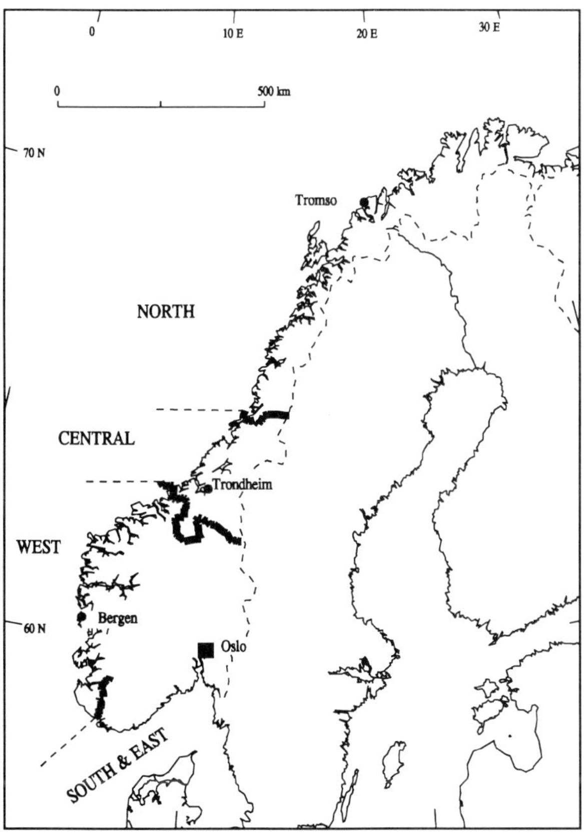

Fig. 2.2 Salmon farming regions in Norway.

Table 2.2 Licences granted and farms in operation by region in Norway: absolute and relative figures, November 1991.

Region	Farms in operation No.	%	Licences granted No.	%
North	162	26	259	31
Central	132	21	139	17
West	321	51	361	44
South and East	17	2	68	8
Total	632	100	827	100

Table 2.3 Licences granted and hatcheries in production by region in Norway: absolute and relative figures, November 1991.

Region	Hatcheries in operation No.	%	Licences granted No.	%
North	42	17	125	22
Central	39	16	61	11
West	153	64	266	48
South and East	6	3	106	19
Total	240	100	558	100

Table 2.2 gives the number of licences granted and farms in operation in November 1991. Of a total of 827 licences, 44% are granted in West Norway. 31% of the licences were granted to North Norway and 17% to Central Norway. The number of farms in operation were 632. 51% of these are located in West Norway, compared to 26% and 21% in North Norway and Central Norway respectively. South and East Norway have an insignificant part of the number of farms – only 2%.

As seen from Table 2.2, nearly 200 of the licences granted were not utilized by November 1991. One reason is the high number of bankruptcies where the farm either temporarily or permanently went out of business. In other cases, despite acquiring a licence, a number of farms never materialized due to a lack of financing. North Norway has the highest number of licences not in use.

Table 2.3 gives the licences granted and hatcheries in production by region in Norway in November 1991. The pattern of distribution for licences by region is basically the same as for farms. 48% are allocated to West Norway, compared to 44% of fish farm licences. The percentage for North Norway is 22 and 17. Comparing hatcheries and farms in operation, the distribution between the two largest producing regions in Norway is more unequal. 64% of the hatcheries in operation are located in West Norway compared to 51% at the operating farms. For North Norway, the figures are 17% and 26%.

Even though South and East Norway were granted 19% of licences for hatcheries, only 3% of the hatcheries in operation are located in this region.

Table 2.3 shows that only 240 hatcheries or 40% were in operation in November 1991, out of a total of 558 licences granted. The downturn in salmon farming and difficulties in financing are some of the reasons that licences failed to become production units. Also, the number of bankruptcies in this part of the aquaculture industry has been high.

Government regulations

In this section the government regulations controlling the number of farms, size, location and ownership are described. The rationale and the impact of these policies on industrial structure and profitability are also discussed.

The first law regulating aquaculture was passed in 1973 and stated that it was necessary to obtain a licence to establish a hatchery and for the feeding of fish. Farms established before the licensing system was introduced were exempt from applying and could continue production at the same site and with the same pen volume as before 1973. Licences were easily obtained and only six out of 97 applicants were rejected between 1973 and 1977. The applications denied were mainly due to risk of disease or excessive pollution in the areas of the proposed farm sites. These applications would probably have been rejected according to other laws (Aarset, 1988).

However, from 1978 the law became the main barrier to entry into the field of salmon aquaculture. The Ministry of Fisheries used the law from that time on to reduce the potential growth rate of the industry primarily due to market concerns. To illustrate: the Ministry granted 54 licences out of a total of 800 applications in 1981 and in 1983, only 100 out of 1200 applications were granted. This demonstrates that the government intended to control the number of farms and thus, the output.

The law regulated the farm size in relation to cage volume and maximum volume was set to 8000 cubic metres. However, up to 1984 new licences awarded had maximum cage volumes of 3000–5000 cubic metres. In 1988, the upper limit was increased to 12 000 cubic metres. The law also regulates ownership of fish farms. The farms should be locally owned, and this is encouraged by limiting controlling interest (51%) to only one farm by an individual or a company.

The rationale of the licensing system relates to controlling industry output and to regional development. The authorities were afraid that production would rise so quickly that the market would not be able to sustain profitable prices for the total output, a fear confirmed in 1989. Secondly, the farming industry has been used as a tool in regional policy. The government wanted the industry spread along the coast in small units where the owner was also the operator to prevent a decline in the population of rural Norway.

The present law regulating the aquaculture industry came into effect in 1985, and still regulates the number and size of farms, location and ownership. According to the new law, the government cannot take market considerations into account in the licensing policy. Another change has been made in respect to ownership. Majority interest should still be held by a local region, but minority interests can be held outside the region. Secondly, one person (or firm) may hold majority interest in only one farm. If a fish farm is to be sold, the transactions have to approved by the authorities. Priority will be given to local (regional) buyers, but the farm will be sold to the highest bidder. In fact, the sale of fish farms has become a ‘free market’ (Aarset, 1988).

The main constraint on development of the Norwegian aquaculture industry up to

1987 was the supply of smolt. Because infected smolt were imported from other countries to Norwegian fish farms, strict regulation of smolt imports was imposed. The government encouraged the development of local smolt production by making it easier to obtain licences for hatcheries. The main policy change in the 1985 law was in respect to hatcheries. There is now in fact free entry into this part of salmon aquaculture, once environmental considerations are satisfied. There is, however, a ceiling of one million smolt to be produced annually by each hatchery. As a result of the 1985 law, the number of hatcheries increased in the second half of the 1980s.

Even though the new law represents a liberalization compared to the 1973 law, there are still political concerns that salmon aquaculture should remain a small-scale decentralized industry and the licensing of farms also provides a means of using salmon farms in regional development.

Geographical and industrial structure

Up to 1977 all applications fulfilling environmental requirements were granted licences which meant the regional location of farms was not important. It was in the well-developed farming district in West Norway that the interest in new licences was largest. The salmon farming industry was slow to develop in North Norway. However, the success in West Norway in the late 1970s led to an explosive interest in the north to enter the industry.

Out of the total of 154 licences granted in 1981 and 1983, 97 were given to applicants in North Norway. (Hallenstvedt *et al.*, 1985) The deliberate willingness to develop aquaculture in North Norway can be seen in Table 2.4.

Examining the three main regions in salmon aquaculture as shown by Table 2.4, indicates the priority given to North Norway. The percentage of total number of farms located in North Norway in 1972 was only 13%. The share increased to 30% in 1985 and to 34% in 1990. The percentage was reduced from 25% to 18% in Central Norway and from 62% to 48% in West Norway in the same time period. To sum up, licences issued in the period of 1973 to 1977 were awarded in already-established aquaculture regions. In the beginning of the 1980s, new licences were issued to applicants in new regions.

The above statement relates to the *number* of licences, not to production. When the maximum pen volume was increased to 12 000 cubic metres in 1988, the production increased dramatically in the established regions in West Norway where

Table 2.4 Distribution of licences for farms by region and year: absolute and relative figures.

	1973		1977		1985		1990	
Region	No.	%	No.	%	No.	%	No.	%
North	24	12	75	21	158	30	253	33
Central	50	25	74	20	103	19	140	18
West	123	62	214	59	280	51	361	48
Total	197	100	363	100	541	100	754	100

farms were able to respond quickly in terms of investment an where they also were able to purchase the necessary smolt to increase production.

As pointed out above, there are legal constraints to ownership of fish farms. The system gives only one licence to each owner or group of owners. Those awarded a licence for farming do not receive one for hatcheries. This limits vertical integration between hatcheries and farms and make smolt a marketable commodity.

Trading companies are not able to integrate backwards into farms. Farmers have the opportunity to integrate forwards. The decentralized pattern of the industry into the hands of local farmers and fishermen has been pointed out. Both groups lack financial resources, competence and motivation for full integration of the industry. The trading companies may have minority interests or trade agreements with fish farms. Such vertical integration is commonly found in the Norwegian industry.

In 1978, the fish farmers' sales organization (FOS) was established by law. FOS had a monopoly for the sale of farmed salmon up to its bankruptcy in 1991. All fresh salmon from the grow-out farms had to be sold through this organization which set the minimum price for sale of salmon from the farms to the trading industry. FOS also authorized purchasers. Of 72 authorized purchasers in 1989, 30 were also farmers.

To export salmon, a fresh fish export licence is required and most companies involved with buying and processing fish have such a licence. However, the farms decide who they will sell to and most of the purchasers are also exporters. This has led to a high degree of concentration in the exportation of salmon. Some 70 companies are exporting salmon but the 20 largest companies have 70% of the total sales, while the ten largest companies have 51%. The concentration has become less over a period of time because of the large increase in volume; everyone has been able to expand at the same time. The organizations and regulations of the industry limit the vertical integration. Therefore, the segments are organizationally separated with strict division among the segments.

The government has, however, so far succeeded in the policy of local ownership of farms. A study by Holm *et al.* 1990 concludes that the majority of farms are indeed locally owned.

To illustrate the size of a Norwegian fish farm in terms of employment, Table 2.5 gives the distribution of farms by number of people employed.

23% of the farms have only 1–2 employed, while 45% employ 3–4. This implies that approximately two-thirds of all farms employ only 1–4 workers which clearly

Table 2.5 Distribution of farms by number employed: relative figures – fully employed and part-time (1988).

Number of employees	Distribution (%)	Farms (No.)
1–2	23	128
3–4	45	244
5–6	14	77
⩾7	19	103
Total	100	552

illustrates the small-scale operation of Norwegian fish farms. 19% or slightly more than 100 farms, however, employ seven workers or more. In addition, an average farm employs three workers seasonally.

In the autumn of 1988 the first serious problems for Norwegian salmon aquaculture emerged. At that time, the salmon produced from the abundant 1986 smolt came onto the market. For the first time in the short history of Norwegian salmon aquaculture, Norway was able to supply sufficient quantities at the peak season, Christmas sales.

Given the minimum price, supply exceeded demand during the winter and through the spring of 1989. The sales organization had not changed the minimum price nor regulated the quantity. Having an excessive minimum price, undermined the minimum price system because the producers sold at lowered prices. The sales organization had tried to reduce the supply in the market by freezing and storing some fish in the spring of 1989, but this lasted only for five weeks.

The situation in the market did not change much during that year. In January 1990 the sales organization decided to freeze a large quantity of salmon in an attempt to maintain a higher market price. All salmon that could not be sold at the minimum price in the fresh market should be frozen. This policy was financed by a 'tax' of 5 kroner for each kilo of fresh salmon sold. The decision to freeze the excess salmon was planned to be a short-run market regulating device to handle overproduction and to secure sufficient price for fresh salmon. However, the result became the opposite. A number of farmers found it profitable to produce for freezing rather than to produce for the fresh market. The quality requirements for the frozen salmon regarding fat content and colour were lower than for the fresh market, and thus producers could produce large quantities of lower quality fish very cheaply. A number of farmers regarded the frozen salmon as a market rather than a short-run market regulation. Loyal farmers producing for the fresh market had to pay 5 kroner for each kilo sold. This was a heavy burden for those producers and resulted in a number of farmers selling illegally outside the sales organization. This again worsened the economic situation for the sales organization.

In addition to the programme of frozen storage, the Farmers Association in cooperation with FOS completed a plan to destroy 12 million smolt in the summer of 1991. In the autumn of the same year, a three-month feeding moratorium was carried out. The result of these two programmes brought a reduction in production for 1991/92 of approximately 30 000 tonnes (Norwegian Salmon Farmers Association, 1992).

Norwegian producers were accused of dumping by both American and Scottish/Irish sides. In Europe, an investigation by the European Commission concluded that dumping had taken place at 11.4% below the cost of production. But no countervailing levy was brought against Norway in 1990. However, in 1991, the Commission decided to temporarily set a minimum price of imported salmon. In addition, a 2% import levy was imposed. In the United States Norwegian salmon was found 'guilty' of both dumping and subsidizing, and a tariff of 26% was imposed on Norwegian salmon exported to the United States market. This terminated the Norwegian export to the US market.

As a result, the sales organization was in a difficult economic situation, and started negotiations with the Government to sell the frozen salmon to non-commercial markets but the Government was not willing to finance such sales and the sales organization went bankrupt. A number of farms went bankrupt too, both as a result of low prices and the bankruptcy of FOS. The Government saw this situation as an opportunity to restructure the aquaculture industry. The number of farms that went bankrupt in 1991 was approximately 200. The number of farms that went bankrupt in 1992 is not known because the information was part of sales organization records which were closed due to bankruptcy. In November 1991, 632 farms of the approximate 800 operating before the crisis started were still operating. The number of farms operating today in Norway is not known, but many of the farms that went bankrupt have been sold to new owners and are in operation today. It is not yet possible to see the consequence of restructuring the Norwegian salmon aquaculture industry. It has changed the structure of selling and today a number of wholesalers and exporters operate rather than the monopoly of the previous sales organization.

Future propects

The future of the Norwegian salmon farming industry is not limited by the availability of suitable sites for production nor lack of infrastructure. The important issues are the regulations of size and ownership and how these regulations influence the profitability of the industry. The concern over the use of the pen volume as a regulatory measure in salmon aquaculture is the consequence of increased stocking densities. Constraint on volume is an incentive to increase the stocking density which in turn increases the risk of disease. Secondly, limiting the size of the pens precludes exploiting economies of scale.

The licensing system may prevent the best qualified people from entering this industry. The people thus prevented from getting a licence at home may emigrate to other countries to establish salmon farms. The substantial investment by Norwegians in salmon aquaculture in other countries, may underline this conclusion. A study by Standal (1987) gives a more complete picture of this. Norwegian investors abroad can be divided into two groups; larger companies invest primarily to gain easy access to markets and secondly, farmers invest abroad who are not able to acquire licences at home.

The question of decentralized structure with locally owned farms has been pondered with respect to developing an economically healthy industry. Many believe that fully integrated enterprises that would include all stages of production and marketing would have better quality control and marketing ability than is presently the case. Such a structure would bring production and consumption more in line with each other.

Iceland by Knut Heen

Even though Iceland has a long tradition in fry and smolt production for enhancement of rivers and sea ranching, salmon farming did not expand until 1980s. At that time Iceland had an annual production capacity of 20 million smolt. In addition to domestic demand, there was a potential market in Norway, Ireland and Scotland where lack of smolt led to high prices. Due to import restrictions in these countries and to the fact that potential buyers believed the smolt quality from Iceland was poor (because of early sexual maturity and slow growth), export virtually ceased. To utilize the surplus production of smolt, Iceland moved into salmon farming.

A special situation exists with the Icelandic salmon farming industry in that a majority of the salmon are grown on land-based farms. In 1992 approximately two-thirds of the salmon was grown on land-based facilities. Because Iceland is mainly a land-based producer, it makes this country of particular interest in the analysis of future prospects for this form of production. Land-based data from Iceland is presented in the analysis of cost structure by technology in Chapter 10.

Table 2.6 gives the number of farms by technology and year. As seen from the table, the number of farms in Iceland is very limited. In 1985, only eight farms were in operation, increasing to 34 farms in 1989. After that time the number of farms has decreased, to 20 in 1992.

The data in Table 2.6 are obtained from the Icelandic Fish Farmers Association. There is uncertainty regarding the total number of farms because of the lack of a licensing system in Iceland. Not all farms are members of the association, particularly some small farms. The figures in the table are therefore an approximation. The small sea-based farms have met great economic problems. In 1989, 25 farms either were bankrupt or were taken out of production. The problems are reflected by the fact that only nine sea-based farms were in operation in 1992.

Table 2.7 gives the production by technology and year, and clearly indicates the

Table 2.6 Number of salmon farms in Iceland, 1985–89.

Year	Net pens	Land-based	Total
1985	5	3	8
1986	5	5	10
1987	16	9	25
1988	15	8	23
1989	16	18*	34
1990	17	12	29
1991	16	12	28
1992	9	11	20

Note: *10 of these 18 farms were recently established and did not harvest salmon that year.

Table 2.7 Farmed salmon production in Iceland by technology and year.

		Production (tonnes)	
Year	Net pens	Land-based	Total
1985	85	7	92
1986	106	17	123
1987	266	223	489
1988	629	424	1053
1989	706	774	1480
1990	977	1759	2716
1991	1490	1195	2685
1992	712	1417	2129

shift from sea to land-based production which took place in the late 1980s. The total production figures were below 500 tonnes up to 1987. From then, the production rose by approximately 500 tonnes annually up to 1990. In 1990, Iceland produced 2716 tonnes of farmed salmon. The production in 1991 fell to 2685 tonnes. This demonstrates the problems the industry has faced and indicates that on the world basis, Iceland is a small producer of farmed salmon with heavy concentration on land-based farming.

The farms are spread around the coast and fjords of Iceland. The Reykjanes Peninsula is the main area for the land-based farms. Aksafjørdur and Austfirdir (East coast) are important locations for sea-based farms. See Fig. 2.3 for location map.

Even though there is free entry into the industry, certain restrictions are imposed on the industry. First of all, only genetic material from Iceland was allowed, which means there was no importation of eggs or smolt. The background for this restriction is related to genetic interaction with domestic stocks and the effort to avoid the import of diseases. There has been an exception to this – a Norwegian-owned farm was allowed to import eggs from Norway. Some other land-based farms have bought eggs from this farm.

The main constraints for the Icelandic fish farming industry are:

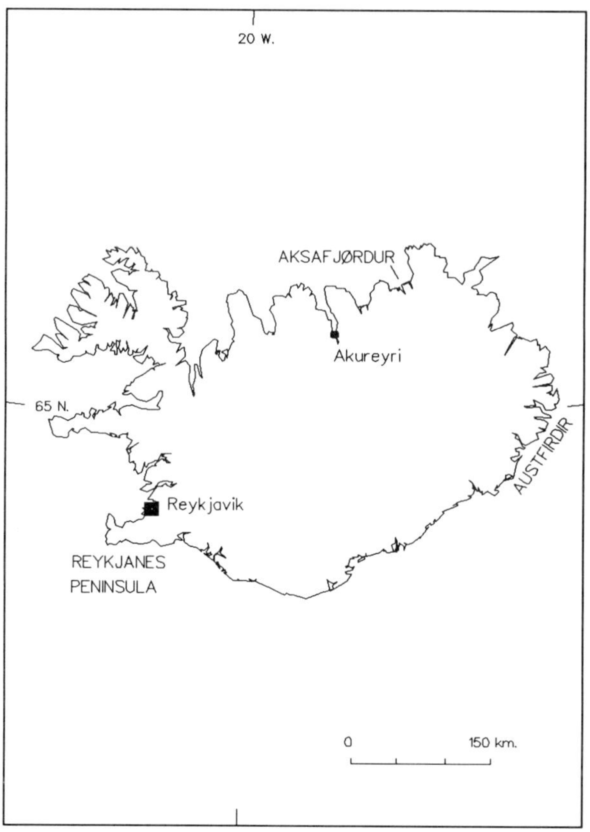

Fig. 2.3 Locations of salmon farming in Iceland.

(1) The genetic material of local species is poor. The problems are early sexual maturity and slow growth rates.
(2) Institutional rules – the import restriction on eggs makes it necessary to develop broodstocks based on local species, a lengthy process.
(3) Lack of sheltered locations for sea-pen farming. The available technology cannot withstand the severe weather conditions and strong winds of the Icelandic winter.
(4) Low temperatures – in the wintertime, the temperature in the sea might reach −1 to 2°C which increases the thermal mortality.

The large number of farms that went out of business at the end of the 1980s and the beginning of the 1990s is an indication of the serious problems the Icelandic industry has been facing.

The land-based industry face many of the same problems as the sea-pen rearing: for example, disease and predators (animals and birds). It seems as if the industry in Iceland has not been able to exploit the potential advantages of land-based farming in controlling environmental factors of production. However, Iceland has two advantages in land-based production compared with most other regions:

(1) Water with a year-round constant temperature and salinity content which is pumped from underground.
(2) Low water pumping costs due to cheap hydroelectric power.

Ireland by Knut Heen

The Irish farming industry has developed in the 1980s. Table 2.8 gives production by year in Ireland. In 1980 the production was only 18 tonnes. Production grew slowly in the beginning of the 1980s and reached 700 tonnes in 1985.

The table shows that the annual increase in production in the latter part of the 1980s was between 1000–2000 tonnes. In 1991, the production was 9000 tonnes.

The Irish industry is characterized by a few large farms with substantial Norwegian investment. In 1985, there were ten farms and in 1986, the number increased to 12. By 1992 there were 35 marine and freshwater salmon farms. They are concentrated in Galway Bay with sites in County Donegal and County Cork. See Fig. 2.4 for

Table 2.8 Farmed salmon production in Ireland 1980–91.

Year	Production (tonnes)
1980	18
1985	700
1986	1250
1987	2232
1988	4200
1989	5214
1990	7170
1991	9000

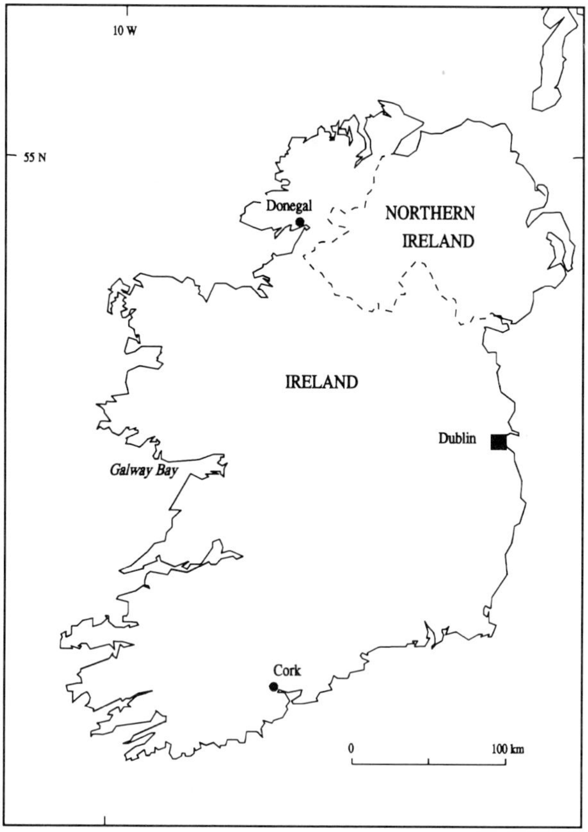

Fig. 2.4 Locations of salmon farming in Ireland.

location map. The Irish government is involved in this industry through research, service support and veterinary services. In addition, the government issues licences for the industry and is also involved in financing new projects.

Like Iceland, site availability is the most important constraint on expansion. All sheltered sites are already occupied. With current technology, there are few possibilities for expanding production through new locations. Future expansion of the industry in Ireland probably depends on development of cage systems which are more resistant to strong winds and waves.

Another problem that the industry in Ireland has faced is high ocean temperatures in the summer which causes disease problems.

Smolt shortage was also a constraint for development of the industry. Smolts were imported from Iceland, Norway and Scotland. In addition, eggs were imported from Norway. By 1990 the problem had been solved through large investments in hatcheries and smolt rearing facilities.

Two diseases have developed in Irish fish farms, furunculosis and a viral disease (infectious pancreatic necrosis).

France is the main market for the products.

Scotland by Knut Heen and John Thorpe

Salmon farming in Scotland was established by Marine Harvest Ltd, a subsidiary of Unilever, in 1968. They were encouraged by the Norwegian efforts and by their own attempts to farm rainbow trout in seawater. In spite of the fact that a number of other companies became involved in the industry at the time, salmon farming grew slowly in the 1970s. However, in the 1980s this growth increased rapidly, most of the expansion occurring in the latter half of the decade. Today Scotland is the second largest producer of farmed Atlantic salmon, with a production of 40 600 tonnes in 1991 (19 600 tonnes as two-sea-winter fish, and 21 000 tonnes as younger fish). Table 2.9 gives production development of the Scottish salmon farming industry.

As Table 2.9 reflects, while totals have increased rapidly, the *rate* of increase has slowed in recent years. The projected plateau at about 38 000 tonnes reflects loss of confidence in the market and difficulties in controlling the bacterial disease furunculosis.

In 1991, 154 companies produced adult salmon from 286 sea-pen sites (6.1 million cubic metres) and 12 seawater tank sites (78 600 cubic metres) employing 1286 staff. Most of the sea sites are located on the west coast, producing 77% of the harvest, and the remaining 23% is produced on the Orkney and Shetland Islands coasts (Fig. 2.5). In general the mainland east coast is too poorly protected for salmon aquaculture. Eighteen sites producing more than 500 tonnes accounted for the largest share of production (31%), with 34 sites producing 200–300 tonnes, accounting for a further 20% (Table 2.10). As in Iceland and Ireland, the Norwegian interest in the industry is substantial, and this interest has increased, particularly during the 1980s.

The production of smolts to supply the farms has increased from 6 million in 1985 to 22 million in 1991. In the event, in 1991, 21.1 million smolts (0.9 million from England) were transferred into the production sites, 96% as one-year-olds. The 20.2 million Scottish smolts had been reared at 143 sites by 80 companies, employing 350 staff. The sites consisted of 86 hatchery units of 52 600 cubic metres capacity, producing 53%, and 57 freshwater cage units of 210 800 cubic metres capacity, producing 47% of the total.

Most of the large companies rear their own broodstock with their own selection improvement programmes. From 27 broodstock sites, 50 million eggs were pro-

Table 2.9 Farmed salmon production in Scotland by year.

Year	Production (tonnes)
1980	600
1985	7 000
1986	10 000
1987	12 700
1988	18 000
1989	28 500
1990	32 000
1991	40 600

Source: The Scottish Office Agriculture & Fisheries Department 1992.

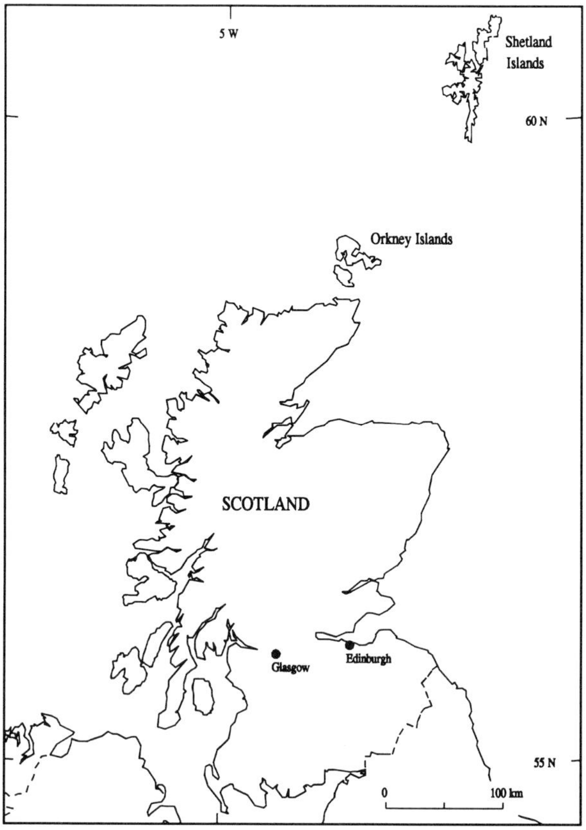

Fig. 2.5 Locations of salmon farming in Scotland.

Table 2.10 Atlantic salmon sea farm sites in Scotland: distribution by tonnage.

Production (tonnes):	<25	26–100	101–200	201–300	301–400	401–500	>500
% share:	1.1	12.2	13.3	20.3	13.5	8.9	30.7

Source: The Scottish Office Agriculture & Fisheries Department 1992.

duced, but a further 0.5 million were obtained from the wild. There is a much higher degree of vertical integration between hatcheries, smolt producers and grow-out farms than is the case in Norway.

The average weight of salmon produced by Scottish farms in 1991 was 3.2 kg for two-sea-winter fish. However, fish harvested in December can weigh 4–5 kg. The average weight of Scottish fish is less than those in Norway. The proportion of the production of a smolt year class harvested as grilse is about 25%, but was a record low at 21% for the 1989 year class.

Table 2.11 Fallowing periods at Scottish salmon cage sites.

Period (weeks)	0	<4	4–8	8–25	26–51	>51
Seawater sites	202	2	24	49	7	2
Freshwater sites	14	7	11	13	9	3

Source: The Scottish Office Agriculture & Fisheries Department 1992.

Major problems faced by the farmers are predators, disease, and severe weather conditions. Seals and diving birds are the principal predators. The most serious disease problems include furunculosis, and parasitic infestations by lice, which together are thought to account for most of the approximately 40% losses from smolt transfer to harvest. Eighty-four of the 286 sea-cage sites, and 54 of the 57 freshwater-cage sites practise fallowing, in an attempt to counter disease and parasite problems and increase production efficiency (Table 2.11). There is also an increasing trend towards reduction in stocking densities, with increasing numbers of farms using more cage space without increasing production.

Scotland does not have a licensing system like Norway's, but does have a compulsory registration system, and farms are required to obtain leases from the Crown Estate Commissioners. Since 1987, farms have been charged a rental of £50 per tonne of fish produced by farms harvesting more than 50 tonnes per annum, and £45 per tonne for smaller farms. The average Scottish salmon farm is larger than its Norwegian counterpart because its size is not legally limited. The UK government has encouraged development of the aquaculture industry particularly through financial assistance from the former Highlands and Island Development Board (HIDB: now Highlands and Islands Enterprise – HIE). This assistance amounted to more than £50 million in grants and loans, and additional smaller sums covering research projects in support of salmon production. Some assistance has also been provided from European Community sources. As a result of the different policies in Scotland and Norway, the industrial structures have developed differently in the two countries. The Norwegian industry is characterized by small farms operated by local owners (fishermen and farmers), while the Scottish industry is dominated by multinational companies. In many cases these companies are integrated, operating their own smolt production facilities and grow-out farms.

The growth in the salmon farming industries in the 1980s has been a combination of increased production in the existing sites and development of new ones. In the mid-1980s smolt supplies were a constraint for the industry, but this problem seems to have been resolved. One constraint on further development of the Scottish industry is the shortage of new sites. Most of the best ones are already occupied and future farms will have to use less well protected places where more expensive equipment and new cage design technology may be necessary.

About half of Scotland's total production of farmed salmon is sold in the United Kingdom, with the European Community and the United States being the other primary markets. As the UK is a member of the European Community, the Scottish exporters do not face any tariffs in the large European markets, unlike non-Community exporters.

Faroe Islands by Knut Heen

In the 1950s, some attempts were made to farm rainbow trout on the Faroe Islands. The efforts were patterned after the Danish farms on land, but the attempts did not succeed. In the 1960s attempts were made to grow rainbow trout in the sea following the Norwegian pattern, but this activity faced serious economic problems. In 1973 the government took over the facilities and the Marine Resource Institute was given the task of operating the Fiskaaling Company and examining the possibility of aquaculture on the islands. Until 1980, only Fiskaaling had sea-based farms. In 1980 the first private farm was established and the industry has grown considerably since the beginning of the 1980s.

Table 2.12 shows that the production increased slowly in the first half of the 1980s. In 1985 the production was only 470 tonnes. In the second part of the 1980s there was a rapid increase in production with the exception of 1988. In 1990 the production was 10 900 tonnes. From 1990 to 1991, there was a dramatic increase in production of 7100 tonnes, leading to a total production of 18 000 tonnes in 1991.

The number of new licences issued in the first part of the 1980s was high especially in 1983 and 1984 when the number of newly issued licences was 16 and 14 respectively. In 1986 and 1987 only one licence was issued each year, but in 1988, the number increased to 16 new licences. Most of the new licences were for a volume of 10 000 to 15 000 cubic metres. The latest established farms are located in areas less protected from wind and waves where salmon farming has not been attempted before.

When the first private farms were established, there was no size limit to the pen volume. Farm size limitations were introduced in the early 1980s. Until 1985, license of 5000 and 7500 cubic metres were the common size. When new licences were issued in the mid-1980s and later, these farms received permission to increase the volume. Therefore, the licensed volume grew faster than the number of new licences would indicate. The licences for expansion were mainly given to the smallest farms, those limited to 5000 cubic metres. Expansion was in most cases made in the additional amount of 2500 cubic metres. Nine of the licences given in 1988 are for farms with the magnitude of 15 000 cubic metres, while six are for 10 000 cubic meters. One can therefore conclude that development has gone in the direction of increasing the size of the farms.

Table 2.12 Farmed salmon production in the Faroe Islands 1980–91.

Year	Production (tonnes)
1980	0
1982	60
1985	470
1986	1 370
1987	3 543
1988	3 420
1989	8 035
1990	10 900
1991	18 000

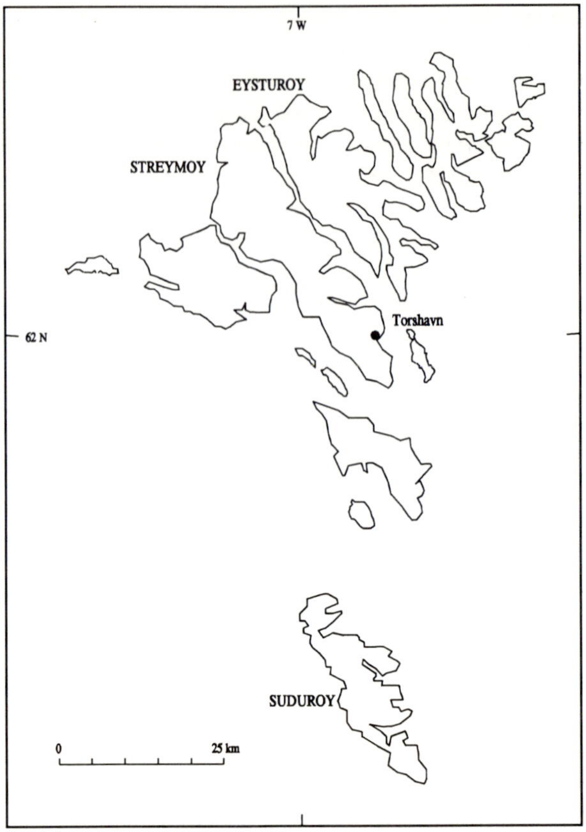

Fig. 2.6 Locations of salmon farming in the Faroe Islands.

By 1990, 70 licences had been granted for grow-out farms in the Faroe Islands with the total pen volume of 750 000 cubic metres. No licences have been given since that time. The licences authorize volumes of 2500 to 15 000 cubic meters, but only one farm has the low level of 2500 cubic metres in licences volume. Over half of all the licences authorize farms for 10 000 cubic metres or larger. Most of the grow-out farms are on the Islands Eysturoy and Streymoy. See location map in Fig. 2.6. Forty-three of the total of 70 licences for farm sites are located on these two islands. Suduroy has ten licences. By 1 January 1990, 63 farms were in operation of the 70 licences that have been awarded. Of those seven licences still not utilized, each authorizes a volume capacity of 15 000 cubic metres.

The salmon farming industry in the Faroe Islands consists of both government and private organizations. In 1985 the local government established a committee (Aliradid) to evaluate all applications for licensing of smolt production, salmon production, shellfish and other aquatic products. The Marine Research Institute and the Veterinarian Institute are members of this committee, as well as institutions involved with sea transportation and insurance. In 1980 a farmers' organization was established to take care of the common interests of farmers.

In some areas the farms are located in close proximity which can increase the problem of disease spread. In the summertime, salmon lice are a serious problem as well as vibriosis, Hitra disease and algae blooms.

Water quality is a problem where farms are in close proximity. Many of the farms have, therefore, two areas available for location, one for the summertime which is not well protected, and one for the wintertime. The number of available sites is an important constraint for further expansion of the salmon farming industry and it seems to be the general consensus that the number of licences have reached the upper limit.

The majority of the product is exported fresh. The main markets are the European Community, Japan and the United States. In the European Community, Denmark is the most important market followed closely by France.

NORTH AMERICA

Salmon aquaculture in North America is located in the coastal regions of the United States and Canada, both on the Atlantic and Pacific coasts. See location map in Fig. 1.1. The Atlantic region consists of the state of Maine in the United States and the Provinces of New Brunswick and Nova Scotia in Canada. On the Pacific coast, the State of Washington in the United States and the Province of British Columbia form the farming region.

Atlantic North America by Neil Ridler

Atlantic salmon farming in the North-East began commercially in the early 1970s, but it failed because of business mismanagement or lethally cold waters (Saunders, 1989). Successful production only began in the late 1970s, more than a decade after Norwegian operations. Handicapped by water temperatures and storm conditions, growth has been steady if slow.

Production and location

Successful salmon farming along North America's eastern coast dates from 1978 when the first commercial farm began operating (Aiken, 1989). Previous attempts had failed because of 'superchill', lethally cold waters of −0.7°C. However, government scientists were convinced that if suitable sites could be found, Atlantic salmon would thrive. Fortunately, the Fundy Isles region of the Bay of Fundy offers such sites, see location map in Fig. 2.7. The Fundy Isles production focuses on Deer Island and Campobello Island to the south and Grand Manan Island located still further to the south and east. On the one hand, they are protected from the severe storms that plague the Bay of Fundy. On the other hand, they are driven by tides with a vertical range of more than eight metres. The Fundy Isles region receives warmer water from the Bay of Fundy. The eight-metre tides also help to control pollution by providing a rigorous flushing action. American and Canadian salmon farming has therefore concentrated on the Fundy Isles region. The first Canadian farm was established in 1978 off Deer Island; while on the American side in Maine, one farm has operated continuously in the same region since 1982.

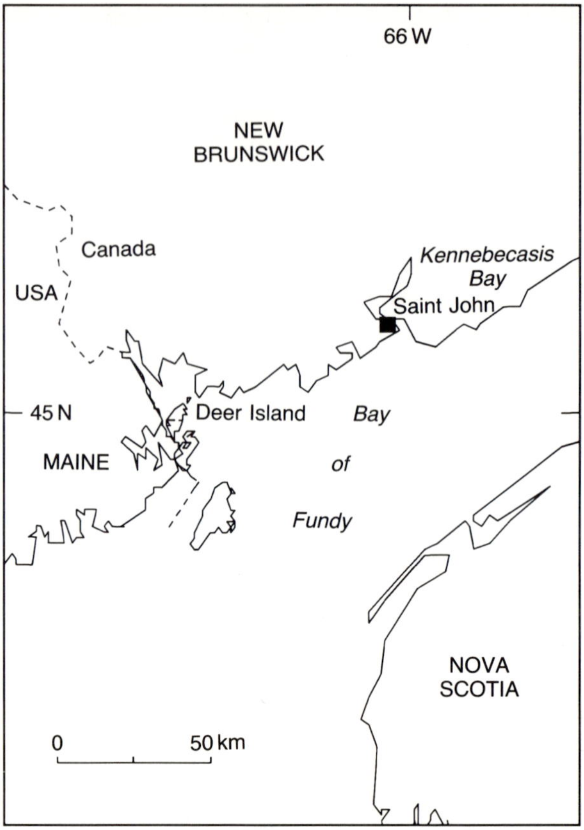

Fig. 2.7 Locations of salmon farming in the Fundy Island region of Atlantic North America.

In 1990 there were approximately 40 cage operations in New Brunswick, and a further 36 finfish sites in Maine. While there has been some loss due to 'winter chill', the 1980s have shown that overwintering is feasible. Regions other than the Fundy Isles are also possible. Kennebecasis Bay, which empties into the Saint John River in New Brunswick, has suitable winter temperatures, and cage and tank techniques have been used to produce Atlantic salmon.

In Nova Scotia, there is some production of Atlantic salmon. Along the Northeast however, New Brunswick remains the principal producer. In Nova Scotia, output of Atlantic salmon in 1988 was approximately 27 tonnes, and in Maine, output was 563 tonnes. As Table 2.13 shows, New Brunswick's output was more than six times that of Nova Scotia and Maine combined.

Table 2.13 illustrates that growth has been steady, if slow. In 1980, 11 tonnes of farmed salmon was produced. In 1985, production increased to 349 tonnes. From 1985 production has about doubled every year reaching 7000 tonnes in 1990. The shortage of sites was not the only reason for the relatively slow growth. From its early days, the industry has been plagued by a shortage of smolts, which not only hampered the establishment of new farms, but it also constrained the expansion of existing farms. To partially meet the industry's needs, the Canadian government

Table 2.13 Atlantic salmon production in New Brunswick 1979–90.

Year	Farm	Tonnes
1979	1	6.3
1980	2	11.3
1981	4	20.8
1982	5	38.1
1983	5	68.0
1984	10	222.2
1985	20	349.2
1986	28	634.9
1987	34	1315.2
1988	34	3600.0
1990	40	7000.0

Sources: Aiken (1989); *Canadian Aquaculture* (1990); Dept. of Fisheries and Oceans (1990).

supplied smolt that it raised for restocking purposes. However, commercial hatcheries have now replaced the government, and the shortage of smolt has eased. A further reason was financial. The industry's infancy and unknown riskiness deterred banks from providing credit. Crop insurance was not initially available for the same reason. Reluctance to lend was compounded by a clause in the Canadian Bank Act which excluded fish as potential collateral for loans, so that farmers could not use their unsold fish as collateral for loans. Credit, therefore, was very difficult to obtain, and in the early years of operation, credit to cover operating expenses is essential. Some assistance, in the form of cash grants and of loan guarantees, was provided, but not enough to finance rapid expansion. Hence the industry has not grown as fast as that on the Pacific Coast.

The slow growth of the industry has resulted in a size distribution and tenure pattern in New Brunswick similar to that of Norway. Unlike Norway, however, this size distribution was not the result of deliberate government licensing policies, but rather of the relatively slow and uncertain growth of the industry in eastern Canada. The risks and uncertainty deterred large investors until the individual entrepreneurs showed that salmon cultivation was feasible and profitable. Only two farms operated in New Brunswick in 1980. The number increased to 20 in 1985 and to 40 in 1990. Limited cage sites have precluded dominance of the industry by a few large companies. A survey of farms showed that in 1987, only 10% had more than 50 cages, (the largest had 70), and that the average farm had 24 cages (Fiander-Good, 1988). Almost half the farms had fewer than eight cages. The large majority of farms are owner-operated.

The industry in Maine has developed differently. It, too, is limited by cagesites, which has kept output small. In 1986, there was only one farm producing, and even by 1988, output was less than 600 tonnes. However, the industry has attracted substantial investment. One farm, for example, has financing (some of it Japanese) of US$20 million. By the turn of the century, output of Atlantic salmon in the North-East United States is forecast to exceed 10 000 tonnes, much of it produced by a few large farms (Porter, 1990). One single farm hopes to be producing 3000 tonnes (Barnett, 1989).

Future prospects

The farm gate prices in the North-East of North America have followed the wordwide trend in 1989 and 1990, and fallen. This has resulted in some consolidation as less competitive farms have been absorbed, but the underlying economic situation of most farms is good. The reason is that most farms have grown slowly, thereby keeping their debt to a minimum. Growth has been slow either because farms have been constrained by shortage of smolt, or because banks have been reluctant to provide credit. High interest rates and low fish prices have therefore been easier to absorb than had the debt/equity ratio been higher.

Over the long term, the market for farmed salmon in North America appears strong, and the principle beneficiary, because of lower transport costs, will be North American producers. Demand is strong because of increased preferences and incomes.

In 1981, Canada was the only exporter of fresh salmon to the United States, but by 1985, Norway had become the major supplier at the expense of Canada, which saw its share fall to 17%. In 1989, Canada regained its pre-eminent position with a 44% share of fresh salmon imports to the USA, compared with 40% for Norway (*Canadian Aquaculture*, 1990). The principal markets for North-Eastern North America are Boston, New York, Montreal and Toronto, but one New Brunswick producer is selling regularly to San Francisco on the Pacific coast.

The potentially buoyant market for Atlantic salmon is being reflected in the reversal of price declines that occurred in 1988 and 1989. Since the beginning of 1990, prices have been firmer. The principal reason for the higher prices has been Norway's reduction of fresh salmon sales, which has left a free and clear opportunity for North American (and Chilean) producers.

Maine and the Atlantic provinces of Canada are among the poorest areas of the US and Canada, with above-average unemployment rates and below-average incomes. The counties where most of Maine and New Brunswick salmon is produced have among the highest non-seasonal unemployment rates within this area. Thus, salmon farming has provided jobs where few alternative opportunities exist.

A corollary of the generation of employment is the survival of isolated rural communities. An explicit criterion in the issuing of licences in Norway has been location with preference given to the three most isolated northern communities. In Canada, no such approach has been adopted. However, with the commercial fisheries facing difficulties, governments are prepared to fund diversification to prevent the collapse of the rural economy, and aquaculture is viewed as one possibility. Already in Maine and New Brunswick, some communities are surviving only because of aquaculture.

The socio-economic benefits of aquaculture have prompted imaginative financing on the part of governments. Originally cash grants and loan guarantees were offered to salmon farmers; without these, the industry would not have developed. Now, to spread the risks, because loan guarantees impose a risk only on taxpayers and not on banks, there is increasing emphasis on subsidized interest rates. A reduction of 7% in interest costs increases the internal rate of return of a hypothetical salmon farm by at least three percentage points (Ridler and Martin, 1990). Subsidized interest rates, therefore, are a means of encouraging the industry without large cash disbursements.

Conclusion

The Atlantic coast of North America produced approximately 5000 tonnes of farmed Atlantic salmon in 1989. An important source of income and employment in a region which has high unemployment, the industry remains small internationally (at less than 5% of world production). Output has been, and remains, limited by the lack of suitable sites, given the cold waters of the Atlantic coast. Sea-cage farming is therefore largely restricted to the Isles of the Bay of Fundy. This constraint could be eased by offshore cages, but these have not yet proved to be economically feasible. As a result Atlantic salmon farming, with present prices and techniques, can expect continued growth up to an approximate maximum of 25 000 tonnes. Even with this output, the Atlantic North-East would remain small internationally, but it would enjoy a comparative advantage in its propinquity to American markets.

Pacific North America by Robert L. Monahan

Salmon farming in British Columbia

Salmon farming was introduced to British Columbia in the 1970s. The industry was aimed at marketing pan-size salmon, but this early attempt to establish salmon farming failed due to lack of financing. The main problems were related to high mortality rates and technical difficulties. The farms used surplus eggs obtained from the Canadian government salmon hatcheries. Up to the mid-1980s, the salmon farming industry in British Columbia remained undeveloped. In 1979 the production was only 38 tonnes. In 1984 the production rose to 100 tonnes.

Norwegian companies began to show interest in the Pacific salmon aquaculture industry about this time. Norwegians were attracted to British Columbia because of the favourable location in relation to the large American market and because the licensing system in Norway restricted their domestic investment. From the mid-1980s, the British Columbia salmon aquaculture industry expanded rapidly. The quick growth was the result of both Norwegian involvement and the positive attitude from the Provincial Government of British Columbia. In 1984 there were only ten marine farms in operation. The annual increase in the coming seven-year period was more than 20 farms, reaching 168 in 1991, see Table 2.14. The table also demonstrates the high number of freshwater operations before salmon farming started. In 1975, there were 46 freshwater sites, increasing to 77 in 1980. At that time, only two marine farms were in operation. The number of freshwater operations nearly doubled up to 1991 mainly due to increased demand for smolt from salmon farms.

The increase in salmon production has a similar growth pattern as the number of farms, but with a two-year time lag.

The increase in the number of farms between 1984 and 1985 is reflected in the great expansion in production from 1986 to 1987, from 391 tonnes to 1739 tonnes (see Table 2.15). From that year, the annual production increased by more than 4000 tonnes. It reached 19 000 tonnes in 1991.

Table 2.13 also illustrates the change over time in species grown. Coho was the major species grown up to 1986. From there, chinook production grows extremely

Table 2.14 Operating Farms in British Columbia – finfish,[1] 1975–91.

Year	Marine	Freshwater
1975	–	46
1976	1	51
1977	1	55
1978	2	57
1979	2	60
1980	2	77
1981	4	89
1982	5	99
1983	8	106
1984	10	107
1985	37	122
1986	69	135
1987	82	135
1988	118	149
1989	135	173
1990	124	88[2]
1991	168	130

Notes:
[1] Includes salmon, steelhead and trout. Salmon is the dominant species farmed, about 98% of total production is salmon.
[2] In 1990, the Ministry of Agriculture and Fisheries took over licensing responsibility for freshwater aquaculture. However many holders of commercial fish culture permits were not in commercial operation and opted not to obtain a fee-based aquaculture licence.

Table 2.15 Farmed salmon production by species in British Columbia 1979–92.

	Tonnes			
Year	Coho	Chinook	Atlantic	Total
1979	41	–	–	41
1980	157	–	–	157
1981	176	–	–	176
1982	230	43	–	273
1983	73	55	–	128
1984	64	43	–	107
1985	66	54	–	120
1986	304	87	–	391
1987	790	949	–	1 739
1988	2 775	3 550	80	6 405
1989	1 815	8 514	1 280	11 609
1990	1 270	10 570	1 600	15 482*
1991	291	15 466	3 396	19 153*
Projected				
1992	400	11 500	6 800	18 700

Note: * Includes smaller amounts of steelhead and rainbow trout.

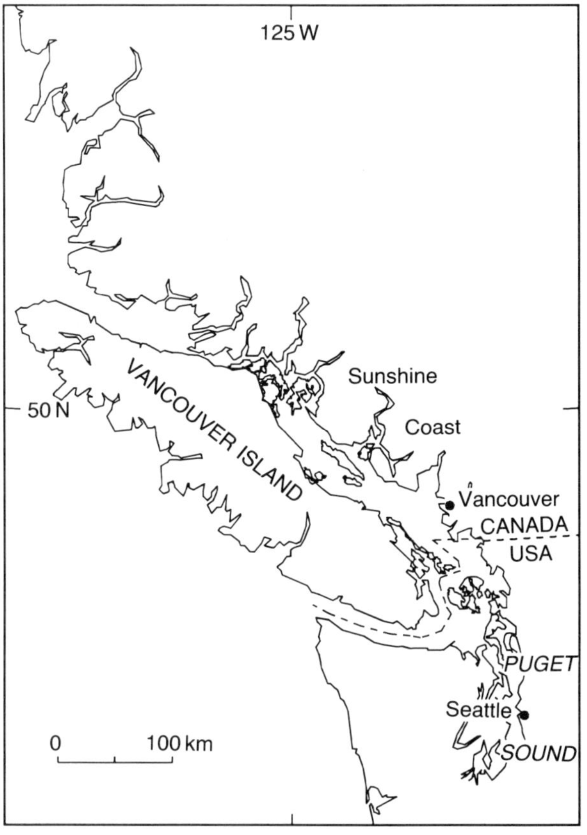

Fig. 2.8 Locations of salmon farming in Pacific North America.

fast, reaching 8514 tonnes in 1989. In that year, the production of chinook was nearly five times that of coho. It is also interesting to note that 1280 tonnes of Atlantic salmon were grown that year.

Most of the farms are located on the Sunshine Coast area and Vancouver Island. See location map in Fig. 2.8. These areas have the best developed infrastructure in the province. However, compared to Norway, the infrastructure in these two areas and particularly in the rest of British Columbia is poor. Transportation to the farms must be done by plane or boat. Many of the sites lack public electrical supply and expensive, farm-owned diesel generation has to be used for the production of electricity. With further development of the industry in British Columbia, one must expect that most new farms will be established in unsettled areas where the cost of establishing a new infrastructure would be substantial.

British Columbia was believed to provide excellent conditions for salmon aquaculture. The background for this was the higher average water temperature than normal that was believed to give fast growth and minimal pollution of the water. However, the salinity of the water shows great variation during the year. The number of islands and fjords protect the farms from strong winds and waves. There is little or no ice during the winter. The great difference in high and low tide insures good water circulation in the farms and it makes possible higher fish stocking rates.

As mentioned above, the Provincial Government of British Columbia has shown great interest in developing the aquaculture industry. Incentives are given in different ways, primarily through financing from the Ministry of Agriculture and Food.

The British Columbia aquaculture industry relies on eggs from the Canadian Department of Fisheries and Oceans (DFO). This department runs 66 hatcheries with the aim of enhancement of the rivers in British Columbia. In 1983, more than 335 million fry/smolt were produced and 2.5 million of these were sold to the farmers. A few private farms were also producing smolt. Their supply of smolt was therefore a critical factor in the development of the industry in the 1980s. Norwegian interests wanted to import Atlantic salmon to the region. In spite of a general prohibition against such import, an exception to the rule was made in 1985. 50 000 eggs were imported from Scotland with a very long quarantine time of 15 to 18 months. All Atlantic salmon eggs and quality are provided by private hatcheries.

Compared to Norway, there are few restrictions to the salmon aquaculture industry in British Columbia. The farmers have to obtain a licence but there is no upper limit to the farm size nor restrictions on foreign investors. Horizontal integration is common at farm level. Most farms have not integrated forward into processing. This activity is mainly done by traditional packing companies as this industry has idle capacity during the off-season of the commercial fisheries.

In the first phase of the development of the industry, feed was imported from Washington State. Since the end of the 1970s, however, British Columbia has had locally produced salmon feed.

The main market for the British Columbia industry is the United States and the Pacific Rim countries, although the USA is the main buyer.

Washington State

The history of salmon farming in Washington State is relatively long. In the late 1960s, farming of pan-sized coho started. The farming was done in marine net pens, ponds and raceways. In the 1980s, production of larger salmon began. Atlantic salmon was introduced from the east coast and Europe. This species is mainly grown in net pens in Puget Sound (see location map in Fig. 2.8) and are marketed at the size of 2–5 kg.

The Pacific Northwest seems to have good conditions for salmon farming. The environment is similar to that of Norway in respect to physical and climatic characteristics. The islands, fjords and coastal inlets provide sheltered areas for sea-pen farming. The water temperatures range from 6–16°C. In addition, the area has a long tradition of salmon enhancement, and substantial infrastructure of hatcheries, feed production, research, education and personnel training to support the salmon farming industry. Table 2.16 gives the production of farmed salmon by year.

As Table 2.16 indicates, the growth has been rather slow: 1000 tonnes produced in 1986 increasing to 2190 tonnes in 1991. Seven farms operated 11 marine sites in 1991. The number of freshwater farms was four, operating five sites. The industry had a major setback in 1990 when production dropped to 1250 tonnes. The reduction in production that particular year was a combination of local, national and international conditions. The local reasons included losses due to plankton blooms

Table 2.16 Farmed salmon production by species in Washington State 1986–92.

Year	Tonnes Pacific	Tonnes Atlantic	Tonnes Total
1986	N.A.	N.A.	1000
1987	N.A.	N.A.	1819
1989	1006	994	2000
1990	230	1020	1250
1991	210	1980	2190
Projected 1992	N.A.	5000	5000

which are the worst problem. Other problems for Washington farms have been infected smolt, primarily by furunculosis, and low levels of dissolved oxygen in the later summer.

However, the main constraint to the industry in the state of Washington is the opposition from owners of property along the shore where farms or potential farms are located, environmentalists, and commercial fishermen. The opposition from fishermen is mainly based on increased competition in domestic and foreign markets that has depressed the salmon prices. Secondly fishermen and environmentalists are concerned with environmental issues such as pollution from fish farms (fish faeces and uneaten food) and possible interaction with the wild species from escapees.

The neighbouring land owners believe cages in the sea reduce property values both through possible environmental pollution and scenic degradation.

JAPAN by Conrad Mahnken

Introduction

Japan is the world's largest consumer of salmon, and in 1985 Japan's supply of 349 000 metric tonnes (catch plus imports) was 39% of world harvest (Nasaka, 1987). Predictable supplies of this important source of quality marine protein are considered vital to meeting the protein needs of this island nation. Two factors have provided impetus for the development of salmon ranching and salmon farming in Japan: growing national demand and the establishment of extended jurisdiction to 200 miles by most countries in the 1970s. Enactment of 200-mile exclusive economic zones by most nations in the 1970s displaced the traditional Japanese high-seas fleets. Therefore, the Japanese have turned to aquaculture to offset the deficit in balance of trade of fish products created by rising imports.

Japan has a long history of salmon culture. The first propagation experiments were carried out in 1876 and the first national chum salmon hatchery was established on Hokkaido one year later. Japan's salmon aquaculture programme presently consists of a combined salmon ranching and farming effort. Japan's salmon ranching programme involves hatcheries that release more than two billion chum salmon fry and lesser numbers of other salmon species juveniles which return as adults for harvest in their high seas and coastal fisheries. The intensive marine farming of coho salmon is carried out in net cages.

Production

Although Japan's salmon ranching programme has been well established, salmon farming is a relatively new phenomenon. The rise of coho salmon farming in Japan had its origins in the 1960s when experimenting with salt-water culture of rainbow trout began.

Because the local market was unfamiliar with large-size rainbow, the farmers could not command a reasonable price. Coho farming technology evolved as a natural outgrowth of marine trout farming. Pilot-scale experimentation with other salmonid species continued in Miyagi and Iwate prefectures and to the north on Hokkaido and chum, pink, chinook, sockeye and coho salmon were all eventually cultured in net pens, tanks or ponds.

In 1973, eyed coho salmon eggs were imported from public hatcheries in the states of Washington and Oregon in the USA and raised to smolt size in a freshwater lake in Shizuoka Prefecture south of Tokyo. The smolts were transferred to net pens in Tokyo Bay in 1974. Encouraged by this success, they began to seek additional farming environments in the more optimal coastal bays of the north-east of Honshu. A farming system began to take shape in 1979 when, after repeated salt-water rearing trials, 350 tonnes of coho salmon was produced with an 80% survival from smolt to adult (Endo, 1985). By 1984 Japanese production had increased to 4400 tonnes.

The Japanese production of pen-farmed coho salmon has increased from 72 tonnes in 1978 to 16 000 tonnes in 1988, a more than 200-fold increase in a decade (Table 2.17). Production failed to increase annually only once, in 1981, as a result of fish losses to a typhoon in December of 1980. The entire production currently is consumed domestically and future increases in production are not expected to be exported. Thus, salmon farming may eventually reduce Japanese reliance on capture

Table 2.17 Farmed Coho salmon production in Japan 1978–91.

Year	Production (tonnes)
1978	72
1979	350
1980	1 850
1981	1 150
1982	2 100
1983	2 900
1984	4 400
1985	7 000
1986	7 700
1987	12 200
1988	16 000
1989	21 000
1990	22 000
1991	N.A.

fishery imports, the majority of which comes from the United States. Japan maintained its dominant position as world leader in the production of farmed coho salmon.

The salmon farming regions in Japan are located at a lower latitude than the other producing regions in the world, see location map in Fig. 2.1. Because of too-high water temperatures in the summer for salmon aquaculture, Japan has developed a unique husbandry practice. The oceanography of the coastal seas surrounding Japan is heavily influenced by two major current systems: the northward-moving warm Pacific Ocean current and the cold south-flowing currents from the Bering Sea. Most coho salmon farming is located on the Pacific coast of Honshu where the oceanographic fronts meet with latitudinal changes with the season. Sea surface temperatures in coastal waters rise in the spring and drop in response to latitudinal shifts at the point of the meeting of the currents. Japanese coastal waters are typical of western boundaries of northern oceans, with more rapid seasonal water temperature changes as well as more extreme highs and lows than are found on the eastern boundaries of northern oceans. The north-eastern Honshu Island (where most of the coho salmon farms are located) has an annual sea surface temperature range from about 9–22°C.

Temperatures for maintaining the maximum growth rate for coho salmon is approximately 12–15°C. Growth declines above a temperature of 15°C and temperatures of 23–24°C are lethal. In Shizugawa Bay, where many Japanese farms are located, maximum growth can be maintained for only about a month of the nine-month grow-out season, and sea surface temperatures may exceed the upper lethal limit for coho salmon for several months of each year. These extreme thermal conditions contrast sharply in comparison to more moderate western North American farming conditions where summer water temperatures usually do not exceed 15°C.

Despite thermal conditions not entirely suitable for Pacific salmon, Japanese farmers have been able to adapt to local conditions. They developed a farming system that produces remarkably rapid growth with production strategies for seawater grow-out that are designed to avoid high summertime water temperatures. Sea surface temperatures usually exceed the growth rate optimum of 15°C by June in Shizugawa Bay, dissolved oxygen may drop, stressing fish and producing disease, and temperatures may exceed the upper lethal limit of 23–24°C (Brett, 1952) by August. Therefore, harvest time is dictated by rapidly rising sea surface temperature. Furthermore, the farmers have an economic incentive to harvest farmed fish early in the spring-summer season before wild-caught Pacific salmon become available and drive down the price of farmed fish.

In order to attain marketable fish weight of 2.5 kg by June, the farmers must transfer large smolts to seawater as soon as seawater temperatures drop below 18°C in late September or early October. Large (150 g) smolts are produced after only about 10 months of accelerated freshwater growth. The saltwater grow-out system is patterned after that developed in the 1960s in Japan for the culture of rainbow trout in seawater where juveniles were transported to the bays and gradually acclimatised to seawater. Very rapid growth occurs and first harvesting begins in May after about eight months in the sea cages and is complete by no later than early July when water temperatures once again exceed 18°C.

The transfer of large smolts to seawater in the autumn is a strategy to avoid unfavourable sea temperature in the coastal bays during the summer. But the Japanese strategy in producing very large (150 g) smolts serves two other purposes as well: only large smolts will attain market size of 2.5 kg in the required nine months of seawater growout; and only large smolts exceed the minimum size for successful osmoregulation in seawater. Fish below about 120 g average size lack salinity tolerance for saltwater transfer at that time of year. Densities of 10–15 kg/cubic metres at harvest are recommended.

Fish harvest usually occurs after six months of rearing in May with 2.0 kg fish and ends early in August with 3–3.5 kg fish. The Japanese system of salmon culture depends on rapid sustained seawater growth to marketable size. Growth stasis due to marine diseases, low or high environmental temperatures, poor physiological condition of smolts or self-pollution will result in late harvest of undersize fish.

Moist pellets manufactured by local farmers are used for feed in coho salmon culture in Miyagi Prefecture. The feed is compounded from a mix of commercial dry diet and raw fish. However, fish farmers generally use a mix containing only 10% dry diet in order to economize (O'Toole, 1988). The use of a high percentage of raw fish in the diet results in a large drop loss and can create local self-pollution problems in the bays. Feed is made fresh daily with pellet size determined as 70% of the maximum gape of the average fish's mouth. State-of-the-art food processing machines are commonly used.

The raw fish constituent is comprised largely of frozen mackerel and sardines, but may also contain saury and juvenile cotids. The proportion of raw fish species will vary from farm to farm and month to month throughout the growing season, depending on availability. For marketability, red-coloured flesh is necessary when cultured salmon are harvested. Flesh colour is examined in March and crustaceans containing the pigment astaxanthin are added to the finishing diet and may comprise 10–20% of the feed from March until harvest (Yoshida, 1985).

Location

Most Japanese coho salmon farms are located along a 120 km stretch of the north-east coast of Honshu Island containing numerous bays, natural harbours, sheltered bays, and inland seas where mariculture operations are safe from storms – primarily in Iwate and Miyagi prefectures (see location map in Fig. 2.9).

Shizugawa Bay in Miyagi Prefecture was the birthplace of commercial coho farming and remains as the primary production site in Japan although several bays to the south, notably Onagawa Bay, also produce farmed coho salmon. These two bays plus some other minor producers also in Miyagi Prefecture, produced more than 85% of the coho salmon farmed in Japan in 1987: increase in production has been conspicuous in southern Miyagi Prefecture particularly Onagawa and total production of farmed salmon in this prefecture alone exceeded 16 000 tonnes in 1989. Southern Iwate Prefecture is the next largest producer contributing 7% of the 1987 production.

In recent years other prefectures have attempted coho salmon cage farming. On the Japan sea coast, several prefectures have produced small quantities of coho salmon in marine environments both warmer and cooler than found in the North-east coast of Honshu.

According to Endo (1985) coho salmon farming has succeeded in the north-east coast of Honshu for the following reasons:

(1) The area has an ideal environment for salmon culture and has a coastline with

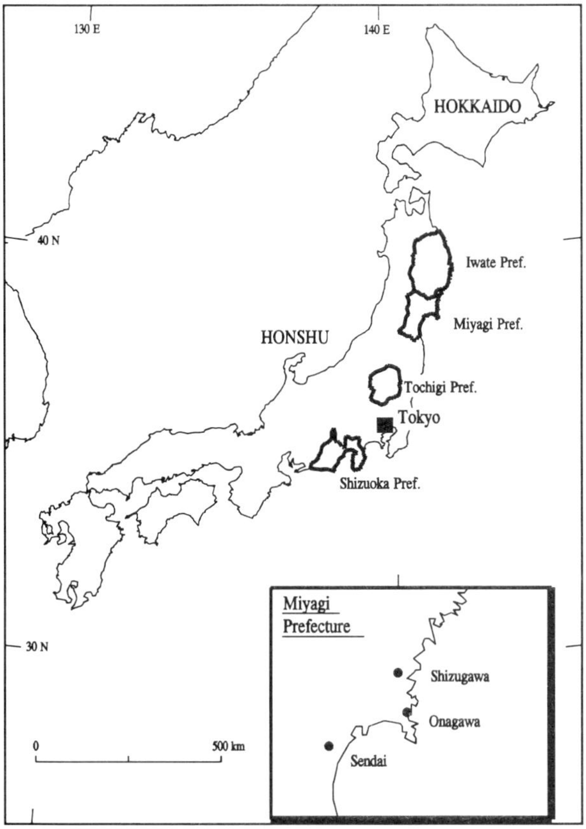

Fig. 2.9 Locations of salmon farming in Japan.

abundant bays and inlets which are well sheltered from storms. The bays are typical drowned river valleys (*rias*) sloping rapidly to the open sea, thereby promoting good circulation. Water temperature ranges from 5–20°C for most of the year. Although summer temperatures from July to October are too high for good growth and survival of Coho salmon, farmers are able to produce 2.5 kg fish for local markets in 9–10 months of salt-water grow-out and still avoid high summer temperature conditions.

(2) Fish can be grown to a harvestable size and supplied to the domestic markets before wild-caught salmon become available.

(3) Farmers and fishermen were seeking new employment opportunities. Prior to 1975, farmers along the coast were engaged in the farming of oysters and seaweeds such as *Undaria* and *Porphyra* but because of limitations in market demand were unable to expand. In the 1970s, many fishermen who had served as crews on high seas fishing fleets, lost their jobs because of the imposition of 200-mile exclusive economic zones and had returned seeking new jobs.

(4) Feed organisms such as mackerel, sardine, and krill for use in coho salmon diets were locally abundant and could be supplied at low cost to the farms.

(5) The larger companies, like Nichiro, were willing to support local fish farmers in

dealing with technical, marketing, and financial problems associated with coho salmon farming.

The total number of Japanese coho salmon smolt farms is reported at between 200 and 300 but their exact numbers and locations are unknown because so many are small-scale operators who produce coho salmon in addition to other traditional species.

Most of the smolts are produced in freshwater hatcheries in northern Tochigi Prefecture (see Fig. 2.9) 230 km south of Shizugawa Bay. By 1987, the 20 smolt producers of the Tochigi Inland Water Fish Aquaculture Cooperative were producing a total of about 250 tonnes of smolts per year (Nasaka, 1987). Individual farmers produce from 1 to 30 tonnes of smolts each year, depending on the capacity of their facility. Tochigi Prefecture is a preferred area for growing coho smolts because abundant groundwater is available with ideal year-round water temperature of 10–15°C, capable of producing the required 150 g smolts between the time of hatch in December and transfer to seawater in October. Some smolts are also produced in Iwate Prefecture.

All of the eggs for production are imported from the United States from the states of Oregon and Washington. The eggs are derived from excess sea-run adults from public hatcheries or sea ranchers. In 1986, the total number of eggs exported to Japan was 28 million. By 1989, egg imports exceeded 120 million.

Regulatory status

A set of regulations cumulatively called the Fishery Rights System (Mottet, 1980) provides the legal framework for salmon aquaculture in Japan. Fishing Rights are authorized by the federal government and granted by prefectural governors (Asada, 1973). These laws consolidate power in the hands of village fishermen and local autonomy in the fishing cooperatives, the parent organization to aquaculture cooperatives. The result is an industry dominated by small family businesses. A major factor contributing to the success of aquaculture projects is the ease with which public waters can be obtained for this use.

The three categories of rights are:

(1) Common fishing rights

These are fishing rights granted to cooperatives of fishermen and are renewable every 10 years. These are monopolistic and exclusive rights to engage in certain fisheries to harvest shellfish, seaweeds, and other sedentary animals in shallow public waters. In theory, the rights must be periodically renewed and can be denied and passed to others at time of renewal; in practice, rights seldom pass to new ownership unless the applicant no longer desires them, or there is adequate room for expansion of the fishery.

(2) Setnet fishing rights

These apply to net fisheries in waters deeper than 27 metres and are operated in a manner similar to common fishing rights. Setnet Fishing Rights are renewable every five years.

(3) Demarcated fishing rights

These authorize the use of a specific area for aquaculture, and are critical laws for aquaculture. The applicant must show that the holders of the common fishing rights and setnet fishing rights have no objections before these demarcated fishing rights can be given. Duration is from 5–10 years.

In 1948, the Fisheries Cooperative Association Law was passed. This law required fishermen to form cooperatives in order to receive priority in obtaining the common and setnet fishery rights. Once established, these cooperatives became all-powerful in allocating demarcated fishery rights for aquaculture purposes to outside fishermen or members of the cooperative. Once obtained, cooperatives must divide demarcated rights among co-op members (usually for a fee) or may conduct aquaculture ventures on behalf of the cooperative, otherwise the government will lease the rights to individuals or companies outside the cooperative. Cooperatives administer these rights as income-producing property with profits divided among members.

Payments to cooperatives derived from fees or leases from farmers are considered compensation for lost fishing areas, but in reality, the revenue generated by the more profitable aquaculture operations often exceeds the foregone fishing income. Cooperatives cannot discriminate on the basis of nationality, sex or religion when allocating sites but are allowed to evaluate applicants on factors such as length of residency in the co-op area and whether an individual has experience in sea farming. From a practical viewpoint, a typical applicant might have to live and fish in a village for five years, have been a voting member of the co-op for three years, and have three years' culture experience in order to have a reasonable chance to receive culture rights.

The Shizugawa Coho Salmon Growers Association, typifies cooperatives that feel that it is important that the use of 'preferred' aquaculture-growing sites within the sea area of the cooperative does not result in large income differentials. Therefore, demarcated areas are periodically redistributed by lottery and individual farmers are allowed to maintain only three cages of the same dimensions, 10 m × 10 m × 7 m deep (2100 cubic metres) (Nasaka, 1987). Shizugawa Bay is limited to 90 farms, each with three cages. Similarly, the 123 pen farmers of the Onagawa Fisheries Cooperative in Miyagi prefecture are also allowed to farm only three cages, but the cages are 13 m × 13 m × 10 m deep (5070 cubic metres) or more than twice the total volume of those in Shizugawa Bay.

Several large fishing corporations have shown a keen interest in the profitable coho farming industry and have contributed considerable technical advice to the farmers. But large fishing companies and other large corporations are neither allowed to participate in the coastal fisheries nor be directly involved in farming the coastal bays unless specifically authorized by the fishermen's cooperatives. Yet the operation of the salmon farming industry by diverse groups of individual farmers, fish farming cooperatives, and large corporations has functioned remarkably well within the framework of the fishing rights system.

The private companies can operate within the framework of the industry through selling eggs and feeds and as fish brokers and commonly remain involved throughout the life cycle of the farm-reared fish. For example, the Nichiro Company has been importing coho and chinook eggs from Oregon and Washington since 1971. They purchase the eggs and distribute them to the freshwater farmers. Marine fish farmers having farming rights to the coastal bays may buy the smolts from the freshwater farmers through intermediation by the private company (Endo, 1985). When the fish are harvested, the companies may also act as brokers by either directly purchasing the product or taking the product on consignment. Several large companies including Nichiro, Taiyo and Nichimo contracted with about 350 farmers in 1986 for their production of coho.

Diseases

Vibrio, furunculosis and BKD are the most frequently encountered diseases during saltwater culture. BKD has been identified as the most serious saltwater disease and is most evident as saltwater temperatures increase in the spring and summer. In recent years, mortalities as high as 30% during salt water culture have been blamed on BKD or BKD-exacerbated diseases. Farmers blame US egg sellers for the rising incidence of BKD in farmed coho. Whether a function of better detection procedures or not, the consensus appears to be that BKD is a relatively recent issue. Furunculosis has also been responsible for some initial losses after transfer. Vibriosis, Costia and gill amoeba were not considered as contributing significantly to seawater mortalities. A nutritionally-related disease ('fatty liver disease') which leads to anaemia is sometimes mentioned as a problem.

Future prospects

The producing areas in Japan are very limited due to sea temperatures. Ocean currents make it too cold in the north (Hokkaido) in the winter, while the summer temperatures along the Honshu coast are too warm. Even though production attempts have been made in regions other than the north-east coast of Honshu, it is reasonable to conclude that the growth potential of the industry in Japan is limited.

SOUTHERN HEMISPHERE

The Southern Hemisphere does not have a natural salmon run. Salmon have therefore been introduced by imports of eggs from the Northern Hemisphere, and is therefore not a traditional food. Many of the nations in the Southern Hemisphere, like Chile, do not have large numbers of relatively affluent citizens who would be the usual consumers of this fish. Markets for high-value salmon products are mostly in the Northern Hemisphere, primarily Europe, the United States and Japan.

A major advantage for salmon farmers in the Southern Hemisphere is, in the competition for Northern Hemisphere markets for fresh salmon, that the southern harvest takes place during the off-season for the northern salmon-capture fishery. The off-season extends from about November to June. Pacific salmon farmers in the Northern Hemisphere also try to harvest in the winter and spring off-season, but this means their salmon must be harvested before they reach the optimum size. In the Southern Hemisphere, on the other hand, salmon harvested in December, January and February are at their peak size just before maturation.

New Zealand by Knut Heen

Sockeye salmon were introduced in New Zealand in the beginning of this century to establish a sport fisheries in the rivers. Later, chinook salmon were introduced to start a commercial fishery. Chinook salmon are widely distributed in the main rivers and

Table 2.18 Farmed salmon production in New Zealand 1982–91.

Year	Production (tonnes)
1982	5
1984	10
1986	500
1987	700
1988	1500
1989	1600
1990	2400
1991	2700

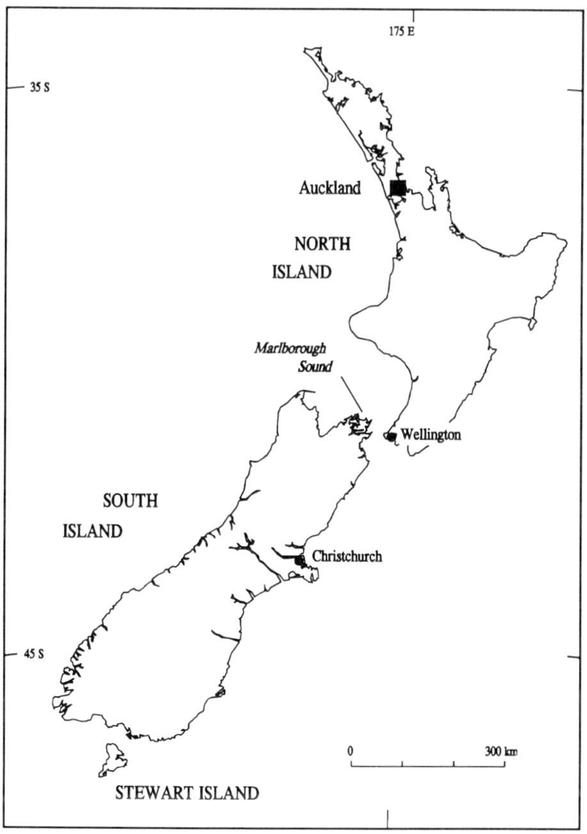

Fig. 2.10 Locations of salmon farming in New Zealand.

lakes of the South Island. Atlantic salmon are also present in New Zealand in various lakes and streams. Three major companies dominate the production of farmed salmon in New Zealand, which started in 1978.

Development of production in shown in Table 2.18. Production grew slowly during the 1980s and appears to continue in slow but steady growth in the 1990s. The farms are located mainly on Stewart Island and in Marlborough Sound, see location map in Fig. 2.10.

The fish are harvested after two years with harvesting taking place mainly during the Southern Hemisphere summer, October through March. Egg supply in the early years came from the ocean ranching programmes. Today farms maintain their own broodstock.

Two feed plants have been established to provide feed for all companies in the various phases of the salmon culture activity.

The advantages that the New Zealand farmers enjoy over the northern producers are:

(1) Excellent supply of unpolluted fresh water.
(2) A salmon stock that is relatively disease-free.
(3) The southern advantage of off-season harvest when wild salmon is not available in the Northern Hemisphere.

New Zealand farmers produce mainly for the Japanese market.

Tasmania by Knut Heen

As early as the 1860s, attempts were made to introduce salmonids into Tasmania. In 1885, the first formal release of Atlantic salmon fry was attempted in the Plenty River in southern Tasmania. Despite a number of attempts in the following 70 years, Atlantic salmon were never established in Tasmania. Attempts to introduce and establish Pacific salmon were not successful either.

Successful attempts in farming rainbow trout in 1980 led the Department of Sea Fisheries to begin investigating the possibility of salmon farming in sea cages. This investigation led to the establishment of the Salmon Enterprise of Tasmania Pty Ltd (Saltas) in 1985. This company has established a hatchery with the capacity of one million smolts per year, which is equivalent to a production of 3000 tonnes dressed weight salmon. Saltas has a monopoly of smolt production until 1995 and it is not believed that Saltas will increase the production of smolt before that time. This will limit the industry capacity to produce salmon. The reason for this monopoly position given by the Government is to have the necessary control in the early development phase of this industry.

The objectives of Saltas are:

(1) Developing techniques for culturing salmonids in salt water as well as good husbandry practices.
(2) Establish a high quality broodstock of Atlantic salmon and trout of known genetic composition and adapted to salt water culture in Tasmania by a programme of selective breeding and genetic research.

(3) Produce certified smolts for commercial culture in salt water.
(4) Training of suitable and qualified staff.
(5) Supervising and financing the operation of research the selective breeding centre and the hatchery which is intended to raise at least one million smolt per year plus an experimental sea farm which is capable of producing approximately 100 tonnes of fish per year.
(6) Conducting and commissioning research directed at assisting the development of all the above areas.
(7) Operating in all the above as commercial ventures.

The monopoly situation for smolt production enables Saltas to create a surplus for financing research and development. Most of the farms are located in the fjord leading to Port Huon and in the D'Entrecasteaux Channel between the main island and Bruny Island and on the Tasman Peninsula in Southeastern Tasmania, see location map in Fig. 2.11. In 1985 a joint venture called Tassal Ltd, between Norway and Tasmania was established. This company now controls about 40% of all salmon production. The first commercial quantities of salmon was successfully marketed in the 1986/87 season. The production in 1989/90 and 1990/91 increased to 3000 tonnes, see Table 2.19.

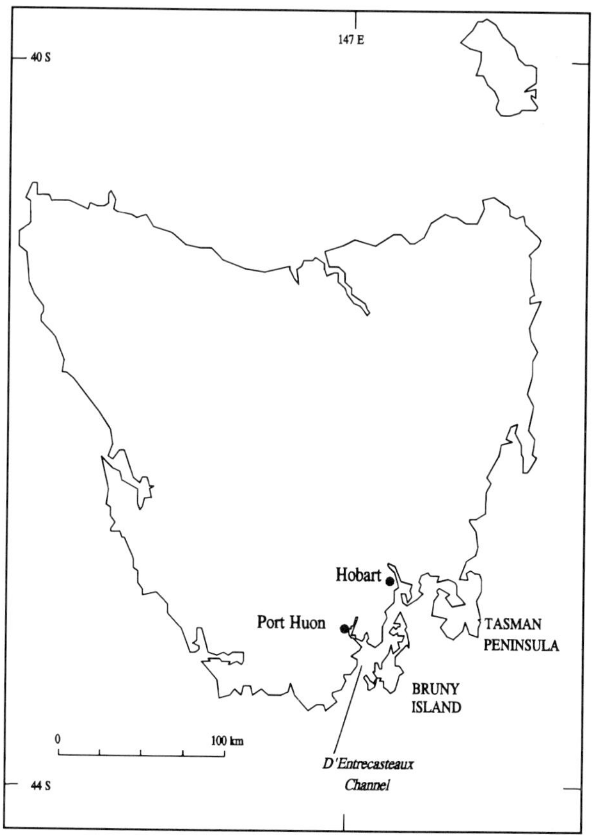

Fig. 2.11 Locations of salmon farming in Tasmania.

Table 2.19 Farmed salmon production in Tasmania 1985/86–90/91.

Year	Production (tonnes)
1985/86	0
1986/87	53
1987/88	300
1988/89	650
1989/90	3000
1990/91	3000

Twenty-one licences were issued by mid-1987 and an additional 15 applications were put forward. Thirteen companies were in operation.

The performance of the fish stock has been successful with very fast growth rates. Salmon of 4–6 kg are harvested before the second autumn in seawater and early sexual maturation is not a problem. The quality of the product has been very good and the product has enjoyed top prices in the market. The favourable conditions are a result of low variation in water temperature (8–18°C), clean water and good infrastructure. The technology in use is very similar to that used in Norway and Scotland but the production cycle is different due to the location in the Southern Hemisphere and to the high growth rates. The time from hatching to harvest for a 4–5 kg salmon is on average less than two and a half years compared to three to four years in Norway. The strict policy for controlling the broodstock and smolt production is to avoid introducing infectious salmonid diseases in Tasmania. The Government feels there is a realistic chance of avoiding the introduction of such diseases by isolating their own stock. Quarantine legislation was introduced as early as 1974. The main problem related to mortality has been amoebic gill disease. This disease can cause up to 90% mortality if not controlled by shifting the fish to brackish water. First summer rearing in brackish water or adoption of a nine-month or a one-year production cycle eliminates this problem. A major constraint for the development of the industry in Tasmania has been the supply of smolt and the aim has been to expand production. There are also limitations on the number of available sites which is another constraint on the further growth of the industry.

Tasmania's location is both a strength and a weakness. This isolation is believed to avoid introducing diseases. The isolation can be interpreted both in terms of geographical isolation and in terms of protective legislation. The geographical isolation means that there is no competition in the Australian market from other producing regions. However, the Australians are traditionally not salmon eaters and only 1000 tonnes annually is expected to be sold in the home market. The rest has to be exported to the United States, Japan and other Far East countries, a marketing handicap due to high transportation costs.

Chile by Jon Lindbergh

Introduction

The topography, oceanography and climate of Chile's western-facing coastline, from Puerto Montt to Cape Horn, is somewhat similar to equivalent latitudes in the North-east Pacific Ocean where some of the world's largest natural Pacific salmon runs exist. While the coast of Norway is at a higher latitude, oceanic conditions are also remarkably similar. Protected Chilean fjords and islands potentially suitable for salmon culture facilities are abundant for 1500 km. The northern segment of this coast, from Puerto Montt to Puerto Aisen, is where most of Chile's expanding salmon farming industry is now located.

Chile's geography has provided extraordinarily favourable locations for salmon culture, but at the same time has often made access difficult. No easy way exists through the Andes to reach vast areas of the coast.

Production

Early salmon farming projects were carried on at an experimental scale. The first significant harvest took place in 1980 with the total production of full-sized salmon being 70 tonnes, see Table 2.20. Production slowly increased to 104 tonnes in 1984. In the early 1980s, salmon growers began to recognize the possibilities offered by good biological conditions in Chile and inviting markets in the Northern Hemisphere. Pioneers in the industry began to expand and dozens of newcomers began applying for permits. In 1980, only two commercial salmon farming programmes existed. By 1991, more than 230 discrete farming sites (concessions) had been authorized. Production reached 1200 tonnes in 1987 and in 1991, the harvest placed Chile behind only Norway and the United Kingdom, but approximately equal to Canada and Japan. Projections for the 1995 harvest are about 80 000 tonnes. Low market prices of salmon worldwide in recent years have slowed salmon farm development in Chile, but production continues to expand. Long-range projections are speculative, but Chile has the potential to equal Norwegian production tonnages.

Most farmed salmon harvested in Chile up to 1988 were coho. However, several large operators began initiating Atlantic salmon farming programmes and by 1991, Atlantic salmon constituted about 45% of the harvest. In 1992, about 50% of Chilean production was Atlantic salmon. Atlantic salmon culture allows Chilean farms to sell large fish in direct competition with North American and European growers. Furthermore, Altantic salmon provide the Chileans a much wider period for harvesting and marketing fresh salmon. The reversed Southern Hemisphere season also allows Chile to produce large numbers of fresh large Atlantic salmon

Table 2.20 Farmed salmon production by species in Chile 1980–92.

Year	Pacific	Tonnes Atlantic	Total
1980	70	0	70
1984	104	0	104
1986	900	0	900
1987	1 200	0	1 200
1988	3 100	200	3 300
1989	6 300	1 770	8 070
1990	10 300	9 500	19 800
1991	15 000	12 500	27 500
1992	19 000	20 000	39 000

during October and November when Northern Hemisphere growers have more difficulty meeting the demand for that size.

Several farms have been carrying out experimental net-cage rearing of chinook salmon. The prevailing opinion of Chilean growers is that chinook are considerably more difficult to raise than coho and Atlantic. Growth rates have been lower, disease and mortality rates higher and market prices have been lower. Chinook production in Chile can be expected to decrease in the next few years.

Some Chilean growers are rearing full-sized steelhead trout in salt water net cages with good success, with production in 1992 being over 11 000 tonnes. Others are beginning to rear commercial numbers of cherry salmon, with an eye towards the Japanese market.

Salmon brokers in the United States have generally given good reports regarding fresh salmon, primarily Atlantic salmon, from Chile. Taste, size, appearance, colour and body form are considered excellent. At least one major broker claims his Chilean imported product is superior to Atlantic salmon imported from Europe.

In 1990, 60% of Chile's salmon harvest, mostly coho, was sold frozen into Japanese markets. Chilean coho have usually received a higher price than Japanese-grown farmed coho and Alaskan troll-caught coho in the market place. Japanese consumers are extremely discriminating in their selection of seafoods, and their acceptance of Chilean salmon products speaks well of their quality.

Financial investment in Chilean salmon farming ventures has come from Chilean national interests as well as from a number of foreign countries, which include the United Kingdom, Japan, Norway, Colombia, New Zealand, Canada, and the United States. Salmon farms with foreign investment frequently include Chilean partners, although Chilean participation is not a legal requirement. Scottish and Japanese interests control some of the largest farms. Many of the larger farms are financed by foreign interests while most of the smaller ones are exclusively Chilean.

Markets for high value salmon products are mostly in the Northern Hemisphere, primarily in Europe, the United States and Japan. Salmon is not a traditional food in South America and many of the nations there do not have the large numbers of relatively affluent citizens who would be the usual consumers of this fish. The first salmon farmers in Chile airshipped fresh coho salmon to the United States to take advantage of high prices available for their product during the off-season of wild-caught Pacific salmon. More recently, most Coho salmon production has been shipped frozen to Japan and most Atlantic salmon has been shipped fresh by air to the United States. Increasing amounts of frozen Atlantic salmon are being sold to Japan and in late 1992, substantial tonnages of frozen Atlantic salmon were being sold to Europe.

A significant amount of Chilean salmon is being smoked prior to being exported. Chileans usually cold-smoke and mild-cure salmon as compared with the hard-smoke technique common in the Pacific Northwest. Smoking equipment and techniques are European in origin. High quality Chilean smoked salmon should be increasingly competitive in the world marketplace.

Location

The first privately-financed salmon farmers in Chile began to develop commercial locations in the late 1970s close to roads and support facilities in the X Region near

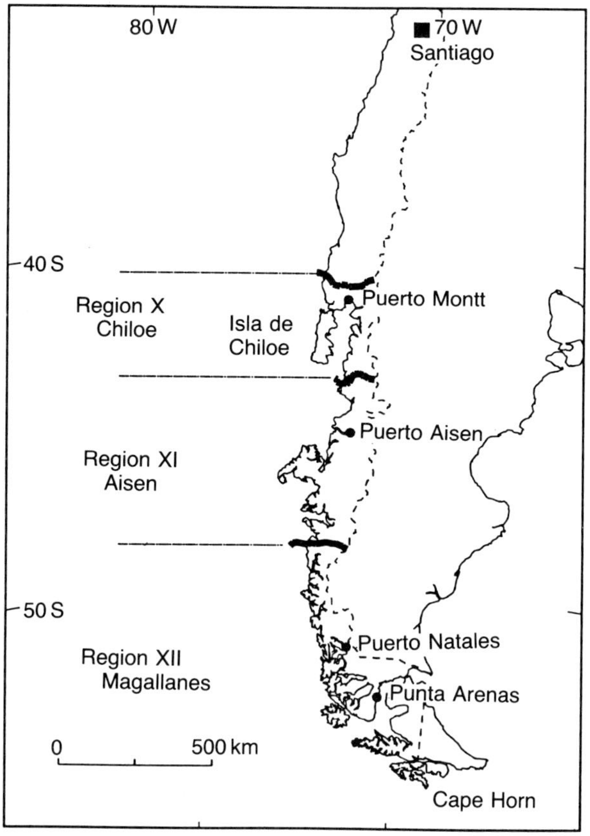

Fig. 2.12 Salmon farming regions in Chile.

Puerto Montt and on the island of Chiloe (see location map in Fig. 2.12). Shortly after, two non-profit groups established projects in the XI Region near Puerto Aisen, a region where the government wished to encourage more jobs and commerce. When the industry began to expand, most of the activity was concentrated in the X Region where transportation was available and labour costs relatively low. In 1991, more than 75% of producing marine salmon farm sites were in the X Region. One large salmon farm and a number of smaller farms operate in the XI Region. Recently, several farms have been established in the XII Region between Puerto Natales and Punta Arenas.

In the X Region, biological conditions are very favourable for marine salmon culture. Seawater temperatures are stable and vary between 10–14°C. Toxic phytoplankton blooms are rare. As a result, growth rates have been usually higher than in the Northern Hemisphere. Coho salmon, the first salmon reared in significant numbers commercially in Chile, have grown very rapidly. Smolts (migratory-sized juvenile salmon) of 30–50 grams in weight, which are transferred to salt water cages in December, reach an average harvest weight of 3 kg or more by the following December. This performance is substantially better than growth rates achieved with coho in Washington State and British Columbia.

Most of the growth experience through 1990 in Chile has been with coho derived from eggs imported from the Northern Hemisphere. Parent northern hemisphere stocks spawn in the fall of the year and their progeny normally take 18 months to reach the critical size necessary within the time period or 'biological window' when they can smolt as year-old salmon. The primary smolting window occurs during the lengthening days of spring in both hemispheres. It can sometimes be extended, with certain risks, well into the summer. Since the seasons in the separate hemisphere are six months out of phase, northern eggs moved to the south must face the critical spring either one half-year early or one-half-year late. Given the good growing conditions in the X Region of Chile, most smolts from imported eggs are ready a half-year early, as large 12-month smolts in December.

First attempts by hatcheries to produce coho juveniles from Chilean broodstock, called 'nacionales', resulted in relatively small smolts, less than a year old, available two or three months later in the season than smolts from imported eggs. The fish performed well, but did not reach as large a size at harvest time as imported stock. By 1989 and 1990, X Region Chilean hatcheries were able to grow a 40 or 50 g S-O 'nacionale' smolt from local broodstock by December. Performance of these large smolts has been very good, with better growth, better survival and larger size at harvest than imported stock. In addition, almost no precocious males have been encountered.

Results with Atlantic salmon in Chile have been extremely good. Farms in the X Region normally produce fish with weights of 3–4 kg after one year in seawater (in most Northern Hemisphere farms, about one and a half years in seawater are required for them to reach an equivalent weight). Relatively few grilse have been encountered.

In any promising new fish farm region, there is a tendency for too many operators to crowd into the most desirable and accessible locations. Overcrowding reduces water quality, decreases growth and survival, and increases disease. The biomass of fish being cultured in certain locations in Chile now may be reaching dangerous levels.

Regulatory status

An application for a concession to establish a marine fish farm must be addressed to the Ministerio de Defensa Nacional, which authorizes the location. An additional application must be addressed to the Subsecretaria de Pesca, which authorizes the particular species to be cultivated. For a freshwater concession, the competent authorities are the Dirección General de Aguas and the Subsecretaria de Pesca. There is no application fee and the annual lease fee for a concession in 1991 was approximately US$1500.

Applications for concessions for salmon farm sites must provide a layout of the proposed facility and a plan of operation. An application can be denied if the Subsecretaria can demonstrate that the location and/or the plan of operation is unsuitable. In order to reduce the opportunity for lateral transmission of disease, farms under separate management must be no less than one and a half nautical miles distant from each other in salt water and 3 km from each other in freshwater. If there is a peninsula or other land mass between farms, the distance can be less. Net cages may be rotated from one place to another within concession boundaries in order to avoid excessive build-up of organic sediment beneath the cages.

Chile has no specific restrictions on fish density in a net cage or on the total weight of salmon to be produced in each concession. There are no restrictions pertaining to current and depth. A company may apply for as many concessions as it wishes. Stated annual production objectives per concession vary from a few dozen metric tonnes to several hundreds. One company applied for more than 60 concessions and has announced an eventual objective of 10 000 metric tonnes of salmon yearly. Concession rights are transferable with the permission of the Ministerio de Defensa Nacional.

The acquisition of permits for salmon farms in Chile is simpler and takes considerably less time than in most other salmon producing nations. Under normal circumstances, four months are all that has been needed to acquire all the necessary permits to begin operations.

Disease, parasites and phytoplankton blooms

No viral diseases of salmonids were known to exist in Chile in 1991. One outbreak of IPN (infectious pancreatic necrosis) occurred in the late 1970s in a trout farm as a result of a shipment of infected eggs from North America. The stock was destroyed, the facility disinfected, and no recurrences have been detected.

BKD (bacterial kidney disease) was first detected in 1970 in two trout hatcheries and the disease has been frequently found in salmon farms in the X, XI, XII Regions. Losses attributed to BKD have been as high as 50% in some farms among coho populations. In most cases, outbreaks have been controlled by medication and better management practices.

A disease caused by a native Chilean *Rickettsia* sp. has been recently identified in salt water cultured salmonids in Chile and is popularly known as 'huito disease', 'U.A.', or 'coho syndrome'. It has been found only in the X Region. This disease appears to be specific to coho and has caused losses as high as 80% in the worst affected farms. The causative agent has been identified in other species, but mortality has been low.

Growers report that coho derived from Chilean broodstock have much higher survival during *Rickettsia* epizootics than coho derived from imported eggs whose parents have not faced the disease. The apparent resistance to *Rickettsia* by coho from Chilean broodstock has provided a strong incentive to rely on Chilean self-sufficiency for coho eggs.

Sea lice infestations have occurred sporadically in coho salmon in net cages in the X Region. Some mortality has been experienced as a result of secondary infections from sea lice lesions. Furunculosis has been reported on at least one occasion, but no epizootics have occurred. Some pathologists question the validity of the identification of furunculosis in Chile.

Only one severe phytoplankton bloom, which occurred in September 1988 in the X Region, has been experienced in salmon culture areas in recent years. The bloom caused about 30% mortality to salt water growers. Some farms were unaffected, several incurred losses in excess of 50% and at least one faced a total loss.

In the past, severe phytoplankton blooms have been unusual in Chile. Should oceanographic conditions change and frequent reoccurrences of the 1988 bloom appear, the damage would be serious.

Future prospects

The coastal geography of southern Chile provides a large number of potential sites for salmon farms which are unusually well protected from storm action. Chilean salmon farmers have generally been able to locate their floating facilities in well-protected bays or channels. The lack of severe wave action allows a farm operator to safely build floating salmon cage facilities with lighter materials than are used in many Northern Hemisphere farms.

Overall, Chile has a very large number of locations which have the possibility of being excellent salmon farming sites. The salmon production potential of the country is roughly equal to that of Norway. A logistic infrastructure must be established before some areas are developed, but the Chilean government has a long-term record of encouraging new activity in isolated regions.

The X and XI Regions of Chile, the location of most of the existing salt water salmon sites, have a temperate maritime climate with high rainfall. Freshwater streams with adequate water flow, good temperatures, and good water quality are abundant. Many have sufficient relief to provide gravity flow to a hatchery and rearing enclosures. Some caution needs to be taken to avoid streams with excessive flooding, extremes of temperature, and very low pH. Lakes have been used to supplement stream-rearing water, but few new lake concessions are being permitted. Although fresh water temperatures in the XI and in particular the XII regions may be low for efficient rearing of Atlantic salmon juveniles, smolts can be moved to the more southerly salt water locations by well-boat. It is unlikely that a shortage of freshwater rearing sites will impede the development of Chile's salmon industry.

Chile has available extremely large volumes of low-cost marine fish protein and fish oils, which are the primary constituents in salmon feed, within the country's territorial sea. The country is a major world producer of these products. Feed is the highest single cost item involved in rearing full-sized salmon in cages, and in an efficient farm in Chile, feed costs are about 60% of the total costs up to harvest time.

Salmon farms in Chile have been operated by Chilean workers supervised by Chilean managers. Advice has been provided by foreign experts, but direct supervision is rarely needed. Labour in Chile is the least expensive of any nation presently raising salmon commercially. Daily costs for an individual to feed fish or maintain facilities is about US$10.00, including fringe benefits. The equivalent cost in the United States would be about US$70.00, and although Chilean salmon farm labour is not yet as efficient as equivalent labour in the Northern Hemisphere, it has proven competent.

In Chile, there has been relatively little mechanical equipment used at salmon farms. Much of the work usually done by mechanical and automatic equipment in the Northern Hemisphere is done manually. Reliance on manual labour in Chile exists for a number of reasons. First, the available labour force is inexpensive and capable of carrying out the tasks at hand. Secondly, there is a shortage of jobs in the regions of Chile suitable for salmon farms and sociological benefits are to be gained by providing more jobs. Finally, most mechanical fish culture equipment is manufactured in the Northern Hemisphere. Imported items are expensive in Chile and repairs to equipment which cannot be done within the country take extended periods of time.

Chilean fish farm managers have, in general, been Chilean nationals and have completed at least some higher education at institutions within the country. A number have had 'on the job' training at foreign salmon culture facilities. Salaries of Chilean managers are lower than for equivalent positions in the Northern Hemisphere, but the differential is not nearly as great as wages for daily workers. As a group, Chilean managers have done an exemplary job and this is reflected in the present success of the industry.

Three institutions of higher education in Chile provide four-year courses of study directed toward careers in aquaculture, including salmon culture.

Most salmon eggs used for farmed salmon in Chile up to 1990 have been imported from the Northern Hemisphere. While a major effort is being mounted in Chile to develop the country's own broodstock, there is still heavy reliance on imported eggs. Events in both the exporting country and in Chile can block the arrival of shipments of eggs and have done so in the past. Many of these problems become apparent only just before planned shipment departure. Finally, accidents and delays in transit can be disastrous to something as fragile as live salmonid eggs. Threats to egg imports include:

- Disease found during broodstock health certification.
- Shortage of eggs in the exporting country. Regulations in the exporting country frequently require bottom priority on foreign sales.
- Political pressures by competitors resulting in an embargo on foreign egg sales.
- Restrictions imposed by the importing country on certain species and stocks for disease and other reasons.
- Accidents and delays in transit.

Chile has an almost ideal situation for the successful farming of full-sized salmon. Biological conditions are good, protected locations are available, government authorities are supportive and costs are generally low. The outstanding drawback Chilean salmon producers face is the country's distance from most market areas. However, it costs little more to air-freight salmon from Chile to the United States than it does from Europe and shipment by sea is relatively inexpensive. In 1991, Chile was in a three-way tie with Canada and Japan as the third largest producer of farmed salmon in the world. In 1992, Chilean production equalled that of the United Kingdom and Chile shared the position of the world's number 2 producer.

References and selected bibliography

Aarset, B. (1988) Strukturen i norsk fiskeoppdrett og endringer over tid, report, FORUT (Foundation of Applied Research), Tromsø (in Norwegian).

Aiken, D. (1989) The economics of salmon farming in the Bay of Fundy, *World Aquaculture*, 20(3), 11–19.

Anon. (1987) Pen-raised chinook, *Bill Atkinson's News Report*, Issue 202. June.

Aqua Noticias International, A quarterly publication published by Fundacion Chile, Santiago, Chile.

Asada, Y. (1973) License limitation regulations: the Japanese system, *J. Fish. Res. Bd. Can.*, 30(12) Part 2, 2085–95.

Askildsen, J.E. and Hersoug, B. (1988) Naeringsstruktur i norsk fiskeoppdrett. Report No. 15/1988, Centre for Applied Research, Norwegian School of Economics and Business Administration, Bergen (in Norwegian).

Barnett, J. (1989) *Canadian Aquaculture*, Ocean Products, Inc. Sept/Oct. 1989, pp. 37–41.

Bjørndal, T. (1988) The Norwegian aquaculture industry. *Marine Policy*, 12(2), 122–41.

Bjørndal, T. (1990) *The Economics of Salmon Aquaculture*. Blackwell Scientific Publications, Oxford.

Bjørndal, T. and Schwindt, R. (1987) Norwegian direct investment in the British Columbia salmon aquaculture industry, Discussion Paper No. 1/87, Institute of Fisheries Economics, Norwegian School of Economics and Business Administration, Bergen.

Brett, J.R. (1952) Temperature tolerance in young Pacific salmon, genus *Oncorhynchus*, *J. Fish. Res. Bd. Can.*, 9(6), 265–323.

Bunya, T. (1983) Report on the inquiry into the strategies to combat self-pollution existing in the silver salmon aquaculture industry, sea water adaptation study. In *Miyagi Prefecture Kesenuma Fisheries Experimental Station Report No. 88*, pp. 57–61 (in Japanese).

Canadian Aquaculture (1990) March/April.

Department of Fisheries, State of Washington (1992) *1990 Preliminary Aquaculture Data*, Fisheries Statistical Report, Olympia, Washington.

Department of Fisheries and Oceans (1990) *Long-term Outlook for the Canadian Aquaculture Industry*, Department of Fisheries and Oceans, Ottawa.

Endo, N. (1985) On the Commercial Coho Salmon Culture, In: *Preliminary Report of the Salmonid Workshop on Biological and Economic Optimization of Smolt Production*, Versailles Summit Working Group, Tokyo, Japan, 22–25 January 1985. Ministry of Agriculture, Forestry and Fisheries, Japan. pp. 146–51.

Fiander-Good Associates (1988) *Economic Assessment of Salmonoid Cage Culture Industry in Southwestern New Brunswick*. Fredericton, New Brunswick, Canada.

Gjøvik, J.A. (1987) The state of salmonid farming in Tasmania, report, Noraqua Technologies Pte, Ltd.

Hallenstvedt, A., Hersoug, B. and Holm, P. (1985) Norsk oppdrettsnæring – organisering for vekst, in *Havbruk, Mulighetenes Næring*. Universitetsforlaget, Oslo, pp. 63–134 (in Norwegian).

Harache, U., Boeuf, G., and Lasserre, P. (1980) Osmotic adaption of coho salmon (*Oncorhynchus kisutch*) Walbaum. III. Survival and growth of juvenile coho salmon transferred to sea water at various times of the year, *Aquaculture*, 19, 253–73.

Hersoug, B., Holm, P., Maurstad, A. and Aarset, B. (1991) *Fra vekst til Krise*. Norwegian College of Fisheries Science, Tromsø (in Norwegian).

Holm, P., Jentoft, S. and Steene, B. (1990) *Norsk oppdrettsnæring ved inngangen til 90-åra*. Kystkompetanseutvalget, Tromsø, pp. 54–71 (in Norwegian).

Hortel, M. (1986) Salmon farming in Tasmania, CSIRO Marine Laboratories, seminar notice, Hobart.

Ireland's Fish Farming Journal (1991), No. 49, Autumn.

Iwata, M. and Clarke, C. (1987) Culturing coho Japanese-style, *Canadian Aquacul-*

ture, 3(1), 28–31.

Jungawalla, P.J. (1989) *The Development of an Integrated Salt-Water Salmonid Farming Industry in Tasmania*, Altas, Hobart.

Kiyama, K. (1983) Report on the inquiry into the strategies to combat self-pollution existing in the silver salmon aquaculture industry. Rate of conversion of feed. In *Miyagi Prefecture Kesenuma Fisheries Experimental Station Report No. 88*, pp. 41–7 (in Japanese).

Mahnken, C. (1982) The application of recent smoltification research to public hatchery releases: an assessment of size/time requirements for Columbia River hatchery coho salmon (*Oncorhynchus kisutch*), *Aquaculture*, 28, 251–68.

Mendes, R.Z. and Munita, O.C. (1989) *La Salmonicultura en Chile*, Fundacion Chile, Santiago, Chile.

Mottet, M.G. (1980) Factors leading to the success of Japanese aquaculture, with an emphasis on Northern Japan. *Washington Department of Fisheries Technical Report 52*, Washington Department of Fisheries, Olympia, Washington.

Nasaka, Y. (1987) Trip report on Japan's coho salmon pen culture, US Embassy Memorandum, Fisheries Attaché, 12 August 1987.

Norweigan Salmon Farmers Association (1992) Annual Report for 1991, Trondheim.

O'Toole, C. (1988) Report of a study trip to Japan. II. Salmonid aquaculture and fisheries. University of California Sea Grant Extension Program, Working Paper No. TT-48.

Perspectivas de la Salmonicultura en Chile (1987) Report of proceedings of 'Seminario International organizado por el Depto. de Recursos Marinos', Fundacion Chile, Santiago, Chile.

Porter, H. (1990) Cobscook Bay Area. Aquaculture Impact Brief, (mimeo).

Province of British Columbia, Ministry of Agriculture and Fisheries (1991) *Fisheries Production*, Statistics for British Columbia, 1989.

Reinert, A. (1985) Akvakultur – Færøerne, paper presented to Twenty-First Nordic Symposium in Aqua Research, P/F Fiskaaling, Torshavn, Faroe Islands.

Ridler, N. and Martin, M. (1990) Salmon farming in an international context, *ASAC Conference Papers*, Whistler, British Columbia.

Saunders, R.L. (1989) Salmonid aquaculture: present status and prospects for the future. In *Cold-water Aquaculture in Atlantic Canada* (Ed. by A. Boghen), Canadian Institute for Research on Regional Development: Moncton, New Brunswick, Canada.

Standal, D. (1987) Internasjonaliseringstendenser i norsk oppdrettsnæring. En studie av norske utenlandsetableringer med hovedvekt på vertslandene Skottland og British Columbia, Kanada. Master's dissertation, Norwegian College of Fisheries Science, Tromsø.

Stefánsson, S.E. (1993) Framleidsla in Islensku Fiskeldi Árid 1992, Eldisdeild Veidimalastofnum, Reykjavik (report, in Icelandic).

Technicas de Cultivo y Manejo del Salmon: Desarrollos Recientes (1988) Report of proceedings of 'Seminario International organizado por el Dept. de Recursos Marinos', Fundacion Chile, Santiago, Chile.

United States Department of Commerce (1988) *Aquaculture and Capture Fisheries: Impacts on U.S. Seafood Markets*, National Marine Fisheries Service, Washington, DC.

Washington Fishgrowers' Association (1992) Country report. Olympia, Washington.

Yoshida, F. (1985) Seawater culture of coho salmon in Japan. In: *Preliminary Report of the Salmonid Workshop on Biological and Economic Optimization of Smolt Production*, Versailles Summit Working Group, Tokyo, Japan, 22–25 January 1985. Ministry of Agriculture, Forestry and Fisheries, Japan. pp. 137–42.

Chapter 3
Developments in Salmon Aquaculture Technology

LUDVIG KARLSEN

Introduction

New developments include improvements in the technology of traditional net cage farms and particularly in designs for net cage farms in more exposed marine locations. The design and commercial scale testing of closed-wall cage farms is continuing. In addition, work is being done on the design and problem resolution for land-based farms supplied with pumped seawater.

The three main reasons behind the need for technological change are the environmental and predator problems faced in many current sites, the desire to improve the efficiency of existing technology and the shortage of good sheltered coastal sites in many regions around the world. Environmental problems for the fish include vulnerability to predators and poor growth conditions which are mainly due to insufficient water exchange, giving rise to low dissolved oxygen and high levels of metabolites and pathogens.

One major factor which contributes to the need for technological development of open sea cages relates to prior claims for the most desirable sheltered cage farming sites. Traditional cage design is inadequate for more exposed localities with respect to structural strength, endurance, safety and working conditions during periods of severe weather. Conventional net cage farms in sheltered localities frequently have faced diverse predator problems which include birds, animals and humans, resulting in substantial loss of fish.

Experience has shown the benefits of deeper nets which allow the salmon to choose the most desirable water depth with respect to temperature, water quality and light intensity. An increased net cage size, typically increasing the diameter from 15 to 30 metres, which corresponds to expanded surface areas from 175 to 700 square metres is also advantageous. The use of more than one farm site and more frequent use of alternate locations within the farm has helped growth rates and reduced disease and pollution problems. Norwegian authorities now allow the use of several localities and encourage rotation between alternate sites.

Current research includes frequent model and full-scale testing of net cage designs and configurations to examine the influence of wind, waves, ocean currents and ice on the shape of net walls, volume changes within the cage, water velocity and pattern within the cage. The total effects of forces on single cages and the entire sea-based farm complex have also been examined. The development and application of new service equipment includes net washers, fish sorting devices and new work boat designs. Investigations of new feeding techniques and operating procedures are also being carried out.

Land-based grow-out farm research includes investigations into water flow patterns as a function of tank form, water inlet location and rate of flow. Oxygen addition systems for closed and semi-closed water circulation farms are being improved. Continuing problems include large-scale cleaning of waste water coming from closed system grow-out farms as well as treatment and possible utilization of farm waste.

Smoltification plant technology

In the construction of facilities for fingerling and smolt production, technological development has been more limited. This appears to be due to the fact that the design of fingerling and smolt production plants tend to be restricted by tradition in countries with programmes for the enhancement of salmon rivers.

Most existing smolt facilities use circular fish tanks, but raceways are favoured by some. Experimental and commercial research has focused primarily on tank design, water inlet and outlet location to improve oxygen distribution, removal of ammonia from the water, and waste water purification. Water flows which encourage swimming exercise are also being evaluated.

To achieve the necessary warm water for the one-year and half-year smolt production, both energy equipment for water heating and recirculation systems are alternatives.

The primary components in the energy plants are the heat exchanger and the heat source, as shown in Fig. 3.1. The use of heat exchangers to utilize the heat capacity of the waste water has long traditions in smolt production, in addition to traditional heat sources such as oil burners, electric heaters, or cooling water from industrial plants.

Use of heat pumps in modern smolt plants has proved economically favourable compared to the more traditional heat sources.

In Norway more than 80% (220 out of 270) of the smolt plants with energy systems are equipped with heat pump and plate heat exchangers (Gebauer *et al.*, 1992). Various heat system layouts are in use, including heating of the waste water

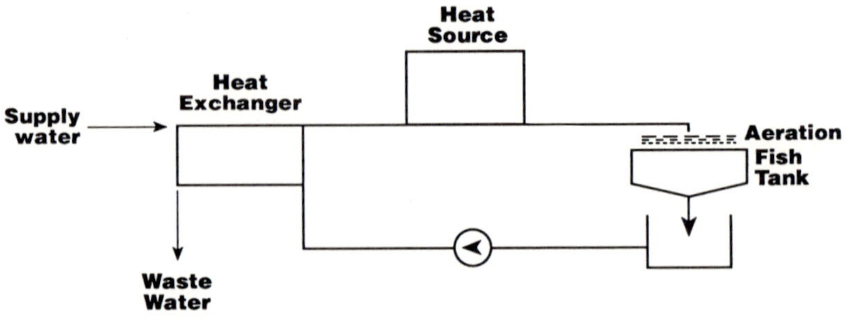

Fig. 3.1 Heat exchanger and heat source.

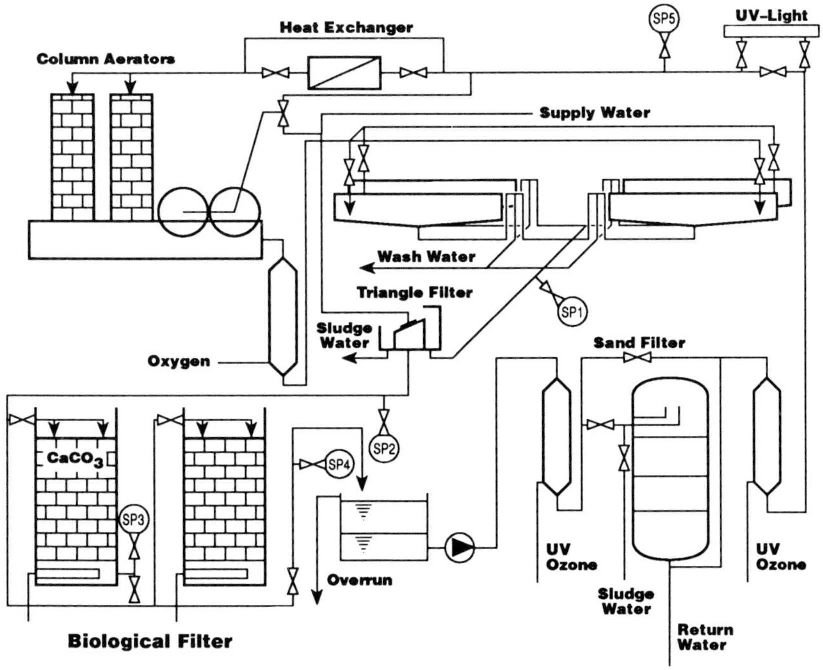

Fig. 3.2 Sketch of pilot scale recirculation equipment for salmon smolt production (from Rusten, 1989).

before the heat exchanger or heating the supply water after the heat exchanger (Fig. 3.2). Other combinations of heat sources such as, for example, heat pump, and electrical heating are in use.

Recirculation systems for smolt rearing have become increasingly important in many countries. In principle, a recirculation system consists of equipment for treatment of the recirculated water, a supply of fresh water, and discharge of waste water.

Recirculation systems have various degrees of water treatment and therefore complexity, dependent on the circulation rate. A conventional system with a low recirculation rate typically includes separation of solid particles and waste water by means of a strainer or (triangular) filter, a sand filter for finer particles, disinfection, addition of fresh water, temperature control/eventual heating and aeration/oxygenation.

A system with a high recirculation rate in addition includes removal of ammonia, either by use of biological filters or by ion exchangers. When using a biological filter it should be placed between the mechanical filter and the fine separation (sand filter). Fig. 3.2 shows a system tested in a research project (Rusten, 1989), which includes:

- A mechanical (triangular) filter for removing large particles.
- Biological filters for removing ammonia.
- Sand filters for removing small particles.
- Disinfection by means of ultraviolet-generated ozone gas and ultraviolet light.
- Column aeration and oxygenation.

Calcium carbonate slurry was used in the high acid incoming water to increase pH-values (Fig. 3.2).

The quality of recirculated water in this facility was found to be very good and intake water flow was only 5% of that of a plant without recirculation (Rusten, 1989).

An alternative to the conventional recirculation systems is to treat and recirculate the water for each tank separately, and also entirely inside the tank. The BIOFISH water recirculation system, developed and tested in Norway, includes a submerged and aerated biofilter, located in the centre of the fish tank, both oxidizes ammonia and adds oxygen to the water. The aeration system creates a lift which maintains the necessary water circulation in the tank while allowing an important reduction in water consumption (Eikebrokk, 1990).

Trials with wheel filters have shown very good performance in both supply and waste water cleaning at smoltification plants. An average of 90% of the feed and faeces particles in the waste water was removed. The wheel filter has also shown good efficiency in removing algae. A special modification of these filters has allowed them to be used for the collection of zooplankton as feed for marine fish larvae (Garmannslund, 1988).

The most recent development in the field of waste water cleaning is to add particle (feed, faeces) traps in the bottom or water outlet for circular fish tanks. By removing the concentrated sludge using only a small portion of the water outflow the cleaning treatment is a more reasonable task than dealing with the total flow. A similar system has also been tried as a dead fish trap, with dead fish which have sunk to the bottom being removed with the water outflow. Research plants by a Norwegian research organization (SINTEF), in the field, have so far shown good results.

An investigation into the effect of double light periods, light intensity and colour within fish tanks on the growing rate and smoltification of salmon parr has been done in Norway. The findings indicate that fish in grey tanks with increased light intensity had a higher growth rate than fish with normal indoor light intensity in dark green tanks. The smoltification process was found to be highly influenced by the light intensity and it was concluded that smoltification would continue during double light periods (Stefansson & Hansen, 1987).

Net cage farms

Net cage farm systems are the dominant production mode in salt water salmon farming. In Norway, more than 99% of grow-out salmon production comes from net cage farms. The main advantages of open-mesh net cage systems designed for sheltered site localities with shore-based service are low capital costs, ease of operation, simple and well-proven technology, and the option for gradual expansion of plant facilities. The latter allows lower capital demands at start-up and hence less complications in financing. A number of different net cage types and constructions are offered by various manufacturers, some of which are shown in Fig. 3.3.

Traditional approaches to Norwegian net cage systems extend back to the 1960s. While improvements have taken place over the years, the basic principle of keeping salmon in net pens attached to a floating unit remains the same.

Recent development trends and research activities dealing with net cage farm

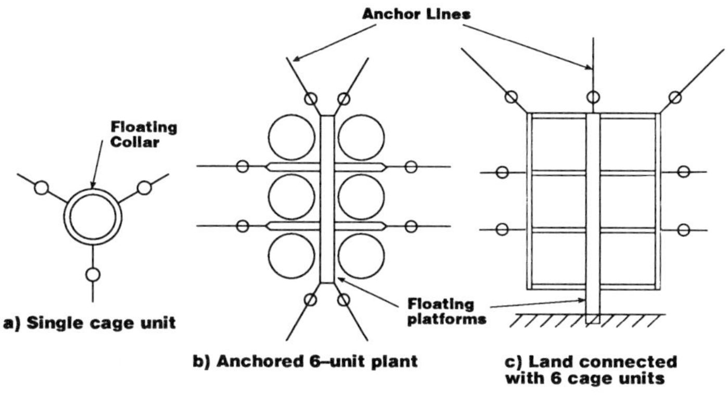

Fig. 3.3 Different types of net cage plants.

systems have concentrated on farm layout and positioning of cages, and on the configuration and size of floating units. Farm system components, including cages, cage materials, moorings and land connections are under investigation. Model testing of nets is also being carried out.

Layout and position

Net cage fish farms may have different layouts in terms of internal positioning of cages, mooring systems and connections to land. Early farms utilized single cages with separate anchoring systems. The principal advantage of this system is minimal influence on the flow and quality of water. Disadvantages include more labour expended in the servicing and husbandry of the separate units.

The converse of a single cage system is a system made up of many cage units secured to a central walkway or working platform anchored by a cohesive mooring complex. The central walkway or working area allows much easier servicing of a large number of individual cage units. Many farms have a shore connection for service, sometimes with motor vehicle access. The impetus for development towards such a compact unit solution is improved working conditions for the farm staff rather than the best environmental conditions for the fish. In many cases, cages down current will have reduced water quality due to shadowing effects from up-current nets. Other negative effects of compact cage systems can be severe pollution problems due to waste and food particles accumulating on the bottom below the cages and concentrated pathogens drifting from one cage to the next.

In Norway, many compactly designed net cage systems have experienced serious problems due to waste accumulation on the bottom beneath the nets. Waste accumulation problems have, in many cases, been solved by avoiding overfeeding and sometimes by the installation of current flow generators. The trend is now towards more spread-out positioning of cage units, often back to the original single cage approach with each cage 50–100 metres from neighbouring units.

Another solution has been a more frequent rotation of cage units to other locations, either within one site or between alternate sites. While results are not well documented by scientific evidence, many farms appear to have increased fish performance and thereby more total salmon production after initiating rotation.

Materials used in construction

Materials used in the floating units of salmon farm facilities include plastic (high density polyethylene), rubber, steel, aluminum, wood and concrete. To date, few controlled tests have been conducted to determine which materials are the best for any specific site. However, each farm operator must select facilities and materials which best allow that farm's salmon product to compete in the market place. Economic factors including price, financing, and efficiency of operation are of crucial importance in the farmer's choice of materials in facility construction.

The physical characteristics of materials used for floating facility construction must match the design requirements of the facility. Plastic and rubber are used for flexible cage frames, while steel and aluminum are used for rigid frames or frame sections. Plastic and rubber cage frames are hollow (mainly pipe structures) and hence self-buoyant, while steel and aluminum must be floated by separate buoyancy materials such as pontoons or polystyrene billets. Cages with separate buoyancy units are somewhat more vulnerable to wave damage, but do allow much higher freeboard than pipe structures.

Shape and size of floating units

The shape of cage frames and hence the internal space can be circular, octagonal, hexagonal, square or rectangular. Cage shapes are usually selected according to the planned layout of the farm. Individually anchored cage units are usually circular, octagonal or hexagonal to maximize volume enclosed within the nets. On the other hand, square cage frames fit more efficiently along a central walkway or working platform servicing several cages.

There are two important factors which should be considered in the selection of the shape of floating frames for rearing units. Firstly, there are distinct differences in the ratio of net area to cage volume (given the same net depth). The circular form is the most efficient, allowing for 12.8% more volume for the same net area than the square form. Calculations for other shapes are shown in Table 3.1.

When comparing the various cage shapes it should be noted that the circular shape allows for better utilization by the fish of the entire cage volume than shapes with corners. Secondly, when placed along a walkway or platform, circular cages will have larger sea water volumes between the cages, allowing for better water flushing than with square pens placed closely side by side.

Table 3.1 Increased net area/volume ratio for various cage shapes (same net depth), compared to the circular shape.

Cage collar form	Increased net area %
Circular	0
Hexagonal	5
Square	12.8
Rectangular, side ratio 4:1	41

Design characteristics and hydrodynamics of single cages: model tests

A recent and important development for net cage farms is the trend towards the use of much deeper nets. The use of deep nets is also associated with a shift to larger cages (surface area) and the return to single moored or widely separated cage units. Net depths have increased from the traditional 4–5 metres to 10 metres and in some instances to as much as 20 metres at many Norwegian farms. Water depth and temperature profiles at a particular site often determine optimum net depth. The increase in net depth has resulted in considerably larger drag forces on nets from currents and therefore more holding power is required from the anchor system.

At MARINTEK in Trondheim, Norway, model tests of different cage shapes and sizes have been carried out. Results have been recorded on video tape and film as well as on paper. These model studies, supported by mathematical calculations, have defined the influence of current velocity on net shape and deflection, including effects on cages masked by up-current cages. When cage nets have small mesh and hence high drag, the current velocity through a cage fourth in line may be reduced to only 15% of the free flow outside. In addition, net deflection for down-current cages is minimal compared to the cage directly facing the flow (Slaattelid, 1990, Aarsnes *et al.*, 1990).

Mathematical model studies of wave forces on sea cage components have been performed and have shown that vertical loadings rather than horizontal force may be the critical factor under some circumstances (Cairns & Linfoot, 1990). Model studies with a Wavemaster cage system in a test tank equipped with wave generators have also focused on cage behaviour and seaworthiness. These studies included mooring failure tests (Whittaker *et al.*, 1990).

A general conclusion from the above-mentioned papers is that model testing gives a realistic and cost-effective alternative to actual field testing in evaluating the effects of forces on net cage systems. Although some uncertainties remain about the influence of scale, tests with relatively large models are presumed to fit closely with full-scale conditions.

Model tests in general are performed with clean cage components. The effects of marine fouling must be considered in view of the dramatic influence fouling exerts under marine conditions, both with respect to vertical forces (weight on floating structures and nets) and the increased hydrodynamic area of net mesh. Beveridge, (1987) refers to extreme net fouling resulting in up to fourfold increases from original twine diameter giving a corresponding increase in the current drag coefficient from 1.42 to 3.99. However, the actual effect of net fouling is dependent on both the fouling growth as well as on net cleaning procedures, and effects may be minimized in practice by good husbandry.

Net cage designs for exposed waters

Net cage salmon farm operators in Norway and elsewhere have faced significant environmental problems when sited in sheltered locations with low water flushing and shallow depth. In addition, finding suitable new sheltered sites for salmon farms has been increasingly difficult. Accordingly, substantial interest has focused on salmon rearing systems that are designed for operation in more exposed marine

Table 3.2 Some net cage designs for salmon farming in exposed waters.

Manufacturer	Type/shape	Material
Aquasystem	Barge	Steel
Bridgestone	Single cage	Rubber
Fishtechnik/Farmocean	Single structure	Steel
Polar Gigante	Single ring	Plastic
Net Systems	Tensioned spars	Steel

locations. After intensive model testing and full-scale field trials of prototypes, several different offshore units are available today, some of which are listed in Table 3.2.

Aquasystem, Bridgestone and Polar Gigante has free hanging nets below the surface components, thus depending on the traditional approach of weights secured to the bottom to reduce volume decrease in strong currents. Another approach, effective in exposed locations, employs a weighted pipe ring secured around the net bottom to help maintain the shape and volume of the net in severe conditions. More recently a new solution has been presented by NET Systems, Inc, in Washington State based on separate tensioned spars which float vertically outside the perimeter of the net. By fastening the bottom as well as the top of the net to the spars, the net cage is better equipped to withstand currents without the volume distortion experienced with free hanging nets.

Other developments

Several other research and development activities related to net cage salmon farms have been carried out. Full scale trials funded through the Norwegian Fisheries Research Council (NFFR) and performed by the Institute of Fishery Technology Research, concluded that a surprisingly small number of Atlantic salmon smolt would escape through a hole in the side of a net pen. On the other hand, Pacific salmon smolt recently transferred to salt water will seek out small holes and escape in large numbers. After a few weeks or months in residency in net cages, however, the urge to escape is greatly reduced. (JML experience in Puget Sound, Washington).

Shade covers over production size salmon net pens have shown a marginal reduction in salmon growth during winter and spring and a marginal growth increase in summer and early autumn. The use of shade covers has been accompanied by a slight reduction of salmon lice infestations and a significant reduction in algal fouling (Huse *et al.*, 1990). A number of 'mortality collectors' have been developed and put in use, with Fig. 3.4 showing one example. The development and testing of sludge bags for waste collection continues, as shown in Fig. 3.5.

The newest development in this field is a combined mortality and feed waste collector. In addition, work continues on current flow generators, water aerators, and the development of bar graders for use within cages (Fig. 3.6).

Closed-wall sea based farm systems

The farming of fish in closed-wall cages or enclosures, floating on the sea surface in a manner similar to traditional net cages, has been regarded as a possible alternative

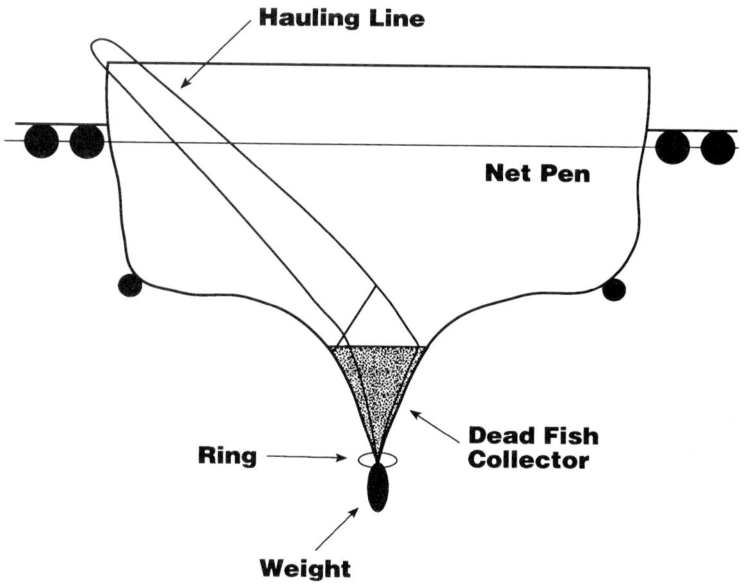

Fig. 3.4 Mortalities collector.

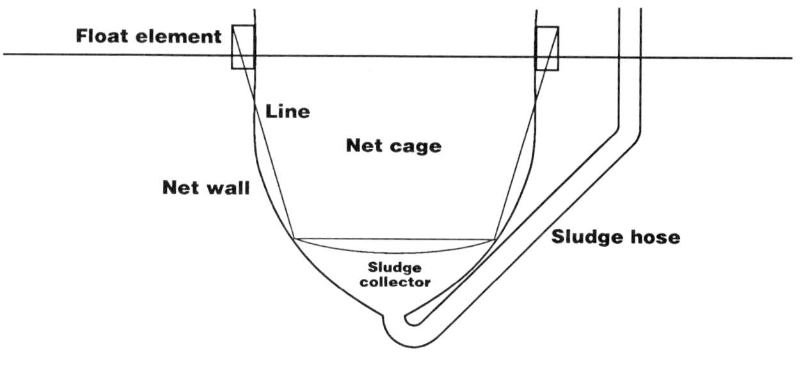

Fig. 3.5 Sludge collector arrangement.

for some years. Closed-wall cages allow much better control of water quality when compared to conventional net cages, as they can use intake water from any part of the water column where noxious phytoplankton, parasites and disease organisms are minimal and temperatures are more optimum. The closed walls also allow easier protection from predators. Disadvantages include higher capital costs, higher energy use and higher threat of sudden catastrophic system failure. Supporters of closed-wall cages believe advantages outweight disadvantages and hope these systems will play an increasing role for future salmon farm development. In Norway, two different designs for closed-wall cages have been developed and field tested under production conditions. Both designs utilize large bags of impermeable fabric to enclose the fish, which are attached to floating collars of the same type used for standard cage units.

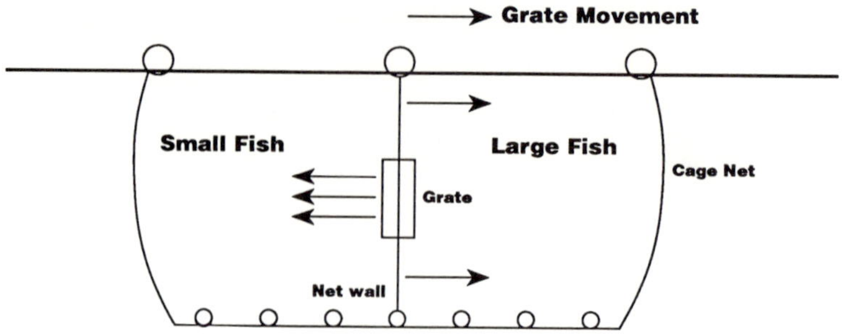

Fig. 3.6 Functional principle of the salmon size grader.

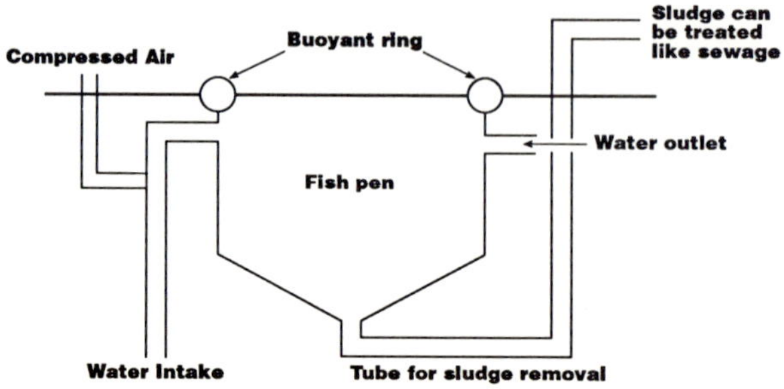

Fig. 3.7 The Marenor closed floating fish pen concept.

In addition, closed-wall cage facilities pump sea water through an adjustable intake, allowing selection from anywhere in the water column. Further development may allow the collection of solid waste for disposal away from the farm site.

The first closed floating cage system was designed by Marenor, a Norwegian commercial company, after extensive research and development work in cooperation with Norwegian universities and assisted by the Norwegian government. The original Marenor system (Fig. 3.7) consists of circular cages of variable sizes. A typical example is a 500 cubic metre unit having a diameter of 11.5 metres and a depth of 5 metres.

A commercial scale farm designed by Oppdrettsservice, A/S, with cages of 8–11 metres using the Marenor concept, was installed in southern Norway to gain practical experience.

A second concept for closed-wall enclosures has been called a floating raceway system. Such a system has been designed by Ocean River Systems (Fig. 3.8). In addition to taking advantage of a seawater supply from selected depths, the water velocity generated in the Ocean River System is claimed to promote better swimming exercise by the fish and hence a better quality product.

The Ocean River Systems design consists of two parallel water compartments, each 50 metres long, 4 metres wide and 5 metres deep holding a total volume of 2000 cubic metres. The walls and the bottoms are made of reinforced impermeable fabric. A portion of the outlet water from one raceway is passed to the inlet to the

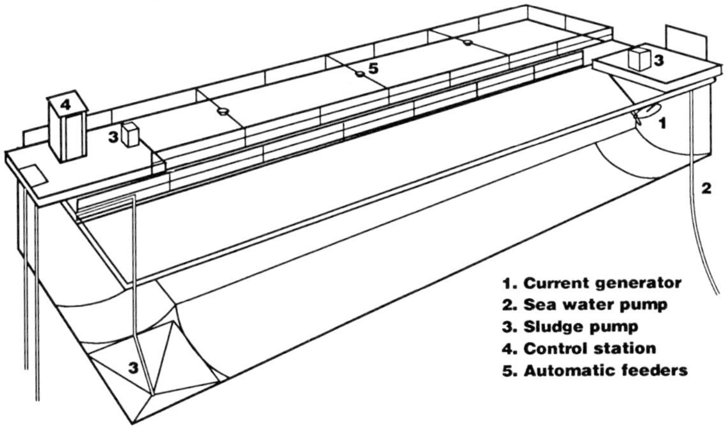

Fig. 3.8 Prototype floating raceway (2000 cubic metres).

other thus providing considerable re-use. Fresh seawater is supplied by two pumping stations with a total capacity of 800 litres per second. In addition, two flow generators keep the desired water velocity within the raceways.

Early trials of the floating raceway concept at a commercial scale have indicated good growth rates as well as good fish quality. However, further development will be needed to bring the system into commercial production.

Designs and materials

A typical closed-wall circular cage (Marenor) is shown in Fig. 3.7. The upper section of the cage walls are vertical to achieve optimum volume while the lower section is conical so as to concentrate waste material at the bottom. The exact configuration of closed-wall units need to be carefully evaluated and tested to determine optimum water flow rates, flow distribution, water exchange rates, settleable solid separation, and fish density to achieve healthy fish at acceptable cost.

Theoretically, the material in cage walls may either be stiff or flexible (impermeable fabric). Only flexible materials have been used in existing Norwegian designs.

The floating collar of a closed-wall circular cage may be of various designs, comparable to standard net cage systems. The original Marenor system used buoyancy elements employing polystyrene. When designing and calculating buoyancy, the comparatively greater water density to be expected within the cage due to higher salinity water pumped from a deep water source must be considered.

$$W = 9.81V(\rho_i - \rho_o)N$$
$$V = \text{cage volume in cubic metres}$$
$$\rho_i = \text{inside water density in kg/m}^3$$
$$\rho_o = \text{outside water density in kg/m}^3$$

To give an example of buoyancy compensation requirements, a water density difference of 5 kg per cubic metre between water inside the cage and outside (i.e.

approx. 0.5%) will require additional buoyancy of 2500 kg for a 500-cubic metre cage.

Water supply

The seawater source and flow distribution are critical factors in the success of floating closed-wall fish culture systems. Water flow and velocity must provide sufficient oxygen to all parts of the system and to carry away wastes. Two different approaches for providing seawater supply have been used. Marenor tried an air-lift pump system and Oppdrettsutvikling along with Ocean River Systems utilizes submerged centrifugal pumps. In both cases incoming water is carried through separate pipes to the various cage units, and then enters the circular cages through a specially designed valve.

Incoming water for closed-wall cages can be drawn from any level of the water column in order to take advantage of the most desirable water conditions available. Consideration should be given in particular to temperature and suspected concentrations of pathogens. Commercial closed-wall systems appear to be feasible only if a deep water supply is available.

Another potential advantage of closed-wall seawater rearing units is that settleable solid waste might be collected and disposed of away from the farm or pumped to a sludge tank for further treatment. Practical means for treating large volumes of waste (sludge) removed from a commercial scale salmon farm have not yet been developed. Research programmes focusing on sludge treatment and possible uses are being started by Norwegian research institutions (SINTEF).

Influence of waves and current

Closed-wall floating sea-based rearing units are far more affected by hydrodynamic forces generated by ocean currents and waves than facilities using open mesh nets. Both theoretical calculations and model test have been performed to study hydrodynamic effects of closed-wall cages. It has been found that wave- and current-generated forces can be approximated by means of simple hydrodynamic formulae. Wave-generated forces are relatively more important for closed-wall enclosures than for open-mesh net pens which have much smaller float units (Oltedal, 1989). This problem has not yet been experienced in field tests of full-scale closed-wall facilities.

Land-based salmon farm technology

The commercial farming of full-sized salmon in land-based pumped seawater facilities is in an advanced stage of development as an alternative to sea cage farming. Several land-based facilities in both Europe and North America have been put into operation, although production still counts for only a small percentage of total farmed salmon production. The magnitude of problems faced by many floating cage farms today relating to the problems of operating at sea, pathogens, environmental contamination and the difficulty in finding suitable locations may encourage more interest in land-based facilities.

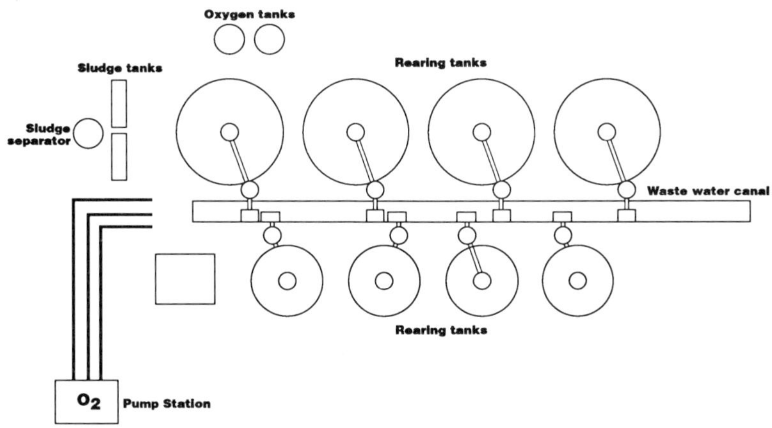

Fig. 3.9 Plant layout of Aquatech Systems' new land-based salmon farm.

Much more research and development activity related to land-based salmon farming has taken place during the last ten years in several countries including Scotland, Iceland, Norway, Canada and Denmark. Facility construction has largely been oriented towards commercial production but opportunities also exist in research and development.

The first full-scale land-based facility in Norway to carry out research and commercial scale development was established in 1985/86 by Maritech (now Aquatech Systems) at Averoy with a total fish tank culture volume of about 2500 cubic metres. Subsequently, four other facilities with fish culture volumes between 2000 and 12 000 cubic metres have been put into operation. Aquatech Systems established a new division of their farm in 1991 with 4500 cubic metres of culture volume (Fig. 3.9). Two other commercial farms with both land-based (500 and 1000 cubic metre volume respectively) and net cage production have begun operation.

Primary research and development activities related to land-based farms include location and farm layout, production systems, fish tank size and shape, water flow systems, inlet and outlet design and supplemental oxygenation.

The main purpose of investigations has been to find out whether land-based farms can compete with sea-based net cages by stocking higher fish densities, increasing growth rates, and reducing disease and fish loss problems through a better water environment and improved predator control. Some recent research has been directed towards determining if fish flesh quality is better in land-based farms due to forced swimming exercise in the tanks. And if flesh quality is better, will the market pay a higher price for these fish?

Types and designs of land-based farms

Designs for fish culture enclosures for land-based farms include circular tanks, raceways and one closed pipe system. Five Norwegian land-based farms using full-scale facilities use round tanks, while one has a raceway in combination with sea-based production. One prototype of a closed pipe system ('Preline') has been under test.

Large circular tanks with spiral flow have shown good performance characteristics in terms of uniform water flow which permits excellent fish distribution, high fish densities, and effective self-cleaning. The cost per unit of volume and area compares favourably with other designs.

The more recently introduced 'Preline' closed pipe system offers a new concept for land-based salmon production. Fish culture takes place in partly buried pipes with a diameter of 2.5 metres, allowing for a volume of about 1000 cubic metres per 200 metres of length. The 'Preline' production unit provides for supplemental oxygen, feeding stations and 'traps' for waste removal at intervals along the pipe.

Location and farm layout

Location and plant layout for land-based fish farms become fixed factors after construction. They are unlike floating sea-cage facilities which can be moved temporarily or permanently, reorganized internally, or divided into parts for multiple locations. A critical prerequisite for land-based pumped sea water farms is easy access to seawater supplies of good quality at 50 to 100 metre depths. The availability of deep water allows the farm to take advantage of higher temperatures during winter (6–8°C compared to 1–3°C for Norwegian coastal surface water). In addition, deep water usually has far less toxic phytoplankton organisms, pathogens and parasites than surface water.

Fish culture enclosures should be at the minimum safe elevation to control pump power expenses and sea water intake pipelines should not exceed 500 metres in length to avoid excessive capital and maintenance costs. The movement of live fish among the various tanks should be considered. The efficiency of a land-based facility can be greatly enhanced if buildings, fish culture enclosures, water supply systems, and auxiliary equipment are laid out to optimize operational procedures as well as capital and operational costs.

Number and size of fish tanks

The number and size of circular rearing tanks in a land-based salmon farm is an important design factor to be considered in the production plan. More recent land-based farms in Norway have reduced the number of tanks in comparison to earlier farms in Scotland and Iceland. New farms have generally from six to eight tanks and only one farm has more than ten. Most newer farms have only one or at most two tank sizes. Total water volume used for fish culture in new farms is less, although salmon production has not been reduced proportionally due to the use of supplemental oxygen and higher fish densities.

Experience has shown that tank diameter should not be too great, as low water input flow can create poor circulation and unreliable self-cleaning problems in very large tanks. This is of particular importance for farms using supplemental oxygen. The most common tank diameter in more recent farm designs in Norway using supplemental oxygen is between 12 and 18 metres. Tank diameters for farms planning no oxygenation are up to 20 metres. Water depth in tanks varies according to the diameter and is usually between 3.5 and 5.5 metres. Diameter/depth ratios less than 2.5 are to be avoided.

By comparison, Scottish and Icelandic land-based farms built several years ago have fish tank diameters of 25, 30, and 42 metres. Tank volumes in recently designed farms are less than 1000 cubic metres per tank while earlier Icelandic and Scottish farms have single tank volumes of 2000, 4000 or 6000 cubic metres (Tvinnereim, 1990).

Shape of tanks

With respect to tank shape, circular and multi-sided tanks (more than 15 side elements) have demonstrated good circulation characteristics with respect to oxygen distribution, water exchange and removal of settleable solids. The slope of the bottom has not been found to be a critical factor for self-cleaning tanks. The design of water inlets and outlets is very important. Hexagonal and octagonal tanks, used by some farms, shows satisfactory water flow characteristics, but square and rectangular tanks should have rounded corners.

Tank materials

Materials used to fabricate tanks include steel, concrete, reinforced plastic and in some cases impermeable fabric supported by an outside wooden frame. For tanks with diameters less than 8 metres, PVC is regarded as a good choice. Larger tanks are often made of prefabricated concrete elements. An advantage of concrete tanks is that they can easily withstand outside loading, a must when fill material is banked against the outside tank wall to establish an optimum working level (about 0.5 metres below the tank rim).

All tanks must have a smooth interior surface for ease of cleaning and to avoid abrasions to the fish. The inside surfaces of concrete tanks must be coated with epoxy paint or covered with PVC fabric.

Sea water supply and distribution

When planning the seawater supply system, pumps must be sized to allow for pressure variations in the supply water distribution system and in a possible waste water treatment system. In particular, excess pressure capacity of 0.3–0.4 bars should be included for possible supplemental oxygen systems.

The number of pumps and supply water pipelines is predicated on the water requirement and planned back-up capacity. Three or four pumps are typical for a farm with 5000 cubic metre fish culture volume. Submersible pumps placed in pump wells installed three or four metres below the lowest tide will usually provide a reliable water supply.

Water oxygenation

The decision to install supplemental oxygenation equipment can have a major effect on the profitability of a land-based salt water salmon farm. Important factors which should influence this decision are general quality, temperature, and minimum oxygen content in the sea water supply. Costs of water delivered to fish culture enclosures, including the capital cost of the total water supply system, pumping head and the safety value of an oxygen back-up should be considered. Recent research and experience from Norwegian land-based facilities show that the use of supplemental oxygen is essential to maintain high fish density and at the same time maintain a satisfactory growth rate.

Recent research has shown that when sufficient dissolved oxygen levels are maintained in culture water, much higher fish densities can be reared at optimal growing conditions in large tanks than previously believed. Fish densities of at least 70 kg per cubic metre are feasible compared to maximum densities of 30–40 kg per cubic metre which can be achieved without added oxygen. It may be possible to supply more oxygen by simply pumping more water, but this is usually more costly than adding supplemental oxygen.

A substantial research programme is in progress at three land-based farms in Norway to define the best methods and equipment for supplemental oxygenation. Early results indicate the deep-shaft system has better efficiency than the cone system.

The technology of fish transport and handling

Transportation of fish and fish products is an important activity in modern salmon aquaculture, and the need increased greatly with the growth of large-scale salmon production. The following transportation requirements are integral to the operation of a salmon aquaculture business:

- Eggs and sperm must be moved to or within hatcheries.
- Fry and fingerling must be moved to and within freshwater facilities, and smolt must be moved from freshwater facilities to grow-out facilities.
- Movement of large fish among or within cage farms.
- Movement of harvest fish from farms to processing plants and the shipment of salmon products to markets.

Smolt almost always must be moved because freshwater rearing facilities are usually separated from salt water grow-out facilities. Frequently it has been necessary to send shipments of smolt more than 1000 kilometres. The transportation of fish to processing plants has become more important because of a tendency towards centralized processing. An emerging development is the transportation of live fish to processing plants or near markets before slaughter.

Implications of live fish transport

All movement of live fish, even internally within a farm, increases stress which may have an adverse effect on both the growth and health of fish. A research project on the handling of smolt by the Landbruksteknisk Institutt (LTI) and Institutt for Akvakulturforskning (AKVAFORSK) is investigating the influence of handling procedures, including transport, on the growth, general health, disease and mortality of salmon smolt. Findings indicate that negative consequences of transportation on fish metabolism are greater than suspected and a stressful movement of fish can reduce appetite and growth rate for several weeks (Lekang, 1989).

Transport systems

Vehicles used for transporting farmed fish and fish products include towed net cages, boats, trucks, airplanes and even helicopters. The method actually used is dependent on several factors including transport distance, required water volume, weight, vehicle availability and cost. Governmental regulations frequently affect the decision.

Net cage transportation

Wild-caught live fish have been moved in net cages for many years in countries around the world. In Norway, live saithe (coal fish) intended for processing are towed long distances in net cages (Fig. 3.10) from offshore purse seine grounds to inshore sheltered areas.

Salmon farm operators have long been towing net cages containing live fish for short distances within farms or to processing plants. In 1988, many farmers were able to avoid toxic algae blooms along the southern coast of Norway by towing cages of fish into algae-free fjords. A new concept for a net cage farm better adapted for towing has recently been designed.

The main advantage of net cage transport is that large quantities of fish and even an entire farm can be transported simultaneously as one unit. The movement can be initiated quickly using only a single towing vessel. If the fish can be transported in rearing cages, which is usually the case, the stress of additional loading and unloading can be avoided. The disadvantages of net cage transportation are very low speed (one knot or less), and movement is limited to relatively short distances. The

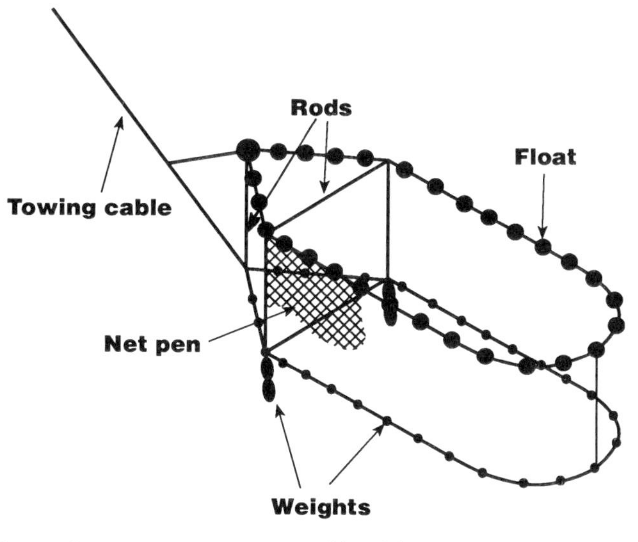

Fig. 3.10 System for net pen transportation of live fish.

risk of fish loss, net damage and other incidents during towing can be quite high in the event of bad weather.

Boat transport

Live fish transport by sea is very important for the salmon farming industry, especially for moving smolt to grow-out facilities. The use of 'well boats' has a long tradition in Norway, primarily for transporting live saithe to processing plants. Farmed salmon producers first adopted this established technology, often using the same type of vessel. However, because live salmon are of much higher value and are subject to more governmental regulation to control disease, advanced well boats have been developed to serve the aquaculture industry. These boats are 20 to 40 metres in length and have a carrying capacity of up to 50 tonnes of live fish. They carry equipment for seawater circulation, closed circulation, aeration, supplemental oxygenation and are able to automatically control dissolved oxygen content, salinity and temperature.

When advanced well boats are in coastal areas with sea water of good quality, water can be exchanged by taking advantage of the hydraulic pressure of the moving vessel. The use of pumps for circulation is necessary when the boat is not underway.

Activating the closed circulation system to prevent water exchange with the ocean environment may be necessary for several reasons. Most are associated with passage through coastal zones in which there are known or suspected disease pathogens, toxic phytoplankton, bad water quality, low salinity or rapidly changing temperatures.

The recommended fish density during boat transport varies with transport time, sea water temperature, fish size and the condition of the fish. For salmon it can vary from about 40 kg per cubic metre for small fish to 70 kg per cubic metre for large fish.

Truck transport

Several systems are in common use for transporting live salmon, usually fry or smolt, by road. Individual fiberglass fish tanks of rectangular configuration may be stacked on the bed of a truck. Another system uses larger elliptical or conical shaped tanks mounted on the truck frame and made from stainless steel or aluminium. The size of transport tanks varies from 200 litres to as much as 22 000 litres. Larger tanks often are subdivided to permit the shipment of different groups of fish. Tanks are insulated and have double walls to avoid heat loss or excessive water temperature variations.

Trucking of salmon requires oxygenation due to high fish densities and heavy oxygen consumption resulting from high stress levels in the fish. Recommended densities vary from about 40 kg fish at 20°C to 100 kg fish at 5°C. Pure oxygen is supplied from standard bottles and is delivered to the water through diffusers at the bottom of the tank. Compressed air added at the tank bottom or from vertical pipes (air-lift pumps) can provide both water circulation and oxygenation.

Fish transport vehicles without the capability of water purification must change water at intervals dependent on temperature and fish density, as frequently as every 10–20 hours. Changing water is stressful for the fish, time consuming and often ineffective in getting rid of sufficient metabolic waste. The purification systems on modern fish transport tankers may have various components, but aeration equipment to remove carbon dioxide, chemical filters for removing ammonia and a particle filter

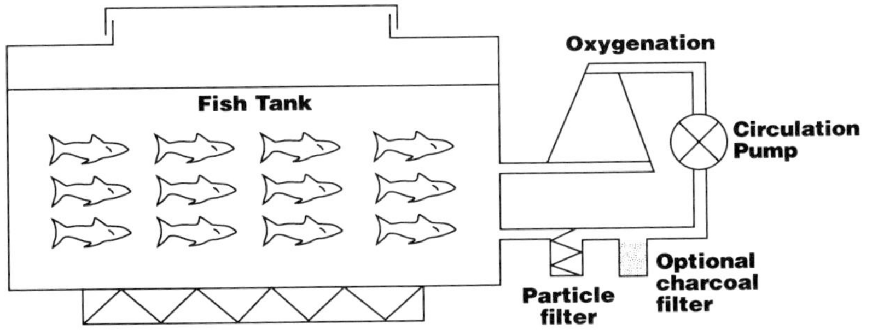

Fig. 3.11 Transport tank with recirculation of water.

for solid waste removal are considered essential (see Fig. 3.11). Many tankers are supplied with cooling systems. Even with such purification equipment, it may be necessary to change water, but at much longer intervals. In Norwegian practice as much as five days' transport without a water change has been possible (Lekang, 1989).

Air transportation

Movement of live fish by fixed wing aircraft is not extensive. Air transport is very costly, but does allow the shipment of fish long distances in a short period of time. It is most feasible for the movement of fry, where large numbers of fish can be shipped in a relatively small weight of water. For long distance air shipment, fish should be starved and placed into strong plastic bags at a density of up to 100 grams per litre. Pure oxygen should be added to each bag to achieve about 50% water and 50% oxygen by volume and the bags sealed and packed in well insulated boxes. If the water is cooled to near 0°C and 15% ice by weight is added, the fish will remain in good condition for more than 24 hours.

Helicopter transport of live salmon smolts is a relatively new development which is attracting growing interest. The fish are carried in special tanks hanging from a helicopter's winch. A standard container for use by smaller helicopters has a volume of 450 litres containing about 130 kg fish at a density of 288 kg per cubic metre (Lekang, 1989). Larger helicopters are able to carry tanks of 1500 to 3000 litre capacity, with as many as 20 000 smolt in one flight.

Other live fish transport and handling equipment

When handling and shipping live salmon, it is important not to take the fish out the water more often than is absolutely necessary. Fish can often be moved by unloading and transferring them through pipes or water-filled troughs. When fish must be lifted, a vacuum pump (Fig. 3.12) assures gentle handling.

Vacuum pumps developed for aquaculture purpose have pipe diameters in the range of 10 to 30 cm and allow for a total lifting height, using both vacuum and pressure forces, of more than 20 metres. The fish can be pumped several hundred metres horizontally with this system.

Other measures of live fish transfer within the rearing facility are fish screw

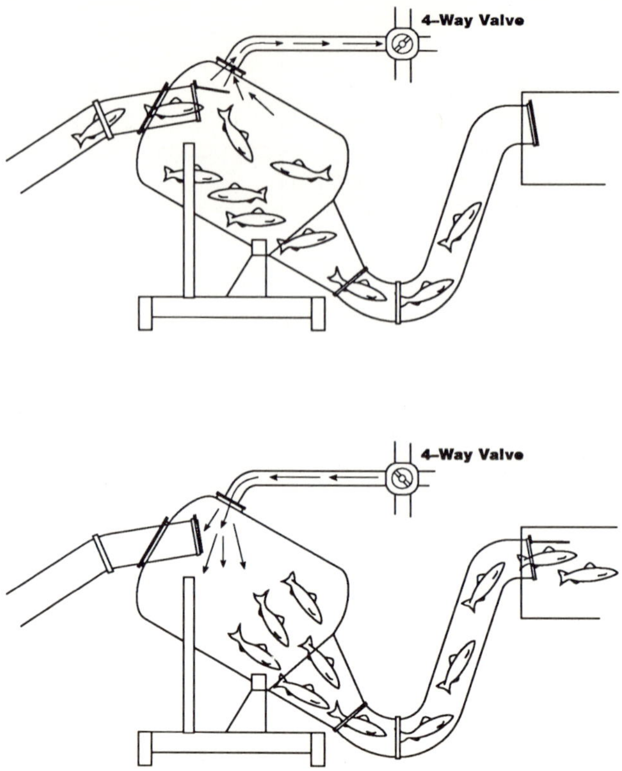

Fig. 3.12 Vacuum fish pump principle.

elevators and brails which should be made of or lined with smooth fabric rather than netting so as to protect fish skin from abrasions and mechanical damage.

Shipment of fresh processed salmon

Air shipment has been the dominant mode of moving salmon products to distant markets. The fish are placed in insulated boxes with a special drainage system and one layer of fish in each box.

A study by SINTEF, in 1989 concluded that:

- The fish should be cooled in chilled seawater before placement in boxes.
- To save weight and volume, the main insulation layer should be placed around a larger shipping package made up of several fish boxes instead of insulating each box.

Other fish handling equipment

New development has been carried out with various types of equipment related to salmon farming, including collection and braillling nets, mechanical size graders, and fish counters. A fish counter based on gravity impact detection, by means of a microphone, of fish fed through a pipe has demonstrated a rate of 10 000 fish per hour (Strachan & Nicholson, 1990). Development continues on containers for treating fish, anaesthetic and bleeding tanks, cleaning and washing tables, slaughtering equipment, and weighing, packing and labelling equipment.

Processing equipment and sanitary routines for the slaughtering and handling of fresh salmon are now subject to stricter governmental regulations and control in Norway. Waste water may no longer be disposed in the sea.

Feeding techniques and equipment

In intensive fish farming, feeding procedures are essential parts of the business and are of critical importance to both technical and financial success. Consequently, extensive and continuing effort has been made towards the development of efficient feeding techniques and equipment. The selection of feeding equipment for a particular situation depends on a number of factors such as:

- Category of facility (smolt culture or grow-out)
- Type of facility (net cage, closed-wall cage, land-based, etc.)
- Type of feed (wet or dry)
- Facility layout
- Degree of technology preferred

Feeding procedures and equipment now available include:

- Manual (feeding by hand or scoop).
- Self-contained automatic feeders (generally electric-powered) with up to 100 litre capacity.
- Self-contained automatic feeders associated with much larger feed bin capacity (400–500 litre capacity).
- Automatic systems with feed pneumatically or hydraulically supplied through pipes from a central service area.
- Semi-automatic systems with feed pneumatically or hydraulically supplied through hoses from service boats.

Automatic systems servicing a large number of fish culture enclosures from a central service area frequently use computers programmed to calculate feed rations according to seawater temperatures and the number and size of fish per enclosure.

While advanced automatic feeding systems are readily available for salmon farmers, manual feeding is still widely used, particularly at grow-out facilities. Many farm managers believe that manual feeding requires much closer observation of feeding behaviour and accordingly provides better fish performance and better feed efficiency. Manual feeding also puts the farm staff in more continuous contact with the fish and hence they have earlier warning of disease and other developing problems.

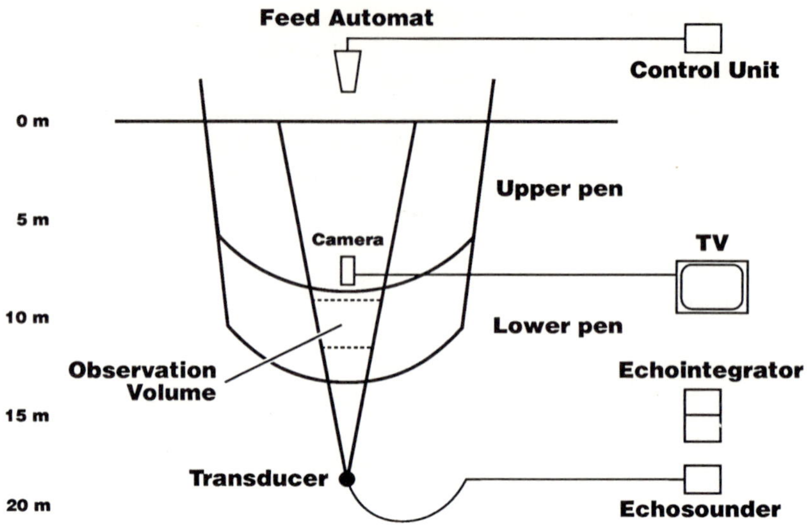

Fig. 3.13 Appetite feeding system (from Juell, 1987).

A system for appetite feeding

An appetite or 'demand' feeding system provides the amount of feed and feeding intervals according to the feeding behaviour and food acceptance of the fish. Thus, manual feeding is often a form of demand system.

Research on the demand feeding of salmon has been done by the Institute of Fishery Technology Research (FTFI) in Bergen. Test equipment included a high frequency echo location device, calibrated to the size of feed pellets, which was installed below a net cage containing fish. The transducer faced upwards to cover a defined observation volume below the fish cage as shown in Fig. 3.13 (Juell, 1987). The transducer was connected to an echo integrator which could measure the echo volume and thereby the amount of food (dry pellets) sinking through the observation volume. An underwater TV camera was also installed to observe the feeding behaviour of the salmon, as shown in the figure. The principle of the system is that the feeder stops dispensing pellets when a certain amount of food, which has escaped through the net bottom, is detected by the echo location device.

This appetite feeding system was found to work effectively and the investigators concluded that such a system could be practical in commercial salmon culture (Juell, 1987).

References

Aarsnes, J.K., Rudi, H. and Løland, G. (1990) Current forces of cage, net deflection, *Engineering for Offshore Farming* (Proceedings of the conference organized by the Institution of Civil Engineers, Glasgow, 17–18 October 1990), pp. 137–52, Tomas Telford, London.

Agrawal, S.C.R. and Bose, N.A. (1989) Sluice design for filling of tide-fed aquacultural farm, *Aquacultural Engineering*, 8, 267–81.

Beveridge, M. (1987) *Cage Aquaculture*. Fishing News Books, Oxford.

Cairns, J. and Linfoot, B.T., Department of Civil Engineering, Heriot-Watt University, UK (1990) Some considerations in the structural engineering of sea-cages for aquaculture, *Engineering for Offshore Farming* (Proceedings of the conference organized by the Institution of Civil Engineers, Glasgow, 17–18 October 1990), pp. 63–7, Tomas Telford, London.

Christie, M. (1987) Flytende Raceway – et nytt og bedre oppdrettssystem, Norsk Fiskeoppdrett No. 5–87, pp. 33–5 (in Norwegian).

Eikebrokk, B. (1990) Design and performance of the BIOFISH water recirculation system, *Aquacultural Engineering*, 9, 285–94.

Garmannslund, H. (1988) Gode erfaringer med hjulfilter, Norsk Fiskeoppdrett No. 10–88, p. 51 (in Norwegian).

Gebauer, R., Eggen, G., Hausen, E. and Eikebrokk, B. (1992) *Oppdrettsteknologi*, Tapir forlag, Trondheim (in Norwegian).

Huse, I. *et al.* (1990) The effect of shading in pen rearing of Atlantic salmon, *Aquacultural Engineering*, 9, 235–44.

Juell, J. (1987) Optimal appetittforing i matfiskanlegg ved akustisk registrering av forspill, Norsk Fiskeoppdrett 5, 38–40 (in Norwegian).

Lekang, O.I. (1989) Transportteknikk ved flytting av fish. NITO/NAFO course 8925: Transport av oppdrettsfisk, muligheter og problemer, 2–3 March, Trondheim (in Norwegian).

Magnussen, O. (1989) Transport av ferskfisk, NITO/NAFO course 8925: Transport av oppdrettsfisk, muligheter og problemer, 2–3 March, Trondheim (in Norwegian).

Oltedal, G. (1989) Lukkede sjoanlegg, NIF-course: Utforming og konstruksjon av akvakulturanlegg, – åpne og lukkede sjøanlegg, 10–12 May, Marin Teknisk Senter, Trondheim, (in Norwegian).

Parker, N.C. and Suttle, M.A. (1987) Design of airlift pumps for water circulation and aeration in aquaculture, *Aquacultural Engineering*, 6, 97–110.

Rusten, B. (1989) Start-feeding of Atlantic salmon in a recycling plant, *Aquacultural Engineering*, 8, 241–56.

Slaattelid, O.H. (1990) Model tests with flexible, circular floats for fish farming, *Engineering for Offshore Farming* (Proceedings of the conference organized by the Institution of Civil Engineers, Glasgow, 17–18 October 1990), pp. 93–106, Tomas Telford, London.

Stefansson, S.O. and Hansen, T. (1987) Bruk av dobbel lysperiode, nye muligheter for smoltproduksjon i Norge, Norsk Fiskeoppdrett No. 5–87, pp. 53–5 (in Norwegian).

Strachan, J.C.C. and Nicholson, F.J. (1990) A novel fish counter for fish farms, *Aquacultural Engineering*, 9, 405–10.

Tvinnereim, K. (1990) Anieggsbeskrivelse, Norske anlegg på land og sjø, NAFO course: Lukkede produksjonsanlegg for laksefisk, 31 October–1 November (in Norwegian).

Watten, B.C. and Todd Beck, L. (1990) Comparative hydraulics and rearing trial performance of a production scale cross-flow rearing unit, *Aquacultural Engineering*, 9, 245–61.

Whittaker, T.J.T., Bell, A.K. and Shaw, M.R. (1990) Model testing of a cage system for the offshore environment, *Engineering for Offshore Farming* (Proceedings of the conference organized by the Institution of Civil Engineers, Glasgow, 17–18 October 1990), pp. 79–92, Tomas Telford, London.

Chapter 4
Nutrition, Diet Formulation and Feeding Practices

MALCOLM JOBLING

Of the species of salmonids farmed worldwide, the rainbow trout (*Oncorhynchus mykiss*) is by far the most widely cultured, both for stocking for recreational fisheries and for food production. When farming is aimed at producing food fish, the trout are reared from the egg until they reach table size. The fish may be reared in tanks, raceways or ponds supplied with constantly running fresh water. While fish reared in ponds may fulfil some of the their food requirements by consuming natural prey organisms found in the ponds, fish grown in tanks and raceways are completely dependent upon the farmer providing food.

In many countries, rainbow trout are harvested when they are portion-sized fish, but in countries such as Norway, the fish may be grown on in seawater cages to larger size (several kilograms) before slaughter. Irrespective of the harvest size of the fish, the farming of rainbow trout is an intensive exercise, and food is provided by the farmer throughout the entire rearing process.

In recent years, there has been a marked increased in the farming of Atlantic salmon (*Salmo salar*). The rearing of salmon, like that of rainbow trout, is intensive, involving both freshwater and seawater growth phases until the fish are harvested at a size of 3–5 kg.

Most of the other salmonids – Pacific salmon (genus *Oncorhynchus*), charrs (genus *Salvelinus*) and trouts (genus *Salmo*) – are raised primarily for stock enhancement and recreational fishery purposes, so most effort has been directed towards developing methods for rearing the young stages of these species.

When fish are reared for stocking, the producer requires start and juvenile feeds formulated to ensure good growth and survival of the fish prior to release. On the other hand, the farming of rainbow trout and Atlantic salmon throughout the entire life cycle necessitates that suitable diets be available for all life history stages. Ideally, information should be available both about the nutritional requirements of first-feeding fry (alevins) and juveniles (parr) in fresh water, and about the requirements of post-smolt and broodstock in seawater.

Studies directed either towards the determination of nutritional requirements or towards the examination of the potential value of various feed ingredients have usually been conducted using juvenile, rather than adult, fish. Juvenile fish generally show more pronounced responses (depressed growth or deficiency symptoms) to unfavourable nutritional conditions than do larger conspecifics, and for this reason are often preferred in nutritional research. In addition, it is much easier and cheaper to carry out practical work on juvenile fish, weighing up to a few grams, than on adults which may weigh several kilograms. Thus, economic constraints on research facilities often govern the feasibility of carrying out nutritional studies on all the life history stages of a given species.

As a consequence, nutritional information is incomplete, and the purpose of this chapter is to give a brief overview of the nutritional requirements of salmonids, and to illustrate how this information can be applied in the formulation of practical feeds.

Feed formulation: aims and constraints

Commercial feed manufacturers aim to develop feed formulations that are as cheap as possible in relation to the levels of fish growth achieved. A good deal of research has been conducted in order to determine to what extent expensive, often protein-rich, food components can be replaced by cheaper ingredients, which supply energy in the form of carbohydrates and lipids. The final goal is to minimize the level of inclusion of the expensive nutrient source in the feed without this having any negative consequences for growth performance, that is, research is often conducted in order to determine the minimum requirements for expensive nutrients. These minimum requirements provide the background for recommended levels of inclusion of the various nutrients in commercially-produced fish feeds.

From this, it follows that a commercially-produced feed, formulated on the basis of minimum nutrient requirements, may differ substantially from the composition of the normal diet of salmonids in the wild. Fish in the wild are not subject to the constraints of economics, but, on the other hand, their dietary composition may be influenced by the vagaries of the production and availability of natural prey organisms.

Discrepancies between nutrient requirement levels established in laboratory growth studies and the nutrient composition of natural prey organisms does not necessarily mean that the recommendations are incorrect, but it should be borne in mind that these recommendations delimit the minimum requirements of the fish. It should also be remembered that the fish used in studies of nutrient requirements will often be held under ideal growth conditions in the laboratory.

In cases where laboratory studies are lacking, and information about nutrient requirements is inadequate, examination of the nutrient composition of the diet of wild fish can often provide an acceptable starting point for the formulation of artificial feeds. The examination of dietary shifts in wild fish can also often give indications that nutrient requirements change during growth and development. Consequently, it is of interest to give a brief description of the diets and feeding behaviour of salmonids in the wild before going on to examine the different nutrient classes, and the requirements of these nutrients for growth.

Feeding behaviour and natural prey

Many salmonids spawn during the autumn, and embryonic development takes place during the course of the winter months. The developing embryo relies upon the nutrients present in the egg yolk both for growth and for use as metabolic fuel. Following hatching in spring, the young fish will be reliant upon the yolk as a source of nutrients for the first few weeks, but as the yolk becomes depleted, the fish will begin to take exogenous food. When the young salmon fry has absorbed most of the nutrients from the yolk-sac, it will swim up from the gravel redd (nesting site where the eggs were laid and the early development took place) at the bottom of the river and begin to search for exogenous food. In the space of the first few hours after emergence from the redd, the fry will attack, in addition to small live prey organisms, air bubbles, leaf debris and other inert material suspended in the water column. It seems that movement is important, and only moving particles appear to attract attention. Results of laboratory studies conducted using hatchery-reared fry and juveniles fed on commercial dry feeds provide additional support for the view that particle movement is important. The young salmon and trout attacked food particles in the water column and particles rolling along the tank bottom, but ignored stationary particles until they were put back into motion. These observations may have been predicted since there is a widely held opinion that salmonids are, primarily, visual feeders, identifying and localizing prey using the sense of sight. Young salmon certainly appear to feed most avidly during daylight hours, and food intake is drastically reduced at night. Thus, feeding opportunity for young salmon is closely linked to the number of daylight hours. This was confirmed in laboratory studies in which groups of young salmon were reared under different conditions of artificially controlled daylength. By the end of the summer, the fish exposed to the largest number of daylight hours, and therefore given the greatest feeding opportunity, had achieved the largest size.

For visually feeding predators, prey colour and shape may also be expected to influence selection. Naive start-feeding fry appear to prefer red-coloured food items, but can soon learn to take particles of other colours, with only one or two training sessions being required. It may not be the colour of the food *per se* that is important, but the degree of contrast provided by the colour of the food against the background colour of the tank or environmental spacelight. A higher degree of contrast between food and background leads both to an increased attack rate and a higher percentage capture success in first-feeding salmonid fry.

In the wild, fry begin to feed on the very small organisms comprising the freshwater zooplankton. At the first-feeding stage, salmonid fry range between 150–500 mg depending upon species, and as they grow, the fry gradually include larger prey items in their diet. The diet of zooplankton (e.g. *Daphnia* and *Diaptomus*) will be expanded to include chironomid and mayfly larvae. As summer progresses, larger insect larvae, such as caddis-fly larvae will be eaten. At the onset of autumn, appetite declines and food consumption remains low throughout the winter months. This is, in part, due to low water temperatures and short daylength, but also reflects the paucity of suitable prey organisms. With the increasing daylength, and a return to somewhat higher water temperatures, in spring, the fish show a marked increase in appetite. At this stage, the diet comprises planktonic crustaceans, insect larvae and a variety of adult insects.

These natural prey organisms differ in a number of respects (movement, size and form) from the types of feed usually offered to young salmonids in commercial hatcheries, and it is often observed that hatchery-reared fish will readily accept natural prey, whereas it is difficult to wean wild-caught fish on to commercially-produced feeds.

Commercial feeds are often produced as rounded particles, whereas many of the prey of wild juvenile salmonids are elongate or worm-like. When the responses of hatchery-reared salmon parr to food particles of different shapes were studied, it was observed that long, thick particles were more frequently attacked than either long, thin or round food particles. Thus, particle shape was important in inducing orientation and attack responses. The young salmon found the long, thick particles more difficult to swallow than the long, thin particles, and a greater proportion of the long, thin particles than the long, thick was eventually consumed. This underlies the importance of supplying food that will both attract attention, induce attack responses and is of a size capable of being handled and swallowed relatively easily by the fish. Some studies have been conducted on the freshwater stages of salmonids in an attempt to determine the optimal size of food particles to be supplied. For Atlantic salmon, in the size range 4–20 cm, growth appears to be maximized if the size of the food is equivalent to 2.2–2.6% of the length of the fish. The smaller-mouthed Arctic charr (*Salvelinus alpinus*) appears to prefer food particles of a somewhat smaller relative size (Table 4.1). These food particle sizes can be considered as giving useful rule-of-thumb guidelines, and commercial manufacturers produce dry feed of different sizes for use in the rearing of the fish at different stages of the life cycle (Table 4.2).

Atlantic salmon undergo the parr–smolt transformation at an age of 1–7 years in the wild (usually 1–2 years in hatcheries) depending upon geographic location. At

Table 4.1 Optimal feed sizes for juvenile salmonids.

Fish species	Length (cm)	Optimal food size (% Fork length)[1]
Atlantic salmon	4–20	2.2–2.6
Chinook salmon	4–5	2.0
	10–11	3.0
Arctic charr	7–11	1.5–1.8
	12–14	2.3–2.4

Note:
[1] Fork length is the length of the fish from the snout to the fork of the tail.

Table 4.2 Approximate sizes of commercially-produced dry feeds for salmonids.

Fish size (g)	Feed diameter (mm)	Feed type
<0.5	<0.6	Starter
0.5–2.0	0.6–0.85	Crumbles
2–5	0.85–1.2	Crumbles
5–10	1.2–2.0	Crumbles
10–20	2.0–3.2	Crumbles
20–100	3.2	Pellet
100–500	4.6	Pellet
>500	6.5	Pellet

the southern end of the range, the parr–smolt transformation will usually occur at a younger age than further north, and this is thought to reflect latitudinal differences in freshwater productivity and length of the annual growth season (which depends both upon water temperature and hours of daylight during the summer months).

On entering seawater, the post-smolt begin to feed on small crustaceans and fish, and during the early part of seawater residence, growth may be rapid. While in the sea, the Atlantic salmon feed predominantly on crustaceans (krill and amphipods), fish (largely herring and capelin), and squid. The relative importance of the different prey types appears to vary with geographic area, and there also appears to be shifts towards larger prey as the fish increase in size. There are, for example, indications that the relative importance of small crustaceans (1–6 cm in length) declines with increasing size of salmon, while large salmon have a greater reliance upon pelagic fish (herring, sprat and capelin) as prey.

Formulated feeds: the options

The types of formulated feeds used in commercial fish farming have characteristics that are radically different from those of the natural prey organisms consumed by salmonids. Commercial feeds differ from natural prey both with respect to moisture and energy content, and nutritional composition. Many natural prey organisms have a relatively high water content (70–90%) and the energy content is generally within the range 5–10 kJ/g wet weight, whereas the majority of commercially-produced dry feeds contain 10% or less moisture and have an energy content of 20–25 kJ/g. These commercially-produced dry feeds can be broadly divided into compressed and extruded pellet types.

Steam pelleting, utilizing compression, produces a dense food pellet that sinks quite rapidly. The steam pelleting process involves the use of moisture, heat and pressure to cause agglomeration of the ingredients into large homogeneous particles. Correct treatment conditions and ingredient composition are important if firmly-bound particles are to be produced. Food ingredients should be finely-ground prior to steam pelleting, and the inclusion of a certain amount of starch from grains is vital in order to ensure good pellet binding. On the other hand, the inclusion of fibrous material and lipids is detrimental to binding, and the amount of lipid included in the pellet mix should not exceed 7%. Consequently, additional lipid must be sprayed onto the food after pelleting, and in this way, pellet lipid levels of 16–20% can be achieved.

Steam is added to the finely-ground dry food ingredients in order to increase the moisture content of the mixture to 15–20%, and to raise the temperature to 70–85°C. The increase in temperature causes partial gelatinization of the starch, which in turn aids the binding together of the food ingredients. The mixture is then passed through a pellet die, and the moisture content reduced by forcing air over the warm pellets immediately after they leave the pelleting machine. Compressed, steam pelleted feeds must be carefully processed in order to achieve satisfactory water stability and to prevent disintegration.

The production of extruded pellets requires more elaborate equipment and the use of more energy for processing than does steam pelleting, so extruded pellets are more expensive to produce than compressed steam pellets. In the extrusion process, the food ingredients are moistened, precooked, extruded and dried. The pellets produced are of relatively low density. Depending upon production conditions, extrusion processes can result in either floating or slowly-sinking pellets.

Usually, the finely-ground food ingredients are pretreated with either steam or water and the resulting mixture, containing 20–25% moisture, may or may not be precooked before being transferred to the extruder barrel. The temperature of the compacted mixture rapidly increases to 105–150°C under pressure and this causes almost complete gelatinization of the starch. As the mixture is forced out of the extrusion barrel and the pellet is subjected to reduced pressure, part of the moisture in the hot mixture vaporizes and this causes expansion of the food pellets. These pellets contain more moisture than compressed pellets and external heating is required in order to reduce the moisture content to a final level of about 10%. Thus, after extrusion, the pellets are usually passed through a drying tunnel.

One advantage of extruded pellets is that they can soak up oil sprayed onto the surface, and lipid contents of 20% and over can be achieved. In addition, the complete gelatinization of the starch produces a firmly bound pellet with good water stability, and there is a lower loss of dust and fines than is the case with steam pelleted feeds.

Both steam pelleting and extrusion lead to destruction of certain heat-sensitive vitamins. These must be either added to the feed mixture in excess prior to processing, or be sprayed onto the surface of the pellets after processing. Ascorbic acid (vitamin C) is particularly heat-sensitive, and 25% may be destroyed in the steam pelleting process. Extrusion may result in a vitamin C loss of 50% or more, and the losses of vitamin A, thiamin and pyridoxine may amount to 30–40%.

The majority of Atlantic salmon producers use dry feeds, but some farmers prefer to rely on wet or moist-pellet feeds manufactured on the farm. Wet- and moist-pellet feeds are made by combining wet and dry meal mixes in different proportions, ranging from 90:10 to 50:50 (wet:dry). The wet ingredients will usually consist of ground fresh or frozen fish (capelin, argentine and small gadoids), squid and shrimp/prawn waste, fish-processing waste and/or acid-preserved hydrolysed fish waste (silage). The dry mixes contain fishmeal, cooked starch, vitamin and mineral premixes and alginate binder. When the proportion of wet material is high (80–90%), the dry mix must contain an effective binder, and alginate (0.5–2.0%) is often the binding agent of choice.

Many commercial feed manufacturers now provide complete dry mixes, containing binder, vitamins and minerals, which when added to wet ingredients in fixed proportions (10, 25 or 40%) will produce a pellet having reasonable water stability. Thus the farmer is spared the effort of manufacturing the dry mix, but the onus is on the farmer to select and blend wet materials that ensure that the fish are provided with a nutritionally adequate diet. These wet- and moist-pellet feeds are made at the fish-rearing site and should be used within a day or two of manufacture. Generally, these types of feed will support good growth of salmonids in sea cages and they are a useful alternative to dry feeds when the cost of wet ingredients is low and the supply stable.

Wet feeds generally contain 50–70% moisture and have energy contents of 10–15 kJ/g, whereas moist pellets contain less moisture (35–40%) and have a somewhat higher energy content (15–20 kJ/g). These differences in physical and other properties between the different feed types may affect both acceptance and the way in which the feeds are handled, digested and absorbed by the fish. For example, some young salmonids start feeding less readily on dry than on moist feeds, with the softer texture of the moist feed appearing to be preferred by the young fish.

Food composition and nutrient availability

The energy content of a dietary ingredient or formulated feed is determined by burning and complete oxidation in an instrument known as an adiabatic bomb calorimeter. This enables the heat released by the oxidation process to be measured, giving the gross energy content of the dietary ingredient or feed.

Fish and other animals require the use of dietary energy for life-sustaining processes such as maintenance of bodily functions and activity. Once these energy requirements are met, any dietary input can be utilized for growth.

Carbohydrates, proteins and lipids can all be used as energy sources by fish, but these organic compounds are not equally well suited for the promotion of tissue synthesis and growth. In order for growth to occur there is a requirement for dietary protein, but if the diet is very rich in protein, the fish will also utilize considerable amounts of the protein as a source of energy. As feed ingredients high in protein content are usually expensive, it is considered disadvantageous that protein is used as an energy source. Thus, attempts are made to spare protein for growth by adding cheaper energy sources (carbohydrates and lipids) to the feed.

As a general rule of thumb, the gross energy contents of carbohydrates and proteins are considered to be 17.5 and 24 kJ/g respectively. The lipids present in terrestrial animals have a gross energy content of approximately 39.5 kJ/g, whereas lipids of aquatic origin tend to have slightly lower gross energy contents. The gross energy content of a specific nutrient, ingredient or formulated fish feed is not, however, a measure of its true energy value to the fish. The difference between the gross energy and the energy available to the fish varies widely, depending both upon the source of the food materials and the processing to which the materials have been subjected.

Differences in digestibilities between different feed ingredients account for most of the differences in available energy between complete feed formulations for fish. Digestible energy is the difference between the gross energy of the food consumed and the energy lost in the faeces. The faeces consists not only of undigested food, but also includes traces of digestive enzymes, cells sloughed from the gut wall and gut microflora, but in practical determinations of the digestibilities of feedstuffs, the contributions of these latter components are usually ignored.

Digestibility can be determined either directly or indirectly. In the direct method, the total amount of food consumed must be known and all the faeces produced collected. Complete collection of faeces is time-consuming and it is often difficult to prevent disintegration of faecal material, so some loss is almost inevitable. The indirect method involves collecting only samples and food and faeces, and digestibilities are calculated on the basis of the ratios of nutrient to indicator in the food and faeces. The indicator is an inert, indigestible, unabsorbed compound that is added to the food in low concentration (usually 0.5–2.0% by weight). In order for the indirect method to be successful, it is clear that the chosen indicator must be truly inert and not interfere with the digestive and absorptive processes. A variety of indicators have been used in digestibility studies with fish, the most popular being indigestible fibre, acid-insoluble ash and chromic oxide. Once the nutrient and indicator ratios are known, digestibility can be calculated according to the formula:

$$100 - \left[\left(\frac{\text{Food nutrient}}{\text{Faecal nutrient}}\right) \times \left(\frac{\text{Faecal indicator}}{\text{Food indicator}}\right) \times 100\right]$$

Chemical analysis and bomb calorimetry can give information about the chemical composition and gross energy contents of feed ingredients and formulations, but digestibility studies are required in order to reveal the availabilities of the various nutrients to the fish. In general, the digestibilities of natural prey, fishmeals and other fish products are high, and fishmeals are usually considered to be an excellent source of protein for farmed salmonids. Digestibilities of meals produced from animal by-products are often lower than those of fishmeal, and the digestibilities of some plant-

Table 4.3 Digestibility of protein and energy contents of feed ingredients used in commercial feeds for salmonids. (Digestibility trials conducted using rainbow trout, *Oncorhynchus mykiss*, as the test species.)

	Digestibility (%)	
	Protein	Energy
Fishmeals		
Anchovy	85	91
Herring	87	92
Animal by-product meals		
Bloodmeal	69	70
Meat meal	75	72
Poultry by-products	70	66
Plant products		
Wheat middlings	68	53
Alfalfa meal	61	45
Soybean meal	75	75
Torula yeast	82	73

Table 4.4 Effects of processing and heat treatment on the digestibility of soybean meal. (Digestibility trials conducted using rainbow trout, *Oncorhynchus mykiss*, as the test species.)

	Digestibility coefficient (%)	
Treatment	Protein	Energy
Heat treatment		
127°C 10 min	40	51
175°C 10 min	70	72
204°C 10 min	78	77
234°C 10 min	72	70
Steam processing		
0.35 kg/cm^2 5 min	37	42
0.70 kg/cm^2 10 min	80	75
1.05 kg/cm^2 10 min	75	69

Table 4.5 Compositions (% dry weight) of some natural prey organisms from freshwater (mixed samples of zooplankton and insect larvae) and marine (prawn, *Pandalus borealis*, and herring, *Clupea harengus*) environments. Compositions of a range of commercial salmonid feeds (pellet) are shown for comparison. (PE:TE = protein energy:total energy = proportion of the dietary energy provided by protein.)

	Freshwater Zooplankton	Insect larvae	Prawn	Herring	Pellet
Protein	48	56	55	47	40–55
Lipid	21	14	17	39	15–23
Carbohydrate	4	9	–	–	10–20
Ash	12	10	14	8	7–12
PE:TE	0.56	0.66	0.58	0.40	0.43–0.58

based meals may be relatively poor (Table 4.3). Digestibility values may also be markedly affected by the treatment to which a feed ingredient has been subjected (Table 4.4).

Nutrient classes, nutritional requirements and sources of nutrients

The nutrient compositions (on dry weight basis) of a number of natural prey and commercial dry feeds are shown in Table 4.5. The most obvious difference is the comparatively high carbohydrate level (10–20%) in the commercial feed.

Carbohydrates

Carbohydrates are not usually considered to be essential nutrients for growth. They may be used, to some extent, by the fish as an energy source. In commercial feeds, the carbohydrates act as binders.

The carbohydrates present in commercial dry feeds are mainly long-chain polysaccharides of plant origin. The two most important are starch and cellulose, both of which are homopolysaccharides consisting of long chains of the monosaccharide glucose held together by glycosidic linkages. The main difference between starch and cellulose is the form of the glycosidic linkage between the glucose units (Fig. 4.1). Cellulose is the major structural component of plant cell walls, is highly insoluble and is almost indigestible for fish. Fish do not produce enzymes capable of cellulose digestion, and any cellulase activity isolated from the alimentary tract of the fish will be derived from the gut microflora. Starches are the main energy storage products in many plants, and two forms of starch may be found in starch granules – amylose and amylopectin. Amylose is a straight-chained molecule, which is soluble in warm water, whereas amylopectin is a more complex molecule with side branches. The relative proportions of the two molecules in the starch granules varies between plant species and this gives rise to differences in digestibility. Fish do produce and secrete starch-digesting enzymes – amylases – into the gut, but there are considerable interspecific differences between fish in the ability to digest these polysaccharides. Raw starches are about 20–30% available for salmonid fish. Steam heating of the starch granules causes gelatinization and digestibility increases to 50%. Complete gelatinization,

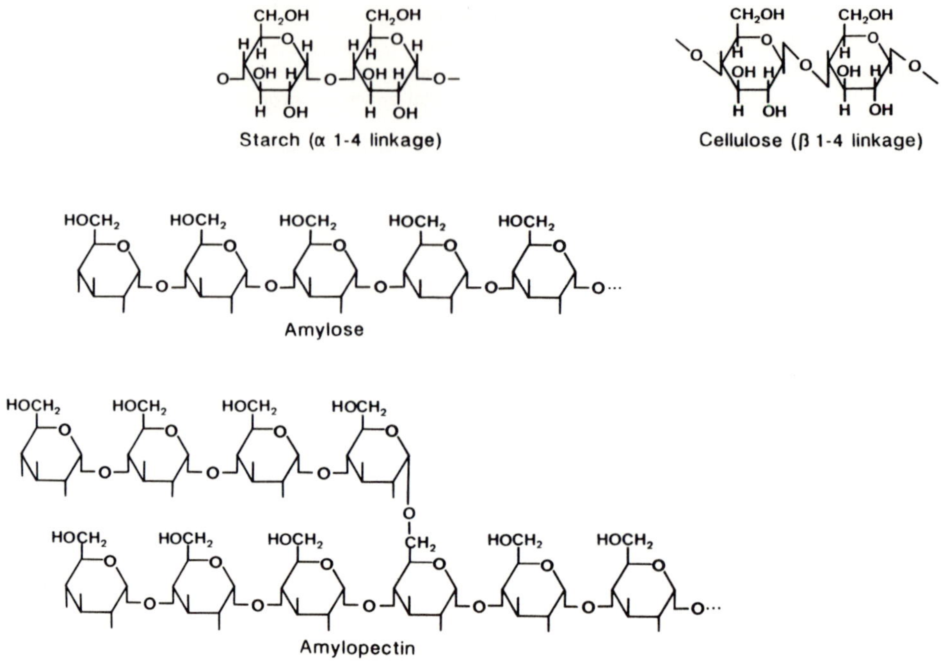

Fig. 4.1 Chemical structures of starch and cellulose, long-chain carbohydrates that are encountered in fish feeds. (Both starch and cellulose are homopolysaccharides, comprising glucose units, but there are different types of glycosidic linkages between the glucose molecules making up the two polysaccharides.)

which occurs during the extrusion process, may further increase digestibility.

Carbohydrates are present in relatively small amounts (1–10%) in the natural prey of salmonids. The storage carbohydrate in animals is glycogen, rather than starch. Glycogen is stored in the liver of most animals, but is also present in relatively large quantities in the muscle tissue of bivalve molluscs. Glycogen is similar to amylopectin, but the side branches are shorter, and glycogen is more water-soluble and digestible than amylopectin.

Partial oxidation of monosaccharides produces sugar acids. For example, partial oxidation of glucose, a hexose sugar, with six carbon atoms, can result in the production of glucuronic acid. The most important of the sugar acids are the uronic acids, especially glucuronic and galacturonic acids, which are components of the heteropolysaccharides. The heteropolysaccharides are mixed polysaccharides, which on hydrolysis yield mixtures of monosaccharides and derived products. A range of heteropolysaccharides, such as gums (e.g. gum arabic), mucilages (e.g. agar) and pectic substances (e.g. pectins) are exploited commercially due to their binding and gelling properties, and may be used to some extent as binding agents in commercial diets for fish.

Alginates, which are components of the cell walls of kelp and other algae, are polysaccharides containing mannuronic acid units. The exact structure and chemical characteristics of the alginates vary between algal species, but as a group, the alginates possess binding properties. Alginates from a range of sources are often employed as binding agents in fish feeds.

Carbohydrate, predominantly starch, has traditionally been added to dry feeds, both because of its binding properties and also as a cheap energy source. Recently, however, some doubts have been raised about the ability of salmonids to fully utilize the energy content of carbohydrates. Long-term feeding with diets containing high levels of carbohydrates (20% or more) can have negative effects on the fish. The reported effects of feeding with high-carbohydrate diets include reduced liver function and metabolic disturbances. Most commercial feed manufacturers limit the maximum carbohydrate contents of their dry feeds to 20%, and many dry feeds now contain carbohydrate levels somewhat lower than this.

Proteins and amino acids

An adequate supply of dietary protein is essential if fish are to thrive and grow well. Results from several laboratory studies with salmonids indicate that optimal growth is achieved when approximately half the food energy is supplied from protein. Findings for other carnivorous fish species are similar and it is generally recognized that protein should provide 40–55% of the dietary energy (i.e. PE:TE within the range 0.40–0.55). The nutritional compositions of a number of the natural prey organisms of salmonids appear to deviate from these values (Table 4.5). In particular, it may appear that the zooplankton and insect larvae diet of the young freshwater stages of salmon is richer in protein than would be required to promote optimal growth. On the other hand, it might be predicted that the protein contents of fish such as capelin and herring (PE:TE = 0.40 or less) would be too low to promote optimal growth rates of large adult salmon in seawater. It appears, however, that these changes reflect real differences in protein requirements of fish at different life stages.

Small young fish in which rates of growth, tissue synthesis and turnover are high seem to have a higher requirement for protein than do larger fish. These ontogenetic changes in protein requirements have been recognized, and commercial feeds have been formulated accordingly. Most start-feeds for swim-up fry contain at least 50% protein, with the protein level in the feed being reduced as the fish grow. Fingerling (parr) diets may contain 40–50% protein, while feeds for the on-growing of larger salmonids sometimes have protein contents as low as 35–40%.

Theoretically, fish do not have a requirement for protein *per se*, but they do require a dietary supply of certain essential (indispensable) amino acids that contribute to protein structure. Amino acids are the structural units of proteins, the proteins consisting of a series of amino acids linked by peptide bonds (Fig. 4.2). All the amino acids have the same basic design, with a carboxyl (–COOH) and an amino ($-NH_2$) group attached to the alpha carbon atom. The remainder of the amino acid molecule (denoted by R in Fig. 4.2) can have different configurations, and amino acids can be divided into a number of series depending upon the nature of the R-group:

(1) Aliphatic series (glycine, alanine, serine, threonine, valine, leucine, isoleucine);
(2) Aromatic series (phenylalanine, tyrosine);
(3) Sulphur-amino acid series (cysteine, cystine, methionine);
(4) Heterocyclic series (tryptophan, proline, hydroxyproline);
(5) Acidic series (aspartic acid, glutamic acid);
(6) Basic series (arginine, histidine, lysine).

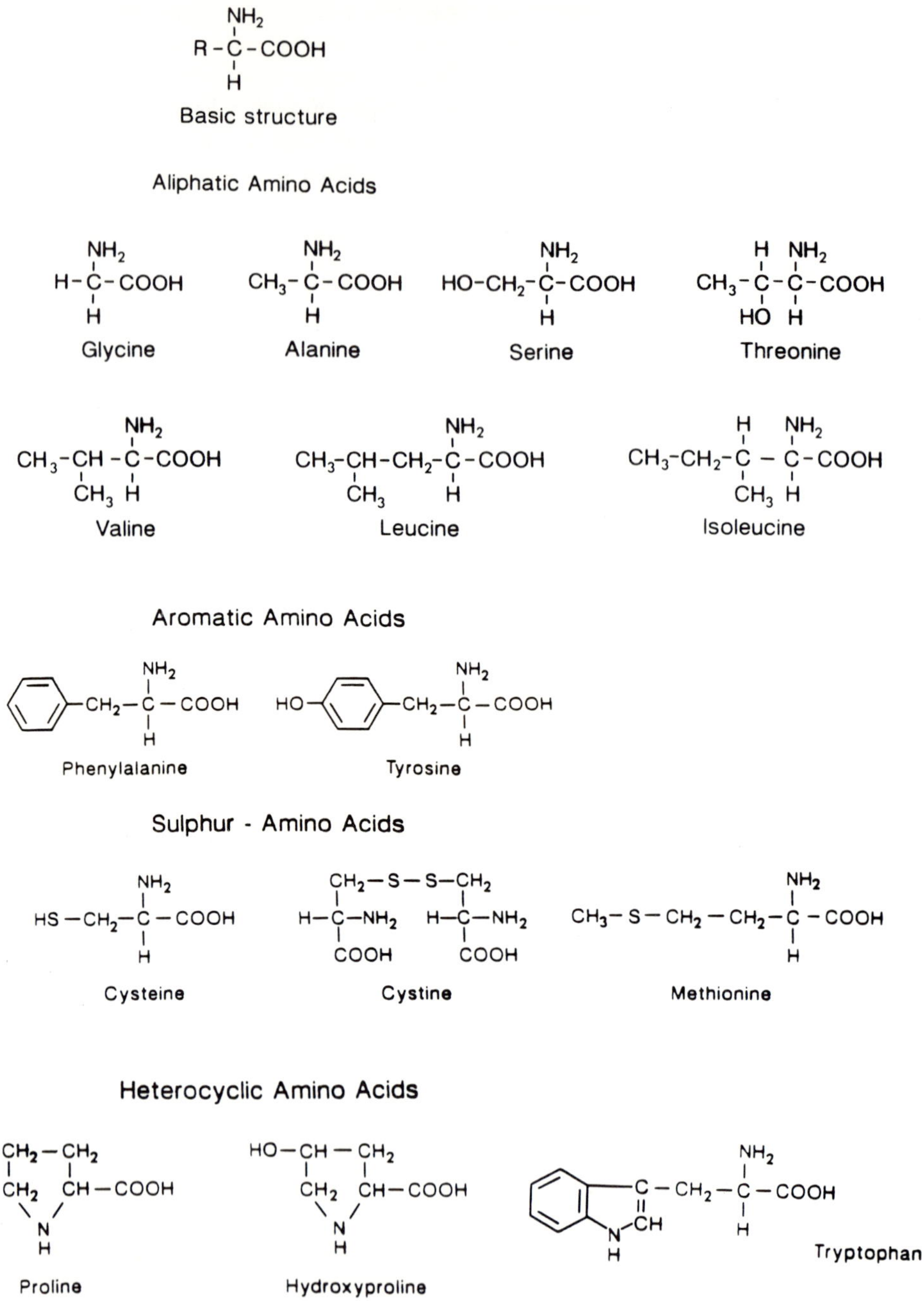

Fig. 4.2a Amino acids and proteins. Chemical structures of the amino acids making up the various amino acid series. (R denoted in the basic structural formula represents the side chain.)

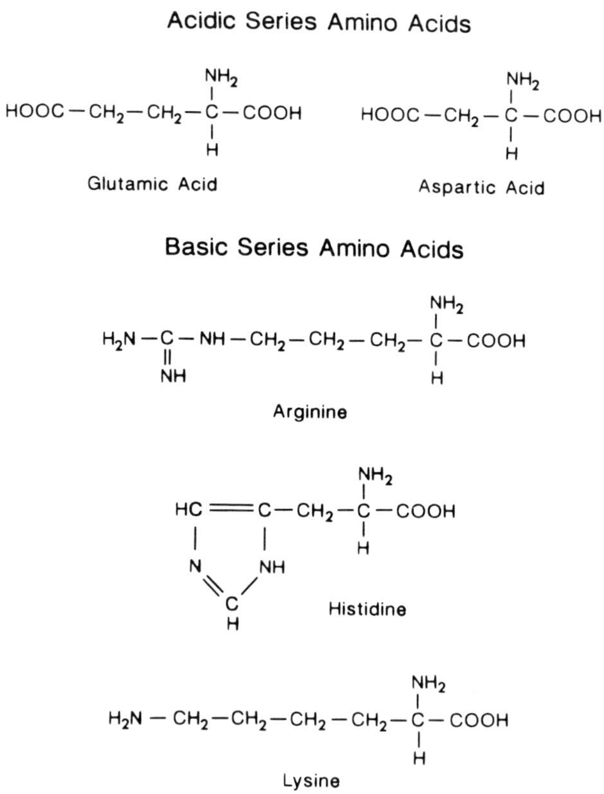

Fig. 4.2a *Continued.* Amino acids and proteins. (a) Chemical structures of the amino acids making up the various amino acid series. (R denoted in the basic structural formula represents the side chain.)

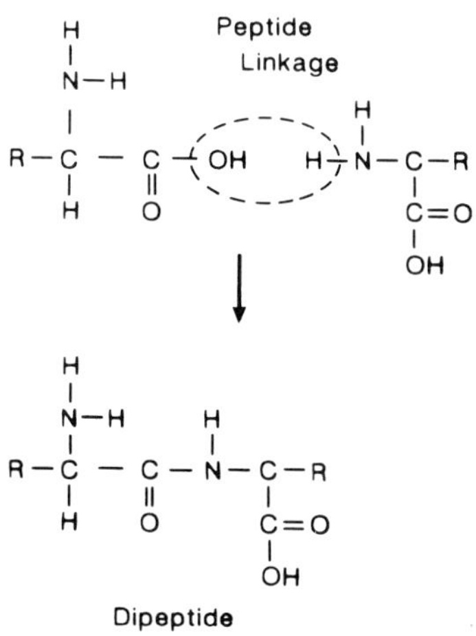

Fig. 4.2b Formation of a peptide linkage between two amino acids giving rise to a dipeptide. (R denotes the side chain of the amino acid.)

Fish and other animals are able to synthesize and interconvert some of the amino acids, but are incapable of *de novo* synthesis of others. Thus, from a nutritional point of view, the amino acids can be divided into non-essential and essential (indispensable) groups. The essential amino acids are those that the fish cannot synthesize, or cannot synthesize in sufficient quantity to enable the maintenance of good rates of growth.

Qualitative requirements for amino acids have been investigated using either growth or labelling studies. In growth studies, the fish are fed a purified diet containing all crystalline amino acids as the control, and other groups of fish are fed test diets similar to the control except that one amino acid has been excluded. Groups of fish that either do not grow, or that grow at rates markedly less than controls, are adjudged to have been fed a test diet deficient in an essential amino acid. In labelling studies, the test fish are injected with a radioactively labelled (usually ^{14}C) substrate, such as acetate or glucose, and it is assumed that the radioactive label will gradually be incorporated into compounds that the fish can synthesize. Thus, amino acids that contain the radioactive label are considered to be non-essential and those that do not contain any trace of label are considered essential.

Results of both types of study have revealed that ten of the amino acids are essential for fish species: arginine, histidine, isoleucine, leucine, lysine, methionine, phenylalanine, threonine, tryptophan and valine.

While these types of study indicate which of the amino acids are essential, they give no information about quantitative requirements, such as the levels of dietary inclusion necessary for the promotion of good rates of growth. Data about quantitative essential amino acid requirements are usually obtained by conducting a series of growth studies in which diets are formulated to contain graded levels of the particular amino acid of interest. Growth responses of the fish are recorded and dose-response curves are constructed by equating growth with the amount of amino acid in the diet.

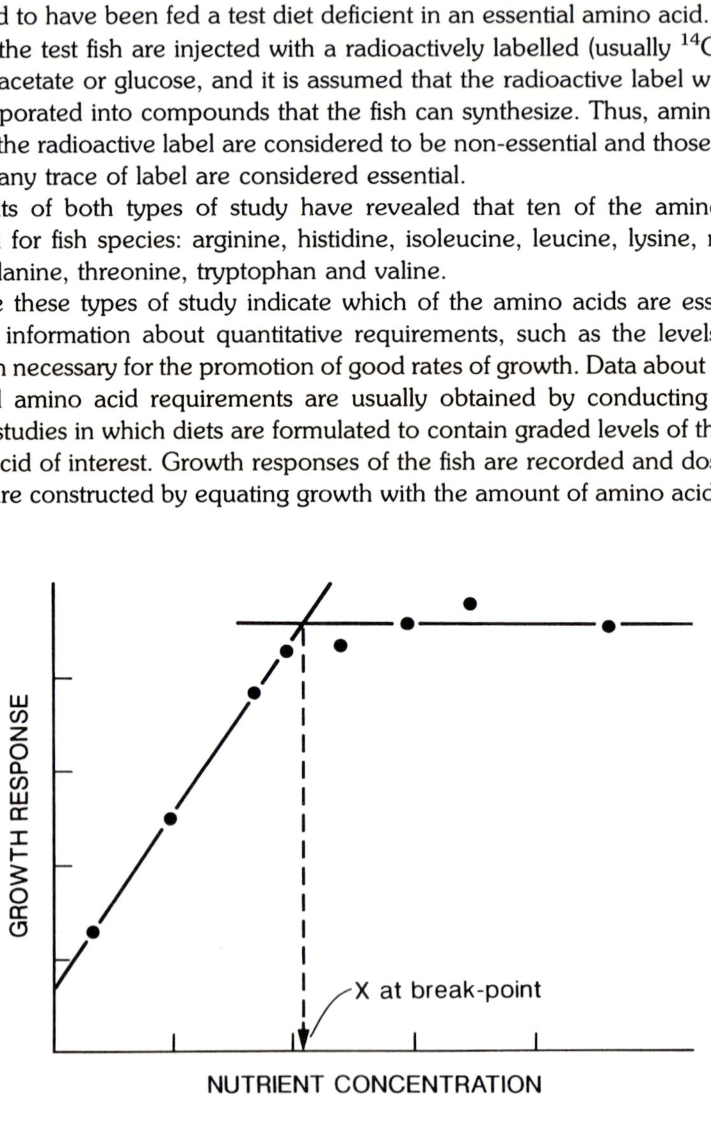

Fig. 4.3 Determination of the dietary requirement for an essential nutrient using analysis of the broken line response. (The intersection of the two regression lines indicates the break-point.)

The requirement is determined by calculating the break-point in the response curve, i.e. the point at which increased inclusion of the amino acid does not promote any further increase in growth response (Fig. 4.3).

It should be appreciated that studies of quantitative amino acid requirements are extremely time- and resource-consuming, and essential amino acid requirements have been examined in detail for only very few salmonid species. There tend to be some differences in the estimated requirements for particular amino acids, both between species and between studies carried out on any given species. In part, these discrepancies may be explained on the basis of methodological differences between studies, but it is also known that there are interactions between amino acids. Thus, the estimated requirement for one amino acid may be influenced by the presence of other amino acids in the diet. For example, phenylalanine and tyrosine are classified as aromatic amino acids. Adequate amounts of both phenylalanine and tyrosine are needed in order for the fish to maintain correct physiological function. Fish can readily convert phenylalanine to tyrosine (i.e. tyrosine is not an essential amino acid) or dietary tyrosine can be used to meet the metabolic needs for this amino acid. Phenylanlanine is an essential amino acid and, from the above, it is clear that the estimate of the requirement for phenylalanine will differ, depending upon whether the test diet contains adequate, or low levels of tyrosine. Similar interactions occur between the sulphur-amino acids. Methionine is an essential amino acid, but cystine can be synthesized from methionine, and if cystine is present in the diet, it reduces the methionine requirement. Because of this, fish are often considered to have a total sulphur-amino acid requirement, rather than a specific methionine requirement, and this is taken into account when amino acid requirements are being determined.

A summary of the essential amino acid requirements of rainbow trout, determined in growth studies, and recommended as dietary inclusion levels, is presented in Table 4.6 (column 1).

Several workers have observed that salmonids show improved growth and feed efficiency when diets are supplemented with essential amino acids to simulate levels

Table 4.6 Comparison of the dietary levels of essential amino acids required for growth and compositions of a range of tissues of rainbow trout, *Oncorhynchus mykiss*. (Values are presented on a percentage of protein basis.)

	Growth requirement	Yolk protein	Yolk-sac fry	Swim-up fry
Arginine	3.5	4.5	6.6	5.7
Histidine	1.6	2.2	2.9	2.5
Isoleucine	2.4	5.7	5.9	4.5
Leucine	4.4	9.1	9.9	7.9
Lysine	5.3	7.1	9.0	7.7
Methionine	1.8	1.2	3.3	2.9
Phenylalanine	3.1	3.8	5.6	4.5
Threonine	3.4	5.4	5.0	4.6
Tryptophan	0.5	–	–	–
Valine	3.1	8.2	7.7	5.9
Cystine	0.9	1.1	1.5	1.0

found in egg yolk or muscle protein. The essential amino acid requirements of certain fish species have also been shown to correlate reasonably well with the essential amino acid profile of whole body protein. Thus, body protein composition data may provide useful background information that can be used for the formulation of diets when the amino acid requirements of the fish have not been adequately established. The essential amino acid compositions of egg yolk and body proteins of young rainbow trout are shown in Table 4.6. Body proteins of some other salmonids have also been examined, and amino acid compositions appear to be remarkably similar. Thus, from the available information, it might be expected that the essential amino acid requirements of the different salmonid species should be similar.

In general, the essential amino acid compositions of the natural prey organisms consumed by salmonids, in both fresh and seawater are adequate to fulfil requirements. Of particular interest is the observation that the amino acid composition of freshwater zooplankton is very similar to yolk protein.

The amino acid profile of hen egg protein is also similar to that of fish egg protein, and hen eggs have been used successfully as a start-feed for a range of fish species. The amino acid content of yolk protein differs in a number of respects from both the amino acid profile of body protein and the requirements determined from growth studies carried out using juvenile fish (Table 4.6). These differences may indicate that the requirements of start-feeding fry differ from those of juveniles, but firm evidence on this point is lacking.

Commercial feed manufacturers endeavour to formulate feeds that contain protein sources to fully meet the essential amino acid requirements of the fish. Feeds containing high levels of good quality fishmeal will usually fulfil the needs, but many feeds also contain other protein sources of single-cell, animal or plant origin.

Fishmeals and oils are produced from industrial fish species (Table 4.7), with yields and compositions of the finished products being influenced by both the species and by the seasonal variations in compositions of the fish used as raw materials. Several types of fishmeal are produced, and it is increasingly common that 'special quality meals' (e.g. NorSeaMink, LT-Meal) are used in the manufacture of trout and salmon feed. 'Special quality meals' can be produced from a variety of fish species, but there are standards required to be met in regard to transport and preservation of the raw materials, and with respect to production procedures. Raw materials destined for the production of 'special quality meals' should be transported under cold

Table 4.7 Chemical composition of some industrial fish species used in the production of fish meals and oils.

Species		Percentage composition			
		Water	Protein	Lipid	Ash
Mallotus villosus	'Winter' Capelin	76.6	13.5	8.4	1.9
	'Summer' Capelin	72.2	12.4	14.6	1.7
Sprattus sprattus	Brisling	68.5	16.1	13.0	2.7
Ammodytes marinus	Sandeel	72.0	16.6	9.5	2.2
Scomber scombrus	Mackerel	62.7	17.4	20.6	1.9
Micromesistius poutassou	Blue whiting	76.3	16.7	3.5	3.5
Trisopterus esmarkii	Norway pout	73.0	16.3	8.5	2.7
Boreogadus saida	Polar cod	76.9	12.9	6.4	2.6

storage, without the addition of chemical preservatives, and should be delivered as fresh as possible. Some autolysis and deterioration, measured as an increase in volatile nitrogen (mostly present as trimethylamine), is permitted. Levels of volatile nitrogen should not exceed 90 mgN and 50 mgN per 100 g in raw material used in the production of NorSeaMink and LT-meal respectively (an upper limit of 30 mgN per 100 g is normally imposed on fish for human consumption).

Fishmeals are produced by 'cooking' the raw materials at a temperature of 70–95°C, followed by pressing, separation and drying. During the 1970s, the LT-(low temperature) process was developed. This allows 'cooking' both at lower temperatures (50–60°C) and for a shorter duration than in the traditional process. The LT-process, nevertheless, gives improved separation of the oil and water fractions, and the milder treatment results in the production of meals of improved quality when compared with those produced using traditional methods. Fishmeals of 'special quality' will usually contain 5–10% moisture, up to 10% lipid and at least 70% protein. Antioxidant is added to the finished product in order to improve storage properties.

While the drying processes which result in the production of fishmeal are convenient for preserving industrial fish, other methods of preservation of fish and fish waste are also available. Commercial fish farmers who manufacture their own wet- or moist-feeds on site may store wet ingredients in frozen block form. A freeze-storage facility is quite expensive to run, and the length of time the ingredients can be held under such conditions without undergoing deterioration is limited to a few weeks or, at most, months. An alternative method of preserving fish and fish waste is by means of treatment with acid. When fish is acid-preserved prior to hydrolysis, the product is known as fish silage.

Production of fish silage involves the thorough mixing of the fish raw material with the acid preservative to create conditions unsuitable for the growth of spoilage bacteria and other micro-organisms. Either mineral (e.g. sulphuric) or organic (e.g. propionic, formic) acids can be used as preservative agents so long as the pH of the ensiled mixture remains low (pH approximately 4 or below), and antifungal agents are added. Use of mixtures of propionic and formic acids allow the preservation of fish waste at a pH of about 4.5, and when these acids are used the pH of the silage remains stable for prolonged periods of time. The silage produced using these organic acids has a fluid consistency that enables easy pumping and tank transport of the ensiled mass.

During acid preservation, the proteolytic enzymes present in the fish or fish waste hydrolyse the proteins resulting in a liquid product. The ensiled mass will tend to form three fractions: an aqueous fraction containing the hydrolysed proteins, a lipid fraction and a bottom sediment. This product can be used as it is, the different fractions may be separated, or the silage may be concentrated by partial drying to remove some of the water content.

Fish silage is usually mixed with other ingredients, such as fishmeal, vitamins, minerals and binder before being fed to the fish in the form of a moist pellet. While ensiling provides a cheap and effective method for the preservation of industrial fish and the waste from fish-processing plants, there may be some limitations to the use of fish silage in the production of feeds for salmonids. First, the fact that the silage is liquid means that a number of dry ingredients must be added in order to form a moist pellet having reasonable properties of water stability. Second, the process of acid-preservation can lead to the destruction of some essential nutrients, notably the amino acid methionine, and, finally, the acids added to the ensiled mixture may affect palatability of the feed leading to reduced acceptance by the fish.

Oilseed meals (derived from peanut, sunflower and cottonseed) are deficient in the amino acids, lysine and methionine, so are not particularly suitable as major protein sources in fish feed. Soybean and canola meals have better essential amino acid profiles than other oilseed meals, but other factors may limit the levels at which these meals can be included in fish feeds.

Soybeans contain several anti-nutritional factors. Some, such as the trypsin inhibitors, may be destroyed or inactivated by the heating and drying procedures involved in the production of soybean meal. Other anti-nutritional factors such as phytic acid, are, however, generally unaffected by processing. Phytic acid can both reduce the availability of minerals and reduce the protein digestibility of salmonid diets. Thus, the potential for the use of soybean meal in salmonid diets may be somewhat limited, and soybean protein concentrate may prove to be a better alternative. The use of protein concentrates surmounts many of the problems associated with the presence of anti-nutritional factors, and provides the fish with a source of protein having an essential amino acid profile that, with the exception of methionine, would seem to fulfil the known requirements for growth.

Meals derived from grains and grain by-products are generally rich in starch, and some grains are included in commercial feeds in order to improve binding properties. The proteins present in most grains are, however, poorly balanced and are deficient in lysine. It is now common practice for grains to be partially precooked or extruded before they are added to the remaining ingredients comprising a commercial fish feed. This pretreatment serves to increase the gelatinization of the starch, leading to improved binding properties, and also increasing the biological availability of the carbohydrate. Pretreatment may not, however, only result in positive effects since starch and non-reducing sugars, such as sucrose, present in the grains may be hydrolysed to form reducing sugars. Reducing sugars (glucose, fructose, lactose and maltose) may react with a free amino group on an amino acid and this chemical reaction can have important nutritional consequences. The essential amino acid lysine is particularly prone to this type of reaction, known as non-enzymatic browning (Maillard reaction), so that considerable care must be taken in order to keep lysine losses to a minimum during the course of pretreatment and extrusion processes.

Meals derived from various animal by-products may often have a relatively high ash content, and digestibility is often lower than that of fishmeal. Thus, availability of the amino acids may be poorer than a chemical analysis would suggest. Animal by-product meals are often rich sources of the essential amino acids lysine and histidine, but they may be deficient in the sulphur-amino acids.

The published values for the amino acid compositions of ingredients and feeds are based upon chemical analyses of the proteins. The results of these analyses may suggest that the essential amino acid requirement is met, but a chemical analysis will not reveal to what extent the different nutrients and amino acids are available to the fish. Consequently, biological tests and growth trials are an important feature of feed development.

There is relatively little information about the availabilities of specific amino acids comprising the various protein sources, but it is known that protein digestibilities vary widely from feedstuff to feedstuff (Tables 4.3 and 4.4). As a rule-of-thumb, when the digestibility of the individual amino acids is not known, the digestibility of the whole protein may be assumed when estimating the availability of amino acids in various feedstuffs. When estimating availability of nutrients and amino acids to fish, the type of processing and storage of feed ingredients must also be taken into consideration. For example, heat treatment (95°C), freezing and drying can result in reduced availability of sulphur-amino acids, and storage of protein mixtures at 37°C for 40 days has been reported to lead to an 80% reduction in lysine availability.

Lipids and fatty acids

Lipids are a large, varied group of water-insoluble organic compounds. In both natural prey and commercial feeds, the lipids represent concentrated energy sources for the fish. Lipids also have other important roles, such as being the transport medium for lipid-soluble compounds (e.g. vitamins A, D and E), as structural

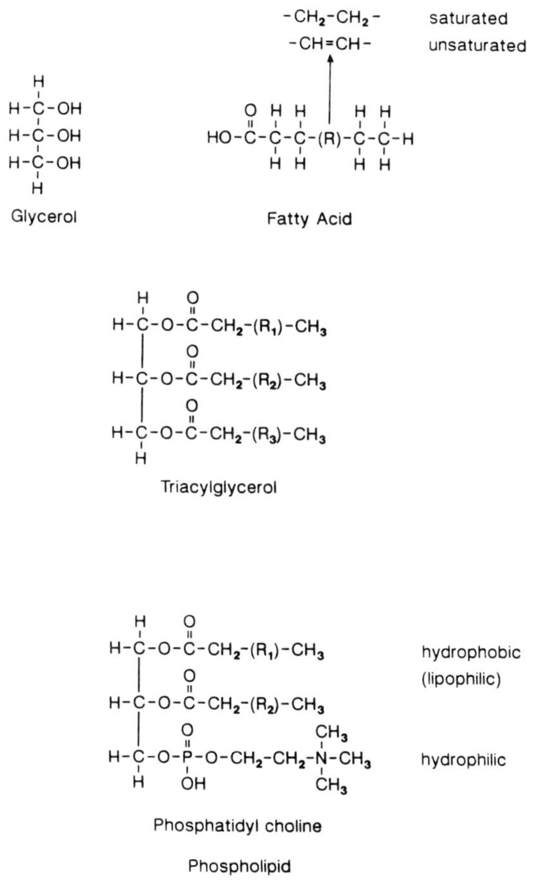

Fig. 4.4 Chemical structures of glycerol, fatty acids, triacylglycerols (or triglycerides) and phospholipids. (R denotes the carbon chain of the fatty acid, which may or may not contain double bonds between carbon atoms. In the triacylglycerols (triglycerides) and phospholipids the fatty acids (indicated by R_1, R_2 and R_3) esterified with the glycerol may be either the same or different acids.)

Shorthand Formula	Structure	Melting Point (°C)
	I Saturated Fatty Acids	
16:0	$CH_3\,(CH_2)_{14}\,COOH$	63
20:0	$CH_3\,(CH_2)_{18}\,COOH$	76.5
24:0	$CH_3\,(CH_2)_{22}\,COOH$	86
	II Unsaturated Fatty Acids	
18:1 (n-9)	$CH_3\,(CH_2)_7\,CH = CH\,(CH_2)_7\,COOH$	13.5
18:2 (n-6)	$CH_3\,(CH_2)_4\,CH = CH\,CH_2\,CH = CH\,(CH_2)_7\,COOH$	-5
18:3 (n-3)	$CH_3\,CH_2\,CH = CH\,CH_2\,CH = CH\,CH_2\,CH = CH\,(CH_2)_7\,COOH$	-11
20:4 (n-6)	$CH_3\,(CH_2)_4\,CH = CH\,CH_2\,CH = CH\,CH_2\,CH = CH\,CH_2\,CH = CH\,(CH_2)_3\,COOH$	-49.5

Fig. 4.5 Examples of fatty acids illustrating methods of classification, and the effects of carbon chain length and degree of unsaturation on melting points.

elements in cell membranes and as precursors for the synthesis of a number of hormones, pigments and growth factors.

The class of lipids most frequently encountered is the triglycerides (or triacylglycerols). Triglycerides represent the main energy storage compounds in animals. Chemically, these are esters of glycerol and fatty acids, with one molecule of glycerol linking with three similar or different fatty acids (Fig. 4.4).

Some marine crustaceans contain large quantities of wax esters, which are esters produced by the linking of a long-chained alcohol with a fatty acid.

Fatty acids differ in numbers of carbon atoms in the molecule, and in the number and positioning of double bonds between carbon atoms. A shorthand system has been devised for classifying the fatty acids (Fig. 4.5). For example, the shorthand formula 16:0 represents a fatty acid with 16 carbon atoms and no double bonds, and 18:3(n–3) represents an 18-carbon fatty acid with 3 double bonds, and (n–3) defines the positioning of the first double bond with respect to the methyl end of the fatty acid chain. Fatty acids found in terrestrial plants and animals generally have carbon chain lengths of 14–18, but longer chain fatty acids, with up to 22 carbons, are commonly encountered in aquatic organisms.

Both the chain length and the number of double bonds (degree of saturation) determine the physical, chemical and nutritional properties of fatty acids. For example, triglycerides, such as pork fat and beef tallow, that contain saturated (no double bonds) fatty acids are solid at room temperature, and have lower digestibility values than triglycerides containing longer-chain unsaturated fatty acids. Triglycerides derived from the seeds of a number of plant species, such as soya, peanut and sunflower, contain quite high levels of unsaturated fatty acids and are fluid oils at room temperature. Marine fish oils, which usually contain a relatively high proportion of long-chain, highly-unsaturated fatty acids have low melting points, are fluid at room temperature, and are well digested and absorbed by fish species. Fish oils are therefore the lipids of choice for inclusion in fish feeds.

Terrestrial and aquatic lipids differ not only in the degrees of saturation of the fatty acids found in the triglycerides, but there are also differences in the unsaturated fatty acids typical of food chains in these environments. The (n–6) fatty acids are typically found in terrestrial and freshwater environments, while in marine systems (n–3) fatty acids tend to dominate. The food chain in fresh water is characterized by 18:2(n−6), 18:3(n−3) and 20:5(n−3) fatty acids, and freshwater fish may obtain substantial amounts of 20:4(n−6) by consuming insects. The fatty acid profiles of marine phyto- and zooplankton tend to be dominated by 18:3(n−3), 20:5(n−3) and 22:6(n−3) and this pattern is generally retained at higher levels in the food chain. Thus, when the total lipid contents of wild-caught freshwater and marine fish are compared with respect to fatty acid composition, large differences are often found. These differences reflect differences in the fatty acid compositions of the prey present in the two environments.

The vast majority of the lipid extracted from a fish will be made up of storage triglycerides (or neutral fats), but the total lipids will also contain smaller quantities of other lipid classes such as phospholipids (Fig. 4.4).

Phospholipids have important functions in the determination of cell membrane structure and fluidity, and may also act as precursors for the synthesis of a series of biologically active compounds. Consequently, recent research has been directed towards the examination of the fatty acid profiles of the phospholipids. The phospholipids, like the triglycerides, are esters of glycerol. Two of the alcohol groups are esterified with fatty acids, but the third is esterified with phosphoric acid, which in turn is linked to a nitrogenous base, the amino acid serine, the sugar alcohol inositol, or glycerol phosphate. The nature of the latter – choline, cholamine (or amino ethyl alcohol), serine, inositol – is used in the naming of the various phospholipids, e.g. phosphatidylcholine, phosphatidylethanolamine, phosphatidylserine, phosphatidylinositol. The phospholipids combine within the same molecule both the hydrophilic (water-loving) phosphate chain, and the hydrophobic fatty acid chains. They are therefore surface active and play important roles as emulsifying agents, and as components of membranes in biological systems. The major membrane phospholipids are phosphatidylcholine and phosphatidylethanolamine.

Phosphatidylinositol, which is present in much smaller quantities, appears to be involved as a mediating agent in a number of metabolic processes. The sphingomyelins do not contain glycerol, but are made up of fatty acids, phosphoric acid, choline and sphingosine. They contain phosphorus, and have similar biological occurrence to the phospholipids, and the sphingomyelins are, therefore, often included in the phospholipid class.

Phospholipid fatty acid profiles tend to be similar, regardless of whether the fish come from marine or freshwater environments, and the long-chained, polyunsaturated fatty acids, particularly 20:5(n−3) and 22:6(n−3), are present in large amounts. The polyunsaturates usually comprise over half the fatty acids of phospholipids, whereas triglycerides are usually dominated by saturates and monounsaturates. Polyunsaturates may only comprise 30% or less of the fatty acids present in the storage lipids.

The polyunsaturated fatty acids present in the phospholipids have important structural and physiological functions, and a dietary supply of polyunsaturates has been found to be essential for both fish and other animals. Terrestrial animals require both (n−6) and (n−3) fatty acids, particularly the former, and these are generally supplied in the diet as 18:2(n−6) and 18:3(n−3). The animals are able to convert and chain elongate these fatty acids into the longer chain polyunsaturates found in the phospholipids (Fig. 4.6). In the earliest work on fish nutrition it was assumed that the essential fatty acid requirements of fish were similar to those of terrestrial animals,

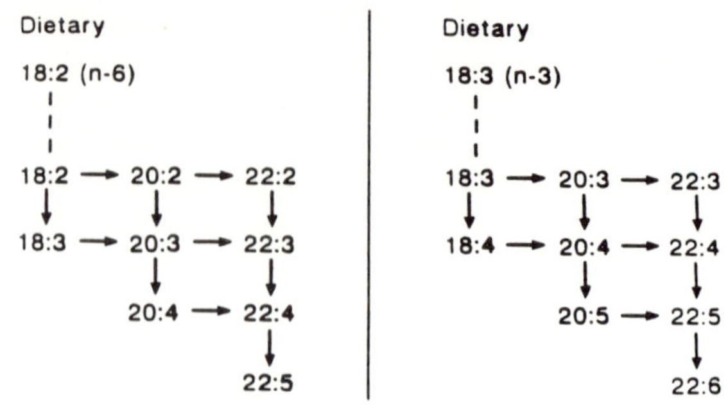

Fig. 4.6 Pathways for interconversion (elongation and desaturation) of dietary fatty acids of the (n−6) and (n−3) series.

and dietary lipids were often supplied in the form of plant oils rich in 18:2(n−6), but poor in the (n−3) series of fatty acids.

Later studies established that most fish species, including the salmonids, have a dietary requirement for (n−3) fatty acids, but can grow well on diets low in (n−6). In a series of experiments carried out on rainbow trout, it was found that the fish were capable of completing the full life cycle on diets containing (n−3) fatty acids, but could not grow when fed diets containing only (n−6) series fatty acids. Thus, the impression was gained that (n−3), but not (n−6), fatty acids were essential for rainbow trout. However, when the fatty acid profiles of the tissue and egg phospholipids of the trout fed the supposedly (n−6)-free diet were analysed, it was found that small amounts of (n−6) fatty acids, particularly 20:4(n−6), were present. The fatty acid 20:4(n−6) may act as a precursor for prostaglandin synthesis, and this group of compounds is known to have important physiological roles in the regulation of respiratory and reproductive functions. It may also be noted that while the major phospholipids, phosphatidylcholine and phosphatidylethanolamine, are rich in the (n−3) polyunsaturated fatty acids, phosphatidylinositol (PI) contains large amounts of 20:4(n−6). PI is metabolically very active, being involved in the transduction of hormone signals through membranes, and in fish, PI appears to be intimately involved in the regulation of gill function. Thus, the dietary requirement of fish for (n−6) fatty acids, particularly 20:4(n−6), may be small, but is nevertheless important.

In practical dry feed, (n−6) fatty acids will be supplied via the inclusion of some plant oils in the diet, and fish oils also contain some 20:4(n−6). Pond-reared trout would be expected to obtain quantities of 20:4(n−6) via consumption of natural prey such as insects and earthworms present in the pond system. For sea-cage farmers formulating their own wet- or moist-feeds, ingredients such as krill and squid can be used to supply a dietary source of 20:4(n−6).

The essential fatty acid requirements of salmonids may generally be met by the inclusion of 1−2% 18:3(n−3) and traces of (n−6) fatty acids in the diet. High dietary levels of 18:2(n−6) can, however, lead to growth depression. Since some of the 18:3(n−3) is converted to longer chain fatty acids before being incorporated in phospholipids, it might be expected that the longer chain acids would be more effective than 18:3(n−3) in meeting essential fatty acid requirements. This has been confirmed in studies with trout, in which 20:5(n−3) and 22:6(n−3) were found to give positive growth benefits at levels of dietary inclusion as low as 0.25% of each of these long-chain fatty acids. Many marine species, unlike salmonids, are unable to elongate 18:3(n−3) and the essential fatty acids must, therefore, be provided as the long-chained 20:5(n−3) and 22:6(n−3) forms. While the inclusion of 1−2% 18:3(n−3) in the diets of rainbow trout are usually sufficient to promote good growth, these levels may be inadequate if the diet also contains large amounts of other fatty acids, especially those of the (n−6) series. Consequently, the recommendation has been amended to 20% of dietary lipid as 18:3(n−3) or 10% of dietary lipid as the longer chained polyunsaturates, 20:5(n−3) or 22:6(n−3). This modification has been made in order to ensure that a dietary balance is maintained between the (n−3) and (n−6) series fatty acids, when lipid-rich diets are fed. The dietary requirements for long-chained fatty acids are usually met by formulating diets to include fish oils of marine origin.

The recent upsurge of interest in lipid nutrition of fish has not only been concerned with ensuring that the diet provides sufficient essential fatty acids to promote good growth of the fish. The possible advantages of fish lipids in human nutrition have attracted increasing attention, as has the realization that there are potential marketing benefits in producing farmed fish with fatty acid profiles similar to their wild-caught counterparts.

During the course of the past two decades there has accumulated considerable evidence of links between nutrition and certain human diseases, for example, the association between high fat diets rich in saturates and the incidence of heart disease. The fact that the flesh of wild-caught fish is composed of high-quality proteins and lipids that are rich in monounsaturated and polyunsaturated fatty acids, should give fish a nutritional advantage over animal flesh that is higher in saturates. The longer chain polyunsaturates – 20:5(n−3) and 22:6(n−3) – are deemed to be especially important for human health, both with respect to aiding in the prevention of heart disease and in the reduction of inflammatory disorders. Fish are the only source of these long-chain polyunsaturated fatty acids that are regularly consumed by humans.

Most of the lipid present in the edible part (fillet) of salmonids will be storage triglyceride, and, as has been stated earlier, the fatty acid profile of the triglycerides in fish tends to reflect the composition of the diet. Wild salmonids consume the fauna found in freshwater and marine environments – prey naturally high in (n−3) fatty-acids – so the storage triglycerides of wild fish will be high in the long-chain polyunsaturates. The 20 and 22 carbon fatty acids may account for 15−20% of the total fatty acids. The fatty acid profiles of the fillet of farmed salmonids are governed by the composition of the diet provided. Fish fed diets in which plant oils and animal fats have been used as the main lipid sources will have storage lipids in which the fatty acids are dominated by 18:2(n−6) and saturates, obviously an undesirable profile from a human health point of view. On the other hand, if the farmed fish are fed diets formulated to include marine fish oils as the major lipid source, the storage lipids in the fillet may be expected to contain high levels of the beneficial long-chain polyunsaturates. Results of recent research have demonstrated that, by using fish-oil based diets, it is possible to produce farmed salmonids having similar fatty acid profiles to wild fish. It should, therefore, be possible, via dietary manipulation, to improve the nutritional value of the saleable product.

The long-chain polyunsaturated fatty acids are particularly prone to oxidation and destruction during extraction and long-term storage. If the nutritional benefits of fish oils are to be retained, it is therefore of utmost importance that treatment processes be as mild as possible and that oils are protected from oxidative damage during storage.

At present the world production of marine fish oils would appear to be more than adequate to meet the demands of the aquaculture industry, and there would appear to be few rational arguments for the large-scale inclusion of plant or animal lipids in diets for salmonids. Nor would there seem to be any acute need to search for alternative sources of (n–3) fatty acids.

Vitamins

Vitamins are organic compounds required in the diet in small amounts. They are essential for the maintenance of health and growth. The vitamins are classified as water soluble and lipid soluble. Eight of the water soluble vitamins, the B complex, have coenzyme functions and are required in small quantities. Three other water soluble vitamins, ascorbic acid (Vitamin C), myoinositol and choline, are required in larger quantities and these compounds are sometimes referred to as the macro-vitamins. Vitamins A, D, E and K are the lipid soluble vitamins. Salmonids require all 15 of these vitamins in their diet, but the precise requirements for fish of different sizes reared under various conditions have not been established. Minimum requirements of small fish held under controlled laboratory conditions have been investigated (Table 4.8), but these relate to the levels required to prevent the appearance of deficiency symptoms.

Most vitamins are partially destroyed during pelleting and long-term storage (Table 4.9), so over-fortification of practical diets is usually carried out. This is

Table 4.8 Vitamin requirements, or levels of recommended dietary inclusion, for a range of fish species. Values presented as either mg vitamin per kg dry diet or IU (International Units) of vitamin per kg dry diet.

	Channel catfish *Ictalurus punctatus*	Carp *Cyprinus carpio*	Chinook salmon *Oncorhynchus tshawytscha*	Dietary recommendation (salmonids)
Thiamin	1	R	~15	15
Riboflavin	9	~15	~25	30
Pyridoxine	3	~6	~20	20
Pantothenic acid	15	~50	~50	50
Niacin	14	28	~200	200
Folic acid	R	–	~10	10
Cyanocobalamin (B_{12})	R	–	~0.02	0.02
myo-Inositol	NR	440	~400	400
Choline	400	~4000	~800	3000
Biotin	R	1	~1.5	1.5
Ascorbic acid (Vitamin C)	60	R	~150	200
Vitamin A	~2000 IU	10 000 IU	~2500 IU	2500 IU
Vitamin D	~500 IU	–	–	2400 IU
Vitamin E	50	100	30	50
Vitamin K	R	–	–	10

Notes:
R = Vitamin required, but quantitative requirement uncertain.
NR = Vitamin probably not required to be included in the diet.

Table 4.9 Information available about the factors affecting the stability of vitamins.

Vitamin	Heat	Light	Oxygen	Humidity	Acid	Alkali
Water soluble						
Thiamin	+	−	+	+	−	+
Riboflavin	−	+	−		−	+
Pyridoxine	+	+				
Pantothenic acid	+	−	−	+	+	+
Niacin	−	−	−	+	−	−
Folic acid	+	+	+		+	−
Cyanocobalamin (B_{12})	−	+	+		−	−
Biotin	+	−	−		+	+
Ascorbic acid (Vitamin C)	+	+	+		−	+
Lipid soluble						
Vitamin A	+	+	+	+	+	−
Vitamin D	+	+	+		−	+
Vitamin E	+	+	+		−	+
Vitamin K	−	+	−		+	+

Notes:
\+ = Stability affected, with there being some destruction of the vitamin.
− = Stability unaffected or only slightly affected.

generally effective in preventing the development of vitamin deficiency symptoms in farmed fish.

Two water soluble vitamins, thiamin and ascorbic acid (Vitamin C) appear to be the most sensitive to processing and storage procedures. Considerable efforts have been directed towards the improvement of the stability of these vitamins, particularly ascorbic acid (vitamin C). The problem has been investigated by applying two different approaches. One approach involves the testing of relatively stable vitamin derivatives in the hope that a stable compound having similar properties to the unstable vitamin will be found. The alternative is to examine the efficiency of a number of coatings in the protection of the vitamin against destruction. Recent evidence suggests that polymer-coating of ascorbic acid may serve to protect and improve the stability of the vitamin during processing and storage. Furthermore, the coating does not appear to seriously reduce biological availability. Approximately 20% of the polymer-coated ascorbic acid appears to be lost during processing, but 50% or more may remain even after 3–4 months' storage at room temperature.

A number of ascorbate derivatives have been shown to be more resistant to destruction during processing, and more stable than ascorbic acid during long-term storage. Several studies have been conducted in order to determine whether these stable derivatives have vitamin C activity. Most work has been directed towards the investigation of the biological activity of ascorbate-2-sulphate and ascorbate-2-monophosphate, although studies have also been carried out using other ascorbic acid derivatives. Although ascorbate-2-sulphate appears to be much more resistant to destruction than ascorbic acid, the results of feeding trials indicate that the biological activity of this derivative is low. The low activity of ascorbate-2-sulphate may be due to a limited absorption of this derivative from the gut of the fish. On the other hand, when ascorbate-2-monophosphate has been included in the diet, the biological activity has been found to closely resemble that of ascorbic acid. Since this derivative both displays vitamin C activity and is relatively resistant to destruction during processing and storage, it must be considered as being a viable substitute for ascorbic acid in the manufacture of fish feeds. When both stability and vitamin C bioactivity are taken into account, it appears that the phosphate derivatives of ascorbic acid are superior to ascorbic acid, sulphate and palmitate derivatives as vitamin C sources in fish diets.

Minerals

Not all the inorganic elements found in the fish body are required to be included in the diet, but the quantitative mineral requirements of fish have been inadequately studied. Fish are capable of absorbing some dissolved minerals from the water and it is almost impossible to prepare the mineral-free diets needed in order to carry out accurate experimental studies.

Salmonids, especially the juvenile stages in freshwater, are thought to obtain the majority of their minerals via the diet, rather than by direct absorption from the water, and approximate dietary requirement levels for some minerals have been determined (Table 4.10).

Almost all necessary minerals will be available in most practical feed formulations, without there being any need for additional supplementation. Nevertheless, mineral deficiencies do sometimes occur in farmed salmonids. This is almost always linked to reduced mineral bioavailability, rather than to a dietary mineral deficiency *per se*. Reduced bioavailability may be associated with dietary imbalance, presentation of the mineral in unsuitable form, or interactions with other dietary ingredients. For example, phytic acid present in some plant protein sources may bind trace elements such as iron, copper and zinc, and thereby reduce bioavailability. Fishmeals are often rich in ash, and the high calcium levels present in the ash can inhibit absorption of other divalent cations such as zinc. Consequently, it has often been thought to be

Table 4.10 Mineral requirements, or recommended levels of dietary inclusion, for a range of fish species. (Values presented as mg mineral per kg dry diet.)

	Channel catfish *Ictalurus punctatus*	Carp *Cyprinus carpio*	Tilapia *Oreochromis niloticus*	Chinook salmon *Oncorhynchus tshawytscha*	Dietary recommendations (salmonids)
Calcium	4500	–	–	–	3000
Phosphorus	4500	~7000	9000	R	6000
Magnesium	400	~500	~750	–	500
Iron	30	150	–	–	30–60
Copper	5	3	~4	–	~5
Manganese	~2.5	13	12	R	~20
Cobalt	–	0.1	–	–	5–10
Zinc	20	~20	10	R	30–100
Iodine	–	–	–	R	~0.3
Selenium	–	–	–	R	~0.3

Note:
R = Mineral required but quantitative requirements uncertain.

advisable to fortify diets containing high levels of fish and soybean meals with a mixture of trace elements. Extreme overfortification is, however, to be avoided since this may have detrimental effects on health and growth performance of the fish. For example, excess dietary iron may have significant negative impacts upon resistance to infectious diseases and fish health in general.

There may be interactions between the levels of water-borne minerals and the dietary requirement, as exemplified by the requirements of juvenile rainbow trout (approximately 1 g) for magnesium. When the fish were held in water containing high levels (1000 mg/l) of dissolved magnesium, approaching the concentrations found in sea-water, the fish became anorexic, body loadings of magnesium increased and considerable mortality was observed within the space of a few days. This suggests that magnesium was entering the fish from the water and that the high levels of water-borne magnesium resulted in toxicity. On the other hand, when fish were held in water with a relatively low magnesium concentration (1.5 mg/l) anorexia and poor growth was observed in groups of fish fed on diets containing less than about 400 mg/kg diet. In this case magnesium deficiency seems to have been the problem for fish fed diets low in magnesium while being held in freshwater with a low magnesium content. When the fish were held in water containing higher concentrations of magnesium (20–150 mg/l), a dietary inclusion of 100 mg/kg was more than adequate to meet the requirement for good growth, and dietary inclusion levels of 700 mg/kg diet gave rise to increased body loading of magnesium. Thus, it is clear that the fish are able to meet some, if not all, of their requirement for magnesium via water-borne sources, and the level to be included in the diet will depend upon the magnesium content of the water in which the fish are reared. In many studies of mineral requirements, the possible contribution of the water-borne mineral source has not been adequately investigated, leading to there being marked disparity in the published values of the dietary requirements for minerals in fish. For example, the range of dietary magnesium requirement may be 50–1300 mg/kg diet, being about 50 mg/kg diet when the water contains high levels of dissolved magnesium, and possibly up to 1300 mg/kg diet when the fish are held in magnesium-free water.

Other feed additives

In addition to the food ingredients which have direct nutritional properties, commercial fish feeds may contain additives of various types. For the most part, these additives – binders, antioxidants, pigments, drugs and antibiotics, antifungals, flavourings and attractants – have no major nutritional value, but they are of practical importance in a number of ways.

Binders

Gelatinized starch is the major binding agent in dry pellet feeds, but agents such as bentonite and lignin sulphonate may also be added at low levels. Wet- and moist-pellet feeds require the use of both nutritive and non-nutritive binders, the most commonly used non-nutritive binding materials being carboxymethylcellulose, alginates and various gums. There is no evidence that the commonly used organic binders have serious detrimental effects on the growth of the fish, but some binders can disturb digestive functions resulting in reduced feed digestibility.

Antioxidants

Oxidation of fatty acids, particularly the long-chained polyunsaturates, present in the lipids can cause reduction of the nutritional value of the feed, and antioxidants are added to prevent lipid oxidation. Oxidation of unsaturated fatty acids results in the feed going rancid, with the development of off-flavours and odours. The oxidation products may be toxic to the fish, especially if the feed is low or deficient in vitamin E and selenium. Vitamin E can act as a natural antioxidant, but the synthetic antioxidants (BHA) (butylated hydroxyanisole), BHT (butylated hydroxytoluene) and ethoxyquin are usually included in formulated feeds at levels of about 0.015%.

Pigments

The pink or red colour of the flesh of salmonids is a characteristic feature of this group of fishes. This coloration is due to absorption and deposition of carotenoid pigments. Carotenoids are the most numerous and widespread pigments to occur in nature, but only plants and protists are able to synthesize carotenoids. Salmonids, like other animals, do not synthesize carotenoids *de novo*, nor are salmonids capable of extensive oxygenation of carotenoids, and oxygenated carotenoids are often deposited with only limited modification. In wild salmonids, the carotenoids are present as a mixture of astaxanthin and its esters, these carotenoids being obtained, primarily, via the consumption of zooplankton and crustaceans. These carotenoid sources are not generally available to farmed fish. Thus, in order to get the colour the consumer associates with the flesh of the salmonid group of fishes, a pigment source must be provided in the diet. The flesh of farmed salmonids will usually be found to contain some astaxanthin and its esters, but the pigment picture may be more complex for farmed than for wild salmonids because a wide range of food ingredients, containing a variety of pigments, may be used in formulating fish feeds. In addition, alternative forms of carotenoids, such as canthaxanthin, have been added directly to formulated feed as a pigment source for farmed salmonids.

Carotenoid supplementation of diets is achieved either by adding pigment-containing natural products, extracts of natural products or chemically-synthesized pigments. The pigment sources most often used in wet and moist feeds are zooplankton, krill and wastes from shrimp/prawn/crab processing. Synthetic pigments, predominantly canthaxanthin, have, hitherto, usually been added to commercially produced dry feeds, but there is increasing use of astaxanthin in these feeds. Certain algae and yeasts, such as *Phaffia*, contain astaxanthin pigments and some species have been investigated as being potential carotenoid sources for farmed salmonids. Selection experiments carried out with *Phaffia* yeasts have led to the development of strains which are particularly rich in pigments. The exact pigment concentrations and bioavailability of the yeast carotenoids are, however, markedly influenced by factors such as the growth conditions, harvesting and processing of the *Phaffia*. Nevertheless, the pigments of plant and micro-organism origin may have considerable potential as carotenoid sources for fish.

In order to meet the consumer preference for red coloration, salmonid flesh should contain at least 4 mg pigment per kg. Since the carotenoids may fade in colour during processing and storage, compensation must be made for this by producing farmed salmon with flesh carotenoid concentrations elevated above the minimum acceptable level. Pigments supplied in the diet are, however, relatively poorly retained (2–10%) in the flesh of salmonids, and pigments are included at levels of 40–100 mg per kg dry feed. Since the cost of carotenoid pigments is high, dietary inclusion at these levels leads to pigments representing a considerable proportion of the overall cost of the feed.

Antimicrobial agents

Unless stored frozen, diets containing 15% moisture or more are subject to attack by bacteria and moulds. Microbial growth can be rapid at room temperature, and a number of moulds produce substances that are toxic to fish. If moist feeds are stored frozen, or are fed almost immediately after manufacture, there is little danger of microbial deterioration. If longer-term storage is anticipated, antimicrobial agents must be added to the feed in order to prevent spoilage. Many compounds are used in the feed industry to inhibit fungal and microbial growth, and these include benzoates, propionates and formic acid.

Flavourings and attractants

Even though a dietary formulation may appear to be nutritionally complete and well-balanced, the diet will not produce the desired growth of the fish if palatability is poor. Fish appear to be very sensitive to certain tastes in their food, and salmonids can lose appetite for several days if diet is changed abruptly. Several salmonids are sensitive to low levels of soybean in the diet, and other flavouring substances must be included in order to mask the taste of the soybean. Squid and prawn/shrimp have been reported to stimulate appetite, and inclusion of high levels of fishmeal is considered to improve feed acceptance in many fish species.

Attempts to isolate the active ingredients which stimulate appetite have not yet been completely successful. Research on feeding attractants has been concentrated on the investigation of the attractant effects of mixtures of amino acids and other organic substances (betaine-glycine, inositol, urea and trimethylamine) that are found in the body fluids and tissues of natural prey species. In some cases, certain amino acids, such as alanine and arginine, together with betaine-glycine have proved to be effective stimulants, but in other cases the addition of mixtures of this type seems to have had little effect on food acceptance. Certain derivatives of inosine are commonly used as flavour enhancers in the preparation of foods for human consumption, and there are some reports that the addition of inosine monophosphate to feed mixtures may promote increased appetite of fish under some circumstances. Tertiary sulphonium compounds, such as 2-carboxyethyl dimethyl sulphonium bromide (DMPT), have also been reported to have appetite and growth stimulating effects in some species, but the modes of action of these feed additives remain unknown.

In the majority of the studies involving potential attractants, the tests have involved the feeding of fish with neutral-tasting gels of agar or gelatin, and then with similar gels to which the chemical mixtures have been added. Alternatively, the potential attractants have been added to diets that were known to be of poor quality and had low acceptability. Increased consumption of a gel or poor quality diet to which a chemical mixture had been added has usually been considered to confirm the attractant properties of the mix. However, when similar mixtures are added to 'high quality' diets that are known to promote good rates of growth under normal rearing conditions, the fish seldom respond to the attractant mixtures by increasing food consumption and displaying more rapid growth. Thus, the efficacy, and necessity, of using such mixtures in practical feeding remains to be verified.

Drugs

Drugs and antibiotics are not routinely included in fish feeds, but medicated feeds are used to treat and mitigate disease.

Nutrient interactions

Interactions between different nutrients may affect bioavailability and metabolism in both a positive and negative fashion. For example, binding of nutrients during preparation of feeds may reduce the ease with which they are digested and absorbed, and hence lower bioavailability. Competition between closely related nutrients for intestinal transport sites may lead to reduced uptake, but interactions can also result in there being stimulatory effects on absorption from the gut. Some examples of the possible forms of nutrient interactions involve those between amino acids, vitamin–vitamin interactions, vitamin–mineral interactions, mineral–mineral interactions, and those of a more complex nature in which nutrients of several different classes are involved.

Amino acid interactions

There are known to be four major transport systems for amino acids in the intestine, these being classified as:

(1) A system transporting most neutral amino acids;
(2) A system specific for the transport of basic amino acids and cystine;
(3) A system transporting acidic amino acids;
(4) A system transporting imino acids (proline and hydroxyproline), glycine and related molecules (e.g. betaine).

From this it follows that several amino acids may be transported by the same system, leading to the possibility of competition between amino acids for binding to transport sites. For example, methionine, valine and threonine have transport systems in common, and absorption of methionine can occur to the detriment of transport of valine and threonine. Some amino acids can be transported by more than one system and, in addition, there are known to be independent transport systems for some di- and tri-peptides, with absorption of the di- and tri-peptides being more effective than that of the free amino acids.

Many observations over a long period have provided evidence that amino acid mixtures simulating proteins may be nutritionally inferior to the proteins themselves, and it has been suggested that the superiority of a protein to an equivalent amino acid mixture can be related to temporal factors in the absorption of the amino acids. It has long been recognized that simultaneous presentation of the amino acids to the tissues is essential for effective protein synthesis. There is also evidence indicating that during the absorption of amino acid mixtures, presentation of the amino acids to the tissues is less simultaneous than when the products of protein or protein partial hydrolysate digestion are absorbed. This is probably due to the fact that when amino acid mixtures are presented, there may be competition between amino acids for intestinal transport sites and the different free amino acids are absorbed from mixtures at very different rates. Thus, when free amino acids are absorbed from mixtures some reach early and high peak concentrations in the blood, while rates of absorption and increases in blood concentrations of other amino acids tend to lag behind. This could lead to an imbalanced supply of amino acids to the tissues and, in turn, to inefficiencies in protein synthesis.

On the other hand, there are a multiplicity of transport systems available for the absorption of the products (e.g. free amino acids, di- and tri-peptides, larger peptides) of whole protein digestion. Thus, both the temporal displacement occurring during the absorption of free amino acids from mixtures, and any disturbance to tissue uptake might be largely or completely obviated when amino acids are presented in the form of whole proteins, rather than free amino acid mixtures. Thus, a 'deranged' pattern of absorption of amino acids into the body seems to be, at least partially, responsible for the nutritional inferiority of amino acid mixtures when compared with the same amino acids presented in peptide or whole protein form. Bearing this is mind, a number of questions will have to be asked. Most notably: 'Can it be regarded as satisfactory to determine amino acid requirements by use of mixtures of free amino acids?' and 'Is it reasonable to expect that additions of free amino acids will prove to be completely effective in compensating for deficiencies when "poor quality" proteins are used in the formulation of fish feeds?'

Vitamin–vitamin interactions

The vitamin folic acid (or folacin) is relatively abundant in natural feedstuffs and may also be synthesized by the intestinal microflora. Folic acid is converted into its active form tetrahydrofolic acid in the liver, with the active form of the vitamin being required for the transfer of methyl groups ($-CH_3$) between organic molecules. It plays a role in the synthesis of purines, pyrimidines, glycine, serine and creatine. Purines are required for the synthesis of nucleic acids (DNA and RNA), and are also necessary for the normal formation of red blood cells. Thus, a deficiency of folic acid may lead both to the fish becoming anaemic and to there being abnormalities in the red blood cells. In many respects the roles of the vitamin cyanocobalamin seem to complement those of folic acid. Cyanocobalamin is, for example, involved in the synthesis of labile methyl groups, and is also required for the synthesis of nucleic acids. This vitamin is needed for the development and maturation of red blood cells, and also plays a role in the recycling of tetrahydrofolic acid. Thus, the two vitamins, folic acid and cyanocobalamin, may closely interact and they may be able to substitute for each other in some metabolic functions.

Vitamin–mineral interactions

Vitamin D is a general term used to describe several lipid-soluble sterol derivatives. Of these ergocalciferol (found in plants) and cholecalciferol (found in animals) are the most important from a nutritional standpoint. Dietary vitamin D is absorbed from the intestine along with the lipids and is transported to the liver where it is converted to 25-hydroxycalciferol. Conversion to the active form, 1,25-dihydroxycalciferol, probably occurs in the kidney tissue. The primary function of vitamin D appears to be related to the absorption of calcium from the intestine, with the vitamin being required for the synthesis of a calcium-binding protein. Some vitamin D metabolites appear to be actively involved in the endocrine regulation of calcium uptake and metabolism, with dietary deficiencies of this vitamin resulting in serious disturbances of calcium metabolism. Vitamin D may also be involved in the metabolism of inorganic phosphate, and be required for the effective functioning of alkaline phosphatase. Thus, vitamin D is intimately linked to calcium and phosphate metabolism, providing an example of vitamin–mineral interactions.

Several other vitamin–mineral interactions exist, most notably those involving the influences of ascorbic acid upon mineral bioavailability. Ascorbic acid has been reported to affect both the toxicity and tissue retention of water-borne copper, but apparently has few effects upon the uptake and metabolism of dietary copper. Ascorbic acid also has a central role to play in the metabolism of iron, with fish fed diets deficient in vitamin C becoming anaemic even though tissue iron stores and dietary iron levels would appear to be adequate.

Mineral–mineral interactions

The bioavailability of zinc can be adversely affected by diet composition, with increased levels of dietary calcium and phytate seeming to have detrimental effects on availability. It would seem that a calcium–zinc–phytate complex forms in the intestinal tract of the fish and this reduces the bioavailability of the zinc. High levels of phosphorus in the diet may also have negative effects, possibly due to the formation of an insoluble precipitate between zinc, calcium and phosphorus that would reduce or prevent absorption from the intestine.

Complex interactions

Since there are a number of factors that can influence oxidative destruction of dietary ingredients, and the oxidative damage to fish tissues, there may be complex interactions between the requirements for the 'natural antioxidants' and dietary composition. Lipid peroxidation is a chemical process that results in the oxidative deterioration of polyunsaturated fatty acids, and thereby will lead to destructive changes in both the dietary lipids and the lipids of the cell membranes of the fish tissues. Fish fed diets rich in polyunsaturates, and depositing large quantities of these fatty acids in their tissues, may be expected to be most at risk of suffering damage due to this form of oxidative deterioration. Iron is one of the metals primarily involved in the induction of lipid oxidation, and ferrous iron is a more potent catalyst of lipid peroxidation than ferric iron. Thus, a combination of high dietary levels of polyunsaturates coupled with high concentrations of ferrous iron might be expected

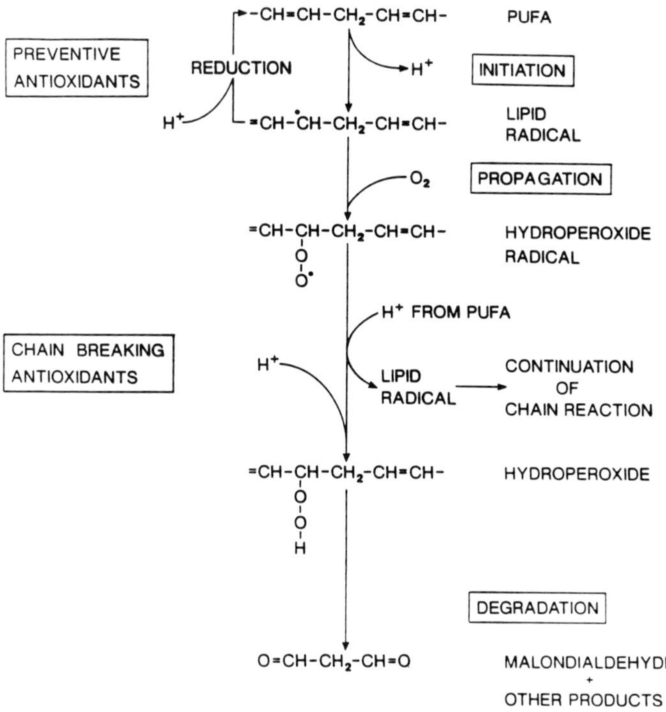

Fig. 4.7 Destruction of polyunsaturated fatty acids (PUFAs) by the lipid peroxidation pathway. (Note the points of intervention of the two types of natural antioxidants – the preventive antioxidants and the chain-breaking antioxidants.)

to have deleterious effects on the stability of the dietary lipids and upon the growth, health and survival of the fish.

In biological systems, lipid peroxidation can be prevented, or at least hindered, by several complementary antioxidant systems. The lipid peroxidation chain reaction is initiated by the removal of a hydrogen atom from a polyunsaturated fatty acid, giving rise to a lipid radical. In the presence of oxygen there will be a rapid reaction leading to the formation of a hydroperoxide radical. Hydroperoxide radicals, in turn, remove hydrogen atoms from other fatty acids leading to a continuation of the chain reaction. Protection of the lipids against peroxidation can be provided by two broad classes of antioxidants: the preventive and the chain-breaking antioxidants (Fig. 4.7). Preventive antioxidants reduce the rate at which lipid radicals are formed, that is, they hinder the initiation of the lipid peroxidation chain reaction. The most important preventive antioxidants are enzymes, such as catalase and glutathione peroxidase, that exert their effects via reduction of the hydroperoxide radicals. Glutathione peroxidase activity is dependent upon adequate supplies of selenium, a micronutrient usually obtained via the diet. The chain-breaking antioxidants owe their antioxidant activity to their ability to bind with and trap hydroperoxide radicals. Ascorbic acid (vitamin C) is an important water-soluble chain-breaking antioxidant, and the most important lipid-soluble antioxidants are the tocopherols (vitamin E and related compounds). The antioxidant properties of ascorbic acid and vitamin E appear to be synergistic, with the ascorbic acid seeming to play a major role in the regeneration of vitamin E rather than acting directly to trap the hydroperoxide radicals. Thus, in this instance there is evidence of a series of interactions between selenium, vitamin E and vitamin C. The requirement for these contributors to natural antioxidant activity will, in turn, depend upon the levels of polyunsaturates at risk from peroxidation and the level of exposure of these fatty acids to initiators of the peroxidation reaction, e.g. ferrous iron, heavy metals and various organic pollutants.

Nutrition and disease

The earliest studies of the nutritional influences upon disease in fish were related to the description of deficiency symptoms, especially those associated with vitamin and mineral deficiencies. Vitamin and mineral nutrition attracted considerable attention because, despite the fact that these micronutrients are required in small doses, the consequences of deficiency can be severe. Typical non-specific deficiency symptoms include loss of appetite and growth depression. More specific symptoms indicative of particular micronutrient deficiencies encompass various metabolic disorders, cataracts of the eyes, damage to gills, muscle and connective tissue, increased fragility of red blood cells and anaemia. These will lead to increased mortality if the deficiencies are not remedied.

Studies of micronutrient requirements have been conducted in a benign environment under laboratory conditions in which the fish would be expected to thrive, and these early studies rarely considered the possibility of nutrient interactions. Consequently, requirement values must be considered as delimiting the minima that can be tolerated by the fish, and it should also be realized that the appearance of deficiency symptoms gives a clear indication that there is something seriously at fault. There is an increasing body of evidence demonstrating that very mild forms of deficiency, that may not be reflected in reduced growth or the development of specific pathological symptoms, can have detrimental effects upon the fish if they are exposed to poor environmental conditions or other potential stressors. In other words, while diets for farmed fish have primarily been formulated in order to promote good rates of growth, there is an increasing awareness that diet composition can affect the general health of the fish and influence susceptibility to disease.

For example, while it has been known for some time that the salmonids require a dietary supply of (n–3) fatty acids in order to display good rates of growth, there is now evidence that a correct balance of (n–3):(n–6) fatty acids is required if the health of the fish is not to be compromised. Results of studies with Atlantic salmon have shown that although a lowering of the ratio of (n–3) to (n–6) polyunsaturates may not depress growth, it can lead to the development of pathological changes, such as cardiomyopathy.

Formulated feeds for salmonids usually contain a certain proportion of polysaccharides, which are included in the diet primarily due to their binding properties. Such polysaccharides are not present in large amounts in the diets of wild fish, and it has been postulated that the inclusion of high levels of complex carbohydrates in formulated feeds can have detrimental effects upon the fish via influencing the composition and activity of the gut microflora. It is known that there are differences between wild and farmed fish in gut microflora, both with respect to bacterial species

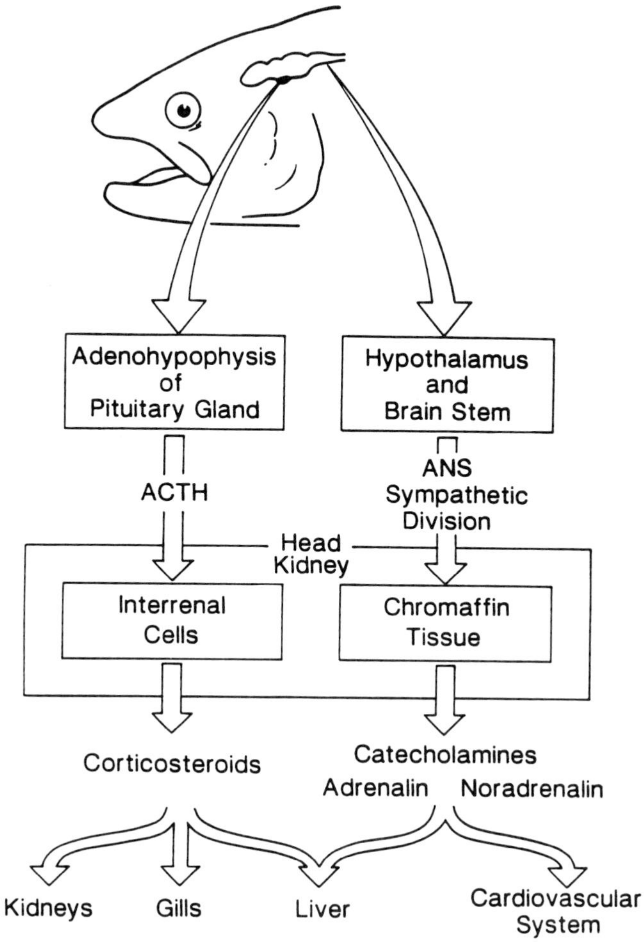

ACTH = Adrenocorticotropic hormone
ANS = Autonomic nervous system

Fig. 4.8 The hypothalamic-pituitary-interrenal (HPI) axis in fish, illustrating nervous and endocrine links between central sites of integration and the periphery.

present and the relative proportions of different groups of micro-organisms, but to what extent these changes either influence susceptibility or can be a direct cause of disease outbreak in farmed fish is still an open question.

Under commercial farming conditions the fish may be subjected to a series of procedures, such as netting, grading and transport, that are potential stressors. Many other factors, related to both the abiotic and biotic environments, for example, pollutants, poor water quality and high stocking densities, may elicit stress responses in fish, and if there is chronic exposure to the stressor the health and survival of the fish may be compromised. Exposure of fish to environmental stressors induces a characteristic series of responses mediated via the hypothalamic–pituitary–interrenal (HPI) axis (Fig. 4.8). Adverse stimuli (stressors) result in the almost immediate release of two types of endocrine factor: catecholamines and cortisol. The catecholamines initiate a series of short-term physiological changes, whereas the effects of cortisol tend to be more long-lasting. Should the exposure to the stressor be of limited duration recovery will generally occur within the space of a few hours, but chronic exposure to stressors may have serious long-term effects on the fish. From a nutritional point of view it is known that exposure of fish to stressors, such as pollutants and poor water quality, leads to a rapid depletion of tissue reserves of ascorbic acid (vitamin C), such that the dietary requirements are elevated in fish held under poor environmental conditions.

Chronic exposure to environmental stressors also increases the susceptibility of the fish to disease via immuno-suppressive effects of repeated or prolonged elevations of plasma cortisol levels. Recent work has been aimed at counteracting these negative effects through the enhancement of the immune response via the manipulation of dietary components. Vitamins, especially vitamins C and E, and fatty acids are considered to be the key dietary components having influences upon the immune system. These nutrients have been implicated in both cell-mediated and humoral immunity. Available evidence suggests that high dietary doses of vitamins C and E can lead to improved protection of the fish against infectious diseases under some conditions, but much more work is required in order to elucidate the potential nutritional links between the immune system and disease resistance.

An alternative approach to enhancing the disease resistance and overall health of the fish may involve the oral delivery of immune potentiators. While these are not nutrients *per se*, some form of dietary manipulation may be required in order to prevent destruction and to ensure adequate absorption of these protective factors from the gut. Various bacterial peptides and glucans are considered to be promising candidates as immune potentiators and experimental trials designed to test the efficacy of these agents are being conducted in several research laboratories.

Special diets

The most common types of feeds used in the commercial farming of salmonids are dry pellet diets prepared either by steam processing or extrusion, but the exact formulation will be changed either as the fish increases in size or in accord with specific production requirements. For example, the diets prepared for fish at the start-feeding stage usually contain approximately 50% protein, with the majority of this coming from special quality fish meals. The protein content of diets fed to larger juveniles (2–50 g) may be reduced to about 45%, and the diets fed to the fish during the grow-out phase in seawater may be formulated to contain 35–45% protein.

Traditionally, starter diets have been prepared as crumbles which may result in particles of somewhat variable size and shape, and can also lead to a certain amount of dust in the final product. It is now more common that starter feeds are prepared using processes that lead to the agglomeration of the dietary ingredients. This results in the production of feed granules of relatively uniform size. These agglomerated diets can be formulated to contain low levels of carbohydrate, since the manufacturing process is not reliant upon the binding properties of the complex polysaccharides.

There is some evidence that the inclusion of low levels (4% by weight) of sodium chloride in the diets of salmonids during the freshwater rearing stage can result in improvements in the ease with which the fish tolerate transfer to seawater. Similar claims have been made for diets formulated to contain betaine-glycine, an agent which is thought to have beneficial effects upon the abilities of the fish to osmo- and iono-regulate. Some commercial manufacturers have developed formulations that include sodium chloride and betaine-glycine. These feeds are fed to the fish in the period prior to the time at which they undergo the parr–smolt transformation and are transferred to seawater.

Salmonids must be supplied with carotenoids in the diet if they are to develop a characteristic pink or red flesh colour, but the fish will not usually be fed with diets containing pigments throughout the entire rearing cycle. Pigmented feed will normally be supplied to the fish during the grow-out phase in seawater, with the most common rearing practice being the introduction of pigmented feed (40–50 mg carotenoid/kg dry diet) approximately one year prior to the projected time of slaughter of the fish.

When salmonids undergo sexual maturation the carotenoids are mobilized from the flesh and are selectively transferred to the skin and gonads. The eggs of wild salmonids contain relatively high levels of carotenoids (0.5–1.0 mg/100 g egg) and this has led to the suggestion that the carotenoids may have important influences on egg viability and larval development. The results from experimental studies are, however, equivocal and there is little evidence to suggest that low levels of egg carotenoids lead to there being detrimental effects on fertility and viability under hatchery conditions. Nevertheless, diets fed to broodstock fish are now often formulated to contain quite high concentrations of carotenoids (100 mg/kg dry diet), primarily in an attempt to minimize the depletion of the carotenoids from the flesh of the fish and to ensure rapid repigmentation of the broodstock following the cessation of spawning.

Broodstock diets are also usually supplemented with higher levels of ascorbic acid than are diets fed to production fish, and there is convincing evidence for the involvement of this vitamin in the reproductive processes. High concentrations of ascorbic acid are found in ovarian tissue and the vitamin appears to be intimately involved in the synthesis and metabolism of the sex steroid hormones. There is also deposition of ascorbic acid in the eggs, and fish fed on diets containing low levels of ascorbic acid produce eggs having poor hatchability. Ascorbic acid functions as a co-factor in the synthesis of collagen, a protein which is abundant in connective tissue and cartilage. Thus, ascorbic acid requirements are likely to be high during embryonic development, and, at this stage the only source available to the developing embryo is that deposited in the egg yolk. Reproduction in salmonids is obviously dependent upon there being a dietary supply of ascorbic acid, but the requirements for optimal broodstock performance, embryonic development and successful post-hatch survival are not known.

While drugs and antibiotics are not routinely supplied to the fish for prophylactic purposes, the commercial producer may be required to use medicated diets in order to treat fish following the outbreak of disease. Diseased fish will usually have a depressed appetite, which makes the oral administration of medicaments difficult, and problems may be further exacerbated by the fact that a number of the drugs and antibiotics used in disease treatment appear to be unpalatable. Thus, medicated diets may need to be specially formulated to contain particularly attractive ingredients in order to ensure consumption by the fish.

Practical diet preparation and feeding practices

In all dietary formulations, limits are placed on the levels of certain feed ingredients regardless of cost. Upper limits are often imposed by anti-nutritional factors or toxicants, or by factors that affect dietary palatability or pelleting properties. For example, untreated soybean meal contains trypsin inhibitors and may also be unpalatable; cottonseed meal contains gossypol which is toxic; and several plant protein sources may contain high levels of phytic acid which reduces bioavailability of nutrients.

The type of feed to be made also determines the limits of levels of the ingredients in the formulation. For example, dry compressed pellets must contain adequate amounts of starch-containing grains in order to ensure effective binding; and in wet and moist feeds the ingredients must be chosen, both to ensure correct nutritional balance, and to guarantee water stability of the pellet, or dough-like mixture.

Once the feed ingredients have been chosen, a series of calculations can be performed in order to define a food containing desired levels of nutrients (see Appendix). Obviously, farmers who rely on commercial dry feeds will not be required to perform such calculations, since the exercise will already have been carried out by the feed producer. Feed producers will usually employ 'least-cost formulation' in which the feed is formulated from the available ingredients using both nutritional and economic criteria. Computer programs are available to assist the manufacturer in carrying out these calculations. Farmers who formulate their own wet and moist feeds should familiarize themselves with the simple calculation procedures required in order to determine nutritional contents of diets, rather than rely on trial-and-error judgements.

Approximately 50% of the production costs in salmonid farming are related to feeds and feeding. The desire of the fish farmer is usually to get the fish to grow rapidly while keeping food costs as low as possible. This can be achieved by employing feeding practices in which feed waste is minimized while at the same time ensuring, as far as possible, that fish are reared under conditions that promote efficient food utilization. In deciding on feeding practices appropriate to given conditions, feeding rate, feeding frequency, pellet type and size, and method of feed delivery should be considered, since all of these factors affect growth rates and efficiency of food utilization. Feeding for maximum growth rate may be desirable in many situations even though it must be realized that food utilization efficiency will be poorer than can be achieved at feeding rates that promote sub-maximal growth. At high rates of feeding, the food may pass rapidly through the gastrointestinal tract of the fish, leading to incomplete digestion and incomplete absorption of nutrient-rich, high-energy dry food. This, in turn, leads to increased nutrient losses in the faeces, and there is often observed to be an inverse relationship between feeding rates and the digestibility of a feed (Fig. 4.9).

Food consumption is markedly affected by fish size and water temperature, and these parameters will account for most of the changes in feed intake observed during the rearing cycle. If allowed unlimited access to food, fish may display quite large day-to-day variations in feed intake, a day of heavy feeding being followed by a day on which the fish feed little or not at all. Nevertheless, feeding charts or tables which account for fish size and water temperature effects on feeding provide useful guidelines as to long-term trends in food consumption, even though they cannot take

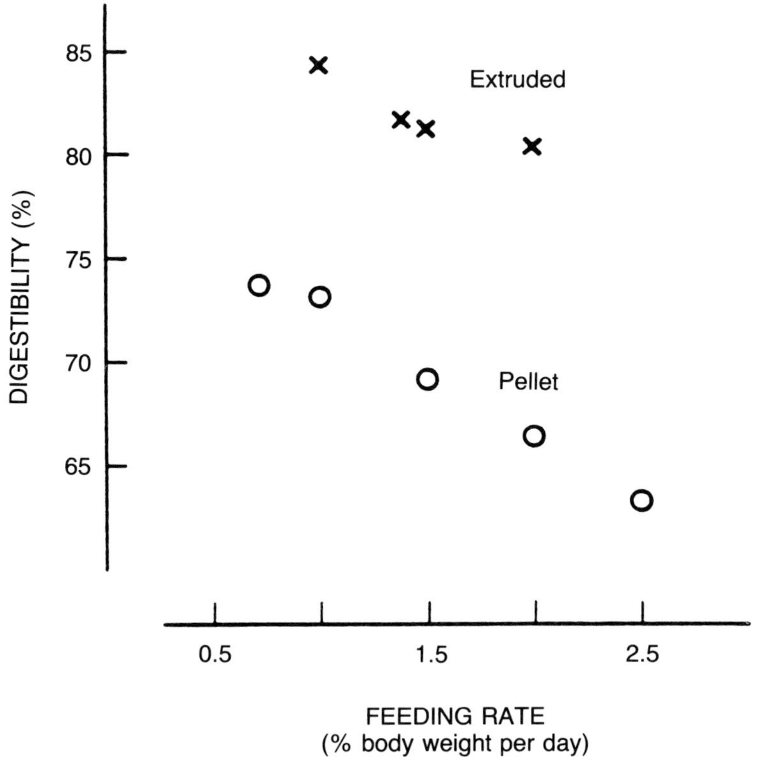

Fig. 4.9 Effects of feeding rate on digestibility of dietary energy in rainbow trout, *Oncorhynchus mykiss*. (Note the higher digestibility of dietary energy of the extruded feed – due to increased availability of dietary carbohydrate. There were no differences in protein digestibility between pelleted and extruded feeds.)

short-term variations into consideration. Food intake is also influenced by season, photoperiod, stocking density, water quality and abrupt changes in rearing conditions, all of which are factors that are not incorporated into feed tables. For example, fish of a given size, held at a given water temperature, will tend to consume more food on increasing photoperiods than on decreasing, so a farmer may observe that the appetite of the fish suffers a decline during early autumn even though water temperature remains relatively high. Reproductive status of the fish will also influence food intake, there being a decline in appetite and growth in the weeks preceding the onset of the spawning season.

High stocking densities and deteriorations in water quality can cause fish to lose appetite, so it is important that feeding activity be monitored carefully. Feed consumption and observed growth rates should be periodically compared to the levels recommended and predicted in the published feed and growth tables. Should large discrepancies arise between the observed and predicted values, causes should be sought and the need for changes in rearing and feeding practices evaluated. The feeding levels given in the published tables are usually based on dry feeds, so when wet and moist feeds are used, the moisture content of the diet should be known, and allowances made in the calculations for the differences in moisture contents between different feed types.

Feeding frequency of the fish is generally varied depending upon size and temperature. Newly-hatched fry are fed almost continuously and young fish in hatcheries are fed at relatively frequent intervals using automatic feeders of various designs. The use of automatic feeders usually demands that dry feed is employed, since most moist feeds tend to stick together and do not have the flow characteristics required for efficient operation of automatic feeders. Some types of automatic feeder, constructed on the conveyor belt or moving band principle, are suitable for use with moist feeds. It must be remembered that moist feeds are very prone to microbial attack and spoilage, so feeders must be loaded at relatively frequent intervals, and any remaining food removed before deterioration occurs.

In many sea-cage operations, fish are fed by hand. This has the advantage that any changes in feeding activity can be monitored quite easily. Hand-feeding has the disadvantage, however, that food may be presented to the fish when their appetite is not at peak level, resulting in sub-maximal food intake.

Large, hopper-style automatic feeders can be fitted to sea cages and these allow fish to be fed many times throughout the day, according to a predetermined programme. Research is currently being directed toward the development of feeders that are controlled via the monitoring of the feeding responses of the fish. Amounts of uneaten food are registered and once a given threshold is reached, the supply of food is automatically terminated. Demand feeding devices have also been developed, and these allow the fish to feed themselves. In theory, this should be the optimal solution since fish should feed when hungry, with minimum food waste. In practice, there are a number of problems involved with getting demand feeders to work efficiently and they are not widely used in sea-cage operations. Demand feeders often require frequent maintenance and servicing for filling with food, and fine adjustment may be required in order to control feed delivery within rigidly defined limits. Many fish within a rearing unit will not learn how to operate the feeder, and there will inevitably be one or two rogues that use the feeder at frequent intervals, whether hungry or not. This can obviously lead to ineffective feeding and food wastage.

Improvements are being made to the design of demand feeders in an attempt to overcome these problems, but it is open to question whether such feeding systems will achieve widespread popularity among commercial producers.

When mechanical methods of food delivery are employed, involving the use of automatic feeders activated by a timer, it may be desirable to ensure that food is broadcast over a relatively wide area, rather than being delivered at a single point source. This can be achieved by fitting several hopper-type feeders to a large sea cage, or by use of a pneumatic feeder that results in the feed pellets being spread over a large area of the water surface. The widespread dispersal of the pellet food is expected to enable the majority of fish within the group to gain access to food, whereas the provision of food at a point source may lead to some fish being prevented from feeding. For example, observations carried out on the feeding behaviour of Atlantic salmon in a sea cage showed that when food was delivered at a point source, the majority of the pellets consumed were eaten by a small proportion (25%) of the fish within the group.

Hand-feeding is often the feeding method of choice in sea-cage operations, and large salmonids will usually be fed once or twice per day. Combinations of automatic- and hand-feeding are also common practice. By distributing 75% of the estimated daily ration via automatic feeders the farmer can ensure that food is supplied to the fish over a prolonged period during the course of the day. Supplementation via hand-feeding, once or twice per day, enables the farmer to monitor the feeding responses of the fish without having to devote a large proportion of the working day to feeding operations. Hand-feeding should, preferably, be carried out either during the early morning and/or during the course of the evening, since feeding response appears to be greatest at these times of the day.

Should water temperature rise excessively during the summer months, fish will lose appetite and feeding can be curtailed for a few days. When water temperatures are low during winter, feed can be provided once per day or less frequently. Fish will not suffer any long-term growth disadvantage even if held on low rations throughout the winter as they tend to show rapid catch-up growth during the spring and early summer, when days become longer and water temperatures begin to rise. There may even be positive benefits to be gained by holding the large fish on low rations during the winter months, since recent evidence suggests that food restriction (feeding and starving on alternative weeks) during winter and early spring can inhibit reproductive development, with the result that the fish direct summer growth towards the soma (including fillet) rather than towards reproductive products (eggs and sperm).

Closing remarks

Fish are ectothermic animals and their energetic requirements for metabolism are much lower than those of endothermic vertebrates (e.g. birds and mammals) of similar size. The metabolic requirements can be met by the use of carbohydrates, lipids or proteins as respiratory substrates. There is no dietary requirement for carbohydrate, with compounds of this nutrient class being used solely as a metabolic energy source. A number of fish species appear to have a relatively limited ability to use carbohydrates as a respiratory substrate, and respiratory metabolism in fish appears to be, predominantly, based upon lipids and proteins.

Both lipids and proteins may be used as respiratory substrates, but there is a dietary requirement for certain fatty acids and amino acids if the fish are to survive and grow. Nutrient requirements have been investigated for a few fish species that are exploited for human consumption, but even for these aquaculture species (e.g. salmonids, cyprinids, tilapias, ictalurids), nutritional information is far from complete.

The same amino acids have been found to be essential dietary constituents for all species examined, but while a qualitative requirement has been demonstrated, the quantitative requirements remain almost unknown.

There appear to be both qualitative and quantitative differences between fish species in their requirements for fatty acids. Salmonids have a greater requirement for the fatty acids of the (n−3) series than those of the (n−6) series, but other fish species (e.g. tilapias) may be able to utilize (n−6) fatty acids effectively. In addition, fish species differ in their abilities to transform and chain-elongate fatty acids and these differences determine whether or not a particular fatty acid will display ‘essential fatty acid’ properties or not. Salmonids have the ability to desaturate and chain-elongate the 18C fatty acids of the (n−3) and (n−6) series. Thus, these fatty acids can fulfil the function of ‘essential fatty acids’ for salmonids, but a number of marine species of fish are unable to transform the 18C fatty acids to long-chained HUFAs, and the long-chained, highly unsaturated fatty acids must be obtained via the diet.

There is an essential requirement for certain vitamins and minerals. While the absolute requirements for these micronutrients are small, a deficiency in one or more of these nutritional factors usually results in there being serious disruptions to metabolism and growth.

Even though the diets fed to farmed salmonids may have been formulated to meet all the known nutritional requirements of the fish, the feeding of such diets will not promote good rates of growth unless the fish are held under suitable environmental conditions. Abiotic factors, such as temperature, salinity and water quality, may have marked influences upon the feeding and growth of farmed fish, and the growth performance of individual fish may also be restrained by biotic factors, such as stocking density, reproductive status and genetic background.

APPENDIX

Practical feed formulation

A farmer wishes to make 100 kg moist feed with an energy content of 13.5 MJ kg^{-1} and a PE:TE of 0.45.

The feed is to be based upon the following ingredients:

	Protein (%)	Energy (MJ kg^{-1})
Capelin(*C*)	14	9
Squid(*S*)	18	5
Fish oil(*F*)	–	35
Fishmeal	70	21
Vitamult/binder	20	13

100 kg feed will contain $13.5 \times 100 \times 0.45/24$ kg protein assuming that protein has an energy content of 24 MJ kg^{-1}.

This gives a protein content of 25.3 kg protein in the total 100 kg feed mix.

The fixed ingredients are predetermined to be 20 kg fishmeal and 10 kg vitamult/binder and these ingredients will contribute the following:

	Weight (kg)	Protein (kg)	Energy (MJ)
Fishmeal	20	14	420
Vitamult/binder	10	2	130
Total	30	16	550
Remainder from other ingredients	70	9.3	800

From this it follows that:

Weight $$S + C + F = 70 \tag{1}$$

Energy $$5S + 9C + 35F = 800 \tag{2}$$

Protein $$0.18S + 0.14C + 0F = 9.3 \tag{3}$$

and from (1) $$S = 70 - C - F$$

(2) becomes $$5(70 - C - F) + 9C + 35F = 800$$

or $$4C + 30F = 450 \tag{4}$$

(3) becomes $$0.18(70 - C - F) + 0.14C + 0F = 9.3$$

or $$0.04C + 0.18F = 3.3 \tag{5}$$

from (4) $$C = 450 - 30F/4 = 112.5 - 7.5F$$

and substituting in (5) gives

$$\begin{aligned} 0.04(112.5 - 7.5F) + 0.18F &= 3.3 \\ 4.5 - 0.3F + 0.18F &= 3.3 \\ 0.12F &= 1.2 \end{aligned} \qquad \underline{F = 10}$$

back substituting to (4) gives

$$C = 450 - (30 \times 10)/4 = 150/4 = 37.5 \qquad \underline{C = 37.5}$$

and finally from (1)

$$S = 70 - 10 - 37.5 = 70 - 47.5 = 22.5 \qquad \underline{S = 22.5}$$

and to check the complete mix:

	Weight (kg)	Protein (kg)	Energy (MJ)
Capelin	37.5	5.25	337.5
Squid	22.5	4.05	112.5
Fish oil	10	–	350
Fishmeal	20	14	420
Vitamult/binder	10	2	130
Totals	100	25.3	1350

The protein energy content (PE) of the mix is

$$25.3 \times 24 = 607.2$$

and this results in a PE:TE of

$$607.2/1350 = 0.45$$

thus satisfying the criteria set for the formulation.

Literature and source material

The scientific literature relating to fish feeding and nutrition is extensive, and several books and monographs have been published on various aspects of the subject. In the preparation of this chapter, material has been collated from over 250 review articles and original publications. Citation of all these works in the main body of the text has been deemed undesirable from the point of view of maintaining continuity. In order to assist readers in finding additional information, some of the more recent literature is presented below.

Selected bibliography

Aquaculture is a scientific journal which regularly contains papers covering aspects of fish growth and nutrition. See, especially, Volumes 79 (1989) and 100 (1992) for review articles.

Camire, M.E., Camire, A. and Krumhar, L. (1990) Chemical and nutritional changes in foods during extrusion, *CRC Rev. Food Sci. Nutri.*, 29, 35–57.

Cowey, C. (1988) The nutrition of fish: The developing scene, *Nutrition Research Reviews*, 1, 255–80.

Cowey, C.B., Mackie, A.M. and Bell, J.G. (Eds) (1985) *Nutrition and Feeding in Fish*, Academic Press, London.

Halver, J.E. (Ed.) (1989) *Fish Nutrition*, Academic Press, London.

Henderson, R.J. and Tocher, D.R. (1987) The lipid composition and biochemistry of freshwater fish, *Prog. Lipid Res.*, 26, 281–347.

Jobling, M. (1983) A short review and critique of methodology used in fish growth and nutrition studies, *J. Fish Biol.*, 23, 685–703.

Lovell, T. (1989) *Nutrition and Feeding in Fish*, Van Nostrand Reinhold, New York.

Sandnes, K. (1991) Vitamin C in fish nutrition: a review, *Fisk Dir Skr. Ser. Ernæring*, 4, 3–32.

Takeda, M. and Watanabe, T. (Eds) (1990) *The Current Status of Fish Nutrition in Aquaculture*, Tokyo University of Fisheries, Tokyo.

Torrissen, O.J., Hardy, R.W. and Shearer, K.D. (1989) Pigmentation of salmonids: carotenoid deposition and metabolism, *CRC Crit. Rev. Aquatic Sci.*, 1, 209–25.

Wilson, R.P. (Ed.) (1991) *Handbook of Nutrient Requirements of Finfish*, CRC Press, Boca Raton, Florida.

Chapter 5
Environmental Considerations

JEFF WALLACE

Introduction

Netpen rearing of salmon has a considerable actual or potential impact upon the environment. Various components of the overall environmental influences that are possible are outlined in Fig. 5.1. The sensory effects are the most obvious. For instance, it is impossible to operate a netpen farm without it being visually apparent in a formerly undisrupted marine area. Such visual modification is an inevitable reality, but is not necessarily 'bad' any more than an agricultural development would be in a formerly wooded area. Other sensory changes, perhaps more intrinsically detrimental, are increased noises inherent in routine operations and enhanced by attracted scavengers or predators, and an increase of offensive odours from decomposition of mortalities, feed, etc.

Of greater importance with regard to their potential for permanent environmental alteration, are those influences that are less immediately apparent. These impacts are the primary considerations of this chapter. Physical and biological concerns related to the immediate environment of the netpen farm are addressed and from a perspective external to the farmed fish themselves.

ENVIRONMENTAL REQUIREMENTS

Introduction and some perspectives

The growth of aquaculture during the last 20 years, together with the general increase in what can be called 'environmental awareness', has resulted in numerous impact studies of various types of aquaculture, and, of course, these have been subject to periodic review. In fact, during the last decade, more or less excellent reviews of the information pertaining to the environmental impact of aquaculture, and especially salmonid aquaculture, have appeared almost yearly. It is not the intention of this author to provide yet another such review; nor is such a review necessary at this point. Rather, this section and its companion 'The question of pollution from salmon farms', provide the reader with a general overview of their topics from the author's perspective. Specific citations within the text are inappropriate in this format, but can be found in a number of relevant articles listed under 'further reading' in both sections. Common to many of these is the conclusion that salmon farming has not, or at least need not have, serious negative impact on the environment. The physiological demands of the fish itself with regard to environmental standards, (described in the following section), ensure that farming cannot be

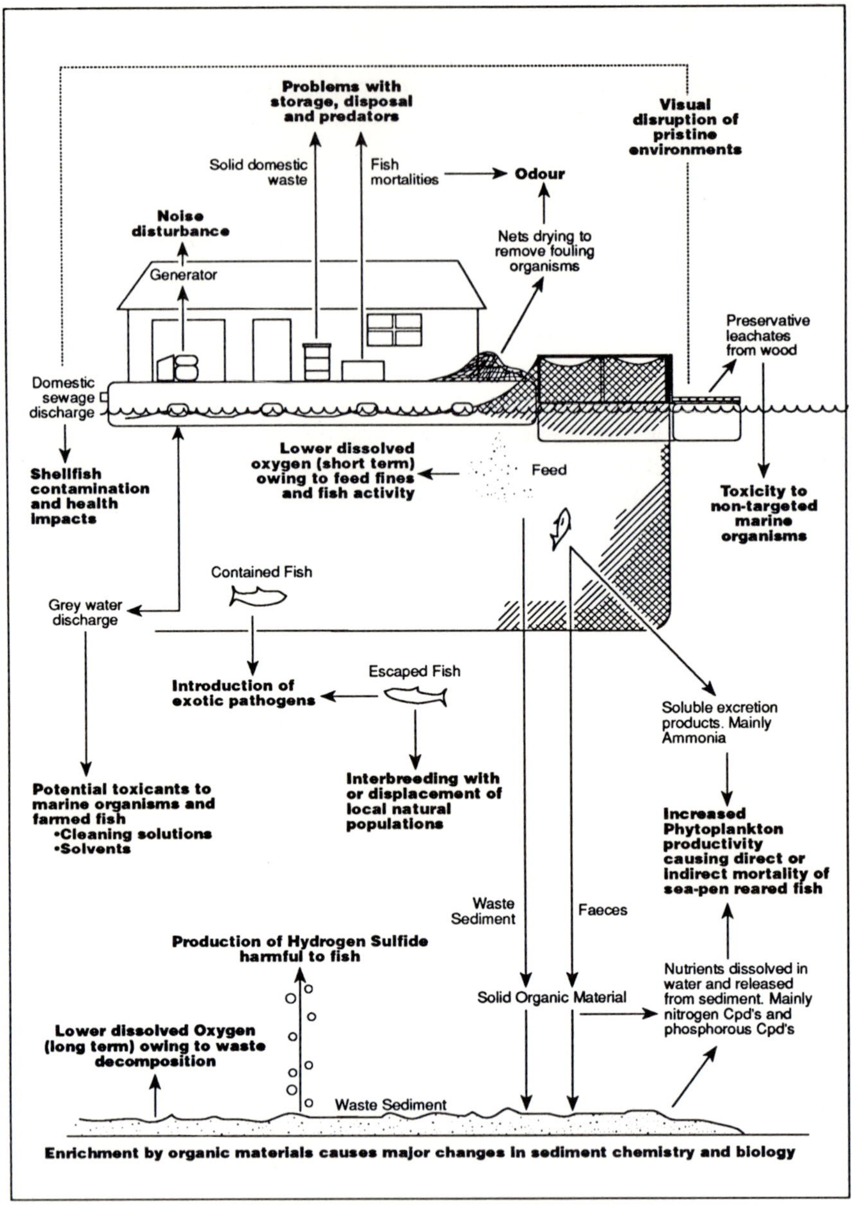

Fig. 5.1 Possible impacts of a salmon farm on the marine environment.

carried out in water of less than very good quality. For this reason, salmon farms are usually situated far from large sources of pollution, human or industrial. Negative effects tend, therefore, with certain exceptions, to be local and limited phenomena. This chapter is intended as an outline of the various ways in which the environment affects the salmon farms and vice versa.

The physical environment of a farmed salmon is usually very different from that which a wild fish encounters. As an egg, instead of lying buried in river bed gravel, bathed in water of low (but varying) temperature, it is incubated in an artificial device, at temperatures usually very much higher than those which could be expected in the wild, and dosed with fungicidal chemicals at regular intervals. As a fry and young fish (parr), instead of establishing and defending a solitary territory on the river bed where it would be well camouflaged and have many hiding places, it is raised in plastic or concrete tanks, in close proximity to thousands of siblings, without cover and under artificial lighting. As a smolt, instead of migrating seawards, it is 'transferred' to seawater at a time not of its own choosing and raised thereafter, also in unnaturally close proximity to other fish, in an enclosure. Throughout its life, it is fed artificial diets. That intensive salmon farming has become a highly successful industry is as much a tribute to the plasticity and tolerance of the species as it is to human ingenuity.

However, salmon are not totally indifferent to the conditions under which they are reared. Tank shape and colour, water depth, lighting conditions and stocking density, can all influence production performance, facts which indicate that the fish may feel more 'comfortable' under some conditions that under others. To put it in more acceptable jargon, the culture environment can be optimized to reduce stress and maximize performance.

One, very essential aspect of the fishes' environment has, however, been studied more than the others, because knowledge about it is absolutely fundamental to the culture process. This is, of course, the water, the quality of which determines not just production performance, but whether or not production is possible at all.

Fish have been described as fluids living within a fluid medium, separated only by a fairly permeable membrane. The point of such a description is to emphasize the very close relationship between a fish and its immediate environment, the water around it. The quality of the water in a given environment will to a great extent determine which species of fish can survive there, and because there exists an enormous variety of aquatic environments in the world, there is also a very large number of fish species, each adapted to one, or in some cases several, of these.

Chemically, pure water is the same everywhere, being composed of molecules consisting of two atoms of hydrogen and one of oxygen (H_2O). However, from the point of view of a fish, even pure water can vary qualitatively, for example in regard to temperature and rate of movement. Moreover, since water is a universal solvent, its quality is changed by nearly everything with which it comes into contact, either naturally – atmosphere, soil, bedrock – or as a result of human interference – pesticides, fertilizers, industrial effluents. These latter factors are often described as 'pollutants', while the former are those which determine the 'natural quality' of water. For a fish, however, there is no such distinction. There is merely water in which it can, or cannot, survive, grow, and reproduce.

Some species of fish are less demanding than others when it comes to water quality. Carp and tilapia are types of farmed fish which can be raised in a wide range of water qualities. The salmonids, however, are not among the less demanding species. (Although many salmonids live in two very different qualities of water, freshwater and seawater, during their lives, the life stages involved require different adaptive physiologies in order to accomplish the change.) This may be at first thought a negative aspect of salmon and trout farming, since it imposes relatively strict water quality criteria upon the farmer. On the other hand, the public image of salmon is that of a fish which lives in pure, clean, cool water, and the sale value is, at least partly, a result of that image. Salmon is a 'high quality' fish, and it requires water of 'high quality'. From the point of view of the fish farmer, water quality requirements may be considered under relatively few main headings.

Oxygen

The quantitative water requirements of fish in culture are in practice defined by the oxygen requirements. The principal gases dissolved in water are the same as those present in air, oxygen, nitrogen and carbon dioxide, but whereas in air the proportions of these, at least of oxygen and nitrogen, are fairly constant, the amounts dissolved in water, as well as being much less, (water contains 20–30 times less oxygen per unit volume that does air), vary greatly with temperature and salinity. At low temperatures, water can contain more dissolved oxygen than it can at higher temperatures, and the higher the salinity of water, the less oxygen it can contain.

Salmonids favour fairly low temperatures, normal culture temperatures usually lying within the range 5–20°C. This means tht the maximum oxygen content of the water in freshwater culture will be between 12.8 mg/l and 9.2 mg/l, assuming that the water is 100% saturated, that is, that it is carrying the maximum amount possible of dissolved oxygen at atmospheric pressure. The corresponding values for seawater will be about 30% lower.

Compared with other farmed species, such as eels and carp, salmonids require relatively more oxygen for normal activity. This is reflected in the levels of dissolved oxygen necessary in the water. Minimum values of 5 mg/l are stated as being necessary over prolonged periods (minimum for eels is about 1 mg/l). At about 15°C, this is around 50% saturation. However, values of no less than 75% oxygen saturation in effluent water must be regarded as a practical minimum in 'safe' planning. Over very short periods, for example under transport lasting only a few hours, salmonids are reported to be able to tolerate oxygen levels as low as 3.5 mg/l. It must be expected, however, that the fish will be stressed by such low oxygen levels.

The basic information necessary for estimation of the production capacity of a farm is, therefore, minimum expected water flow (e.g. in cubic metres per minute) and maximum expected water temperature. From this the minimum oxygen input per minute can be calculated. (In far northern regions, it may not be necessary to take both factors into account simultaneously, since minimum water flow often occurs due to freezing of freshwater sources. However, summer minimum flows due to hot, dry weather will usually coincide with maximum water temperatures). In sea-cage farming, water 'flow' is sometimes referred to as water 'exchange' and the mechanisms involved are coastal and tidal currents.

Production is usually given in tonnes, but this is too imprecise a measure for calculation of water/oxygen requirements, because fish of different sizes have different rates of oxygen consumption. A farm producing pan-sized (or portion-sized) fish will need much more water for each tonne of fish than a farm which produces large salmon. This is because, although fish generally consume more oxygen the larger they are, they actually consume less in proportion to their weight. Based on respiratory measurements, data relating to fish size, water temperature and normal oxygen consumption are available. From these, the oxygen requirements and therefore the water requirements, of any given amount of fish of a given size can be

Table 5.1 Freshwater requirement in litres/min./kg for salmon of various sizes and at various water temperatures. It has been assumed that the inflow water is 95% air saturated and that the outflow water contains 5 mg/l oxygen. (From Liao, 1971.)

	Fish size (grams)								
°C	1	2	4	6	10	20	100	1000	4000
2	0.2	0.2	0.2	0.1	0.1	0.1	0.1	0.1	0.1
4	0.3	0.3	0.2	0.2	0.2	0.2	0.1	0.1	0.1
6	0.4	0.4	0.3	0.3	0.2	0.2	0.2	0.1	0.1
8	0.6	0.5	0.4	0.4	0.3	0.3	0.2	0.1	0.1
10	0.8	0.7	0.6	0.5	0.5	0.4	0.3	0.2	0.2
12	1.0	0.8	0.7	0.7	0.6	0.5	0.4	0.2	0.2
14	1.2	1.0	0.9	0.8	0.8	0.7	0.5	0.3	0.2
16	1.5	1.3	1.1	1.0	0.9	0.8	0.6	0.4	0.3
18	1.8	1.6	1.4	1.2	1.1	0.9	0.7	0.5	0.4

calculated. Often, tables giving approximated figures for practical use are produced (Table 5.1).

Therefore, in calculation of the water/oxygen requirements for a desired production, the numbers, or total weights, of fish of each size/age class present when the farm is holding the maximum quantity of fish need to be known. The total water/oxygen requirement will then be the sum of the requirements of each size/age class. In this connection, it is also important to take into account the fact that the oxygen requirements of fish are greatly increased by feeding, and that many data in the literature which give oxygen requirements as a function of fish size and water temperature only, may underestimate the needs of actively feeding fish.

In general, natural water can be regarded as being fully saturated with oxygen if it is from a fast-flowing mountain stream, or is normal, clean seawater. Water deficient in oxygen may occur, however, at groundwater wells or springs, or under certain coastal sea conditions, and it is usual to measure oxygen levels regularly. When calculating water requirements, it is also possible to introduce safety margins, for example by assuming 95% saturation instead of 100% in the incoming water.

Temperature

While the carrying capacity of a salmon farm is determined by the maximum expected oxygen requirement, and hence the availability of a sufficient flow of oxygenated water, the rate of production is determined by the growth of the fish. The single environmental factor most associated with growth rates in fish farming is, without a doubt, water temperature, since it affects the rates of nearly all metabolic processes.

Another important aspect of temperature, in addition to its effect on the oxygen carrying capacity of water, already mentioned, is that it sets the climatic limits for salmon farming. This is especially true of sea-cage farming, in which technical control of water parameters is usually impossible. In inside hatcheries and smolt production facilities, water temperature can often be manipulated, but the expense of such manipulation can be decisive for the economic viability of the production.

Salmon will die when ice crystals begin to form in the body fluids. Normally, this will occur at about −0.5°C, but under certain conditions the lethal temperature may become slightly lower. Lower lethal temperatures usually occur only under sea-farming conditions, since freshwater normally freezes at 0°C, a phenomenon which would kill the fish anyway.

The upper lethal temperature for post-smolt salmon is around 25°C, or slightly lower. Under farming conditions, high fish densities lead to oxygen lack at high temperatures, because the oxygen requirements of the fish are greatly increased at the same time as the oxygen carrying capacity of the water decreases.

For the early, pre-feeding life stages, fairly low temperatures may be lethal. Newly fertilized eggs should not be exposed to temperatures above 15°C, and even at temperatures in this range, the efficiency of yolk utilization of eyed eggs and yolk sac alevins may be so reduced as to prevent growth and development to the stage capable of feeding, before the yolk supply is exhausted. If this happens, large fry mortalities occur, which are known as 'initial feeding mortalities', a misnomer since the cause of death is starvation. Hatchery water temperatures, where these can be controlled, are therefore determined by a 'compromise' between the farmer's wish for a short incubation time and fast development to initial feeding, achieved by raising the temperature, and the ability of the developing fish to utilize yolk for growth and development to the stage where it can take external food. Yolk is used most efficiently for growth at low temperatures, a fact which is hardly surprising if one considers the likely temperatures in the natural incubation environment of mountain streams during autumn, winter and early spring.

The degree to which salmon will tolerate temperature changes which approach the lethal limits may vary, depending upon the fish's physiological condition. Temperature acclimatization, physiological 'partial adaptation' to a particular temperature, or temperature range, may increase tolerance of temperature changes close to the acclimatization temperature. In other words, fish which have been raised for a period in water of, say 8°C, can be expected to tolerate a sudden temperature fall to 1°C better than fish which have been kept at 15°C. Temperature acclimatization takes time, normally three weeks or more at the acclimatization temperature, to be fully effective. Seasonal changes in natural water temperature are usually slow enough to allow fish to remain more or less acclimatized at all times.

Of course, fish farms are not normally located in areas of the world where water temperatures lethal to the fish occur. The potentially disastrous consequences of mass mortalities aside, production results would be extremely poor under conditions of very low, or high, average temperature. Physiological investigations have shown that fast, efficient growth in salmon is best achieved in water temperatures of 13–17°C. Outside this range, productions becomes less efficient, either due to slower growth or to temperature stress problems. Of course, market factors may allow relatively inefficient production to be profitable, as appears to have been the case during the 1970s in Scandinavia, but such is not the situation today, and at its present northern geographical limits, intensive salmon farming is now beginning to have difficulties as a profit-generating industry, mainly because of the reductions in production efficiency imposed by the temperature conditions.

pH

The pH value of water indicates whether the water is acid, alkaline (basic) or neutral. It is measured on a scale of 1 to 14, where 1 represents maximum acidity and 14 maximum alkalinity. At 20°C, pH 7 represents neutrality. The scale is logarithmic, so that a change of one unit represents a tenfold change in hydrogen ion concentration. It is the ratio of positively charged hydrogen ions (H^+) to negatively charged hydroxyl ions (CH^-) which determine pH. When the two are present in equal numbers, the water is neutral (pH 7).

Substances added to water may influence the pH value by contributing, directly or indirectly, hydrogen or hydroxyl ions. For example, carbon dioxide, mineral and organic acids all have acidifying effects, while algal photosynthesis, by removing carbon dioxide contributes to raise the pH. Calcium salts, especially carbonates, have an alkalizing effect.

Fish, for physiological reasons, need to control their internal acid/base balance. Changes in the external pH can affect the ability to regulate, and most fish species are adapted to particular narrow ranges of environmental pH.

Water used for salmon production should have a pH value between 6 and 8. Slightly lower and higher values can be tolerated, but should be avoided anyway. At low pH values (below 5.5), acidosis occurs, reducing the ability of the blood to transport oxygen, and irritating the skin and gills, so that they produce excess mucus. Chronic acidosis also reduces resistance to bacterial disease. Above pH 8 to pH 9, alkalosis occurs, destroying gill and fin tissue, as well as causing symptoms similar to those of acidosis. Both high and low pH values can also affect the toxicity of other dissolved substances, an aspect to be mentioned later.

Hardness

Water hardness is a measure mainly of the calcium and magnesium salts present. Of these, the calcium salts contribute most. Hydroxides, carbonates and bicarbonates are the main types of salts present. The total hardness, or general hardness (GN) can be divided into temporary or permanent hardness. The former is due mainly to bicarbonate ions. It can be removed by precipitation if the water is boiled, and it is usually the major contributor to the total hardness. Soft water contains, as might be expected, few salts.

There are a number of methods for measuring hardness, each using a different scale. All can, however, be standardized to give mg/l $CaCO_3$ (or ppm $CaCO_3$) as the measure. Soft water contains less than 50 mg/l $CaCO_3$, while hard water has between 300 and 450 mg/l. Hardness is also closely related to the water's conductivity, measured in microSiemens/cm (µS/cm), soft water having values of 10–30 µS/cm and hard water up to several hundred µS/cm. (Seawater gives conductivity values of 40 000–50 000 µS/cm). The temporary hardness of water, being caused mainly by bicarbonates, is closely associated with its alkalinity, which is one reason that hard waters are less affected biologically by acidifying pollution.

As with pH, fish species are generally adapted to particular ranges of hardness. Water hardness is in some ways a type of salinity, and it affects the osmoregulatory systems of freshwater fishes. Soft water species, living in environments relatively devoid of salts, need to have efficient osmoregulatory systems and must do more physiological work to maintain their salt/water balance.

Many Scandinavian and Scottish salmon populations, during the freshwater life stages, inhabit extremely soft waters, due to the granite bedrock of the Scandinavian Shield and the Scottish Highlands. One disadvantage of this is that soft waters, being poorly buffered, are easily affected by acid rainfall, especially where this is accumulated in the form of snow and ice, and then released over a short period of thaw in spring. Soft water appears, nevertheless, to be necessary for some life stages. Salmon eggs, if placed in hard water after fertilization will not swell (or harden), since the salts present in the water reduce the osmotic gradient, between the inside and outside of the egg, necessary for the uptake of water.

pH, hardness and metal ion toxicity

Acid water, if it does not kill fish directly, from acidosis, may be lethal indirectly, in combination with certain metals. In the large areas of the northern hemisphere which are affected by acid rainfall, aluminium may be leached from soils, entering lake and river water. Particulate aluminium, or aluminium which is chemically bound to organic molecules, is not harmful to fish. However, at low pH values, hydroxide salts

Table 5.2 Limit values, taken from various sources (principally EIFAC Technical Papers), of some water quality parameters in freshwater salmon farming. All concentrations are given as mg/l unless noted otherwise.

Oxygen	5–7
Temperature (°C)	>0–20
pH	6–9
Metals	
Aluminium (Al)	0.01 (variable)
Lead (Pb)	0.005–0.05
Iron (Fe)	0.1–0.5
Cadmium (Cd)	0.0003–0.001
Copper (Cu)	0.001–0.006
Chromium (Cr)	0.01–0.1
Mercury (Hg)	0.0002–0.0005
Manganese (Mn)	0.01 (variable)
Nickel (Ni)	0.01–0.05
Zinc (Zn)	0.005–0.06
Silt	25–80
Gases	
Ammonia (NH_3)	0.0125–0.025
Hydrogen sulphide	0.002–0.003
Carbon dioxide	1.15–15.0
Chlorine (as HOCl)	0.002–0.004
Air saturation (%)	105–115
Nitrogen saturation (%)	103

of aluminium can be formed, which are extremely toxic. At pH 5, long-term concentrations of 30–40 µg aluminium per litre water can be lethal to young salmon. Intermittent or 'pulsed' exposure to low pH water containing aluminium, which is a common phenomenon during periods of runoff followed by dry periods, has been shown to weaken, but not kill, salmon fry, making their survival unlikely.

Other metals, especially the heavy metals such as copper, zinc, cadmium, chromium and lead are toxic, even at low concentrations if water conditions allow (see Table 5.2). Generally, the harder the water, the less toxic the metal. Organic matter (humus) in the water can also reduce the effects of toxic metals, but as mentioned previously, the freshwater stages of salmon are mostly raised in soft waters containing little humus, and are thus particularly susceptible to the effects of acid pollution. Acid rain caused by industrial effluent gases and its devastating consequences has, unfortunately, become a long-term problem which will probably take many years to solve.

Silt and gas supersaturation

Certain more natural, but no less dangerous, forms of 'pollution' can occur under certain circumstances.

Spring runoff often contains tiny mineral particles, known as silt. Silt particles are small enough to remain in suspension in moving water, often giving the water a milky appearance. Although non-toxic, silt particles cause irritation to gills of young salmon, which respond by secreting extra mucus. If the fish are kept in silted water long enough the gills can become clogged with mucus, leading to suffocation. Most hatcheries avoid the problem simply by obtaining water through a filtering device.

Gas supersaturation is a phenomenon most associated with the heating of hatchery water. Since the solubility of gases in water is dependent upon temperature and pressure, raising temperature or lowering pressure will cause some of the dissolved gases to come out of solution, causing the water to become supersaturated until equilibrium with the atmosphere is again established, a process which may require considerable time. Technical manipulation of water quality in salmon farming is a wide subject, outside the scope of this chapter. However, 'natural' supersaturation can occur when air is forced under water and thus absorbed under pressure. Air bubbles as little as three metres below the water surface can cause supersaturation. Waterfalls, either natural or under dam gates, hydro-electric turbine outflows, and even seawater close to cliffs subject to heavy wave action, can all become sources of supersaturated water – water containing more gas than can be maintained dissolved at the temperature and pressure involved.

Fish exposed to supersaturated water may die from gas embolism, the blockage of capillary blood flow by gas bubbles. The microscopic bubbles occur because the fish have 'inspired' the pressurized gases from the water. Since neither the water nor the fish blood can hold the gases in solution, they are released as micro-bubbles. In fact, only the biologically inert gases, principally nitrogen, are dangerous in this respect, being in addition, relatively insoluble. Water containing 5–15% more air than can be held dissolved, (that is a total gas saturation of 105–115%) is lethal to salmon (Table 5.2). The fry appear to be particularly susceptible. In terms of nitrogen supersaturation, 103% is the maximum value tolerated. Fortunately, supersaturation is easily removed, by allowing the gases in the water to equilibriate with the natural atmospheric pressure, usually by splashing it through grids, or by aeration.

Salinity

Salmon farming is divided into two distinct phases, the freshwater phase – smolt production – and the seawater phase – growing to market size. The qualitative differences between freshwater and seawater are so great that the fish must make enormous physiological changes in order to survive. These changes are usually described collectively as smoltification, and since the physiology of smoltification is not the subject of this chapter, only a few of the aspects of seawater as a special environment will be mentioned here. The reduced capacity to contain dissolved gases has already been mentioned.

Normal seawater has a salinity of about 35 parts per thousand. That is to say that it contains 35 grams per litre of dissolved mineral salts, the most important being sodium chloride (78%), magnesium chloride (11%), magnesium, calcium and potassium sulphates, calcium carbonate and magnesium bromide. Because the amounts of dissolved salts change the specific gravity of the water, salinity is usually measured by means of a hydrometer. When specific gravity and temperature are known, salinity can be read from tables.

The sheltered coastal or fjord waters often used for cage farming salmon, are usually less saline than full strength seawater, varying from about 20–34‰. However, seawater is in many ways a more stable environment than freshwater. The average pH is about 8.1, the buffer capacity is high, and the high salinity reduces the effect of many pollutants.

To summarize, salmon farming requires plentiful supplies of high quality water, both during smolt production in freshwater and salmon production in seawater. 'High quality', in this respect refers to the physiological demands of this fish species for, essentially, cool, clean, oxygen-rich water. The principal areas of the world in which salmon farming has developed successfully – Norway, Scotland, Ireland, Canada and Chile – are indicative of the kind of environment necessary for this type of aquaculture, as is the present natural distribution of the species. The relatively stringent demands places by the salmon upon its environment have not, however, prevented the growth of the salmon farming industry into a major producer of quality fish, as the present production figures illustrate. In fact, increased competition within the industry, due to high production, are likely in the near future to result in an overall reduction of the areas farmed, as those with less than optimal water conditions, principally with respect to temperature, become competitively marginal.

THE QUESTION OF POLLUTION FROM SALMON FARMS

Implicit in the public image of salmon as quality fish from pure, clear water is the assumption that the water involved in the farming of salmon also remains of the highest quality during and after the production process. That it is of high quality before contact with the fish is a 'requirement' imposed by the fish itself, and the fundamentals of that requirement in terms of oxygen, temperature, pH, dissolved salts and some (but not all) pollutants have already been introduced.

Salmon farms are not normally established where the water quality criteria cannot be met, but changes which affect that quality may occur afterwards. Any changes which are potentially detrimental to the health of the fish may be classified as 'pollution' by the farmer, even though they may be natural occurrences. Fish farms also cause certain changes in the water in and around them, and these may also be detrimental, to the fish, the environment, and to the public image of the industry.

The water passing through a fish farm can be affected by two agents: the fish themselves, through their oxygen consumption and excretions; and the farmer, who adds feed, and possibly antifoulants, disinfectants and medicaments to his or her tanks or cages. Of the two sources of possible pollution, the farmer is by far the most dangerous polluter.

Fish excretions as potential pollutants may pose a limited threat to the environment, and under certain circumstances may reach levels dangerous for the fish themselves. There are three main categories: carbon dioxide, ammonia and fish faeces.

Carbon dioxide (CO_2)

Like all aerobic organisms, fish produce carbon dioxide as an end product of respiration. In addition, there is always a certain amount of carbon dioxide in water entering a farm. The levels in natural water are not high, usually less than 5 mg/l. Carbon dioxide is more soluble in water than the two other principal gases present, nitrogen and oxygen, being 70 times more soluble than nitrogen and twice as soluble as oxygen. It also exists in a complex relationship with bicarbonate and hydrogen ions, such that in hard waters, or sea water, excess carbon dioxide reacts with the alkaline buffers, making its accumulation unlikely. Soft waters, however, do not have high buffering capacities, and since carbon dioxide is toxic, due to the effect on hydrogen ion concentrations, it could pose a threat to salmon fry being raised at high stocking densities. Levels of 20 mg/l are certainly lethal, and levels as high as 12 mg/l are probably so. Fortunately, excess carbon dioxide in water exchanges readily with the atmosphere, because there is much less of the gas in air. Turbulence created by aeration is usually sufficient to reduce dissolved carbon dioxide to acceptable levels in hatchery tanks.

Ammonia (NH_3)

The major end product of protein catabolism in fish is ammonia, which is a very soluble, alkaline gas. It is excreted by fish through the gills, and takes part in the ion regulation system, being exchanged for sodium. In water, ammonia also exists as the ammonium ion (NH_4^+). The ionic form is more prevalent at low pH and low temperature, while the proportion of dissolved free ammonia increases as temperature and pH rise. At 15°C and pH 8, for example, only 5% of the ammonia is free, while at pH 9 that figure is 20%. Water at 25°C can hold five times more free ammonia than can water at 5°C. These facts are important because free ammonia is much more toxic to fish than is the ammonium ion. The toxicity of ammonia is also about 30% less in seawater than in freshwater.

Ammonia is the most toxic nitrogenous compound. Salmon raised in water containing just 0.006 mg/l free ammonia will show a chronic toxic reaction. Levels as low as 0.2–0.5 mg/l are lethal. Eggs and fry are particularly susceptible to this natural toxin. It is, therefore, important in hatchery production to make sure that water conditions and rates of exchange are sufficient to prevent the buildup of high levels of ammonia in tanks. Hygiene is also important in this respect, since ammonia is also produced by bacterial decomposition of organic wastes, such as dead fish and excess feed.

As a potential polluter of the environment however, ammonia excretion by the fish themselves is probably negligible. In the presence of oxygen, ammonia is converted to nitrite (NO_2^-) by *Nitrosomas* bacteria. Nitrite is toxic, but less so than ammonia. Values of toxic concentration vary widely, probably because, like several other toxic substances, nitrite's toxicity varies, being reduced in hard water and in seawater. Nitrite is oxidized further, to nitrate (NO_3^-) by bacteria of the genus *Nitrobacter*. Nitrate is, together with phosphate, often a component of agricultural fertilizers. It is about 2000 times less toxic than nitrite. In seawater, nitrates, nitrites and ammonia are taken up by algae, and they are found at extremely low levels in the water.

Fish faeces

The faeces represents the undigested remains of the fishes' feed. Its composition and consistency will vary somewhat, depending on the quantity and quality of the diets being used. In investigations of sediments under and around fish farms, it has been found that faecal material is not an important component, compared with waste feed. Fish faeces sink slowly, disintegrating as they do so. Sinking rates of about 0.15 cm/second have been estimated. Since fish farms are usually situated in locations with good currents, for water exchange, it follows that the faeces become fairly widely dispersed. The fish also digest most of their feed, so that faeces have a much lower energy content than waste feed.

Fish excretions from salmon farms are unlikely to influence natural phytoplankton production on a large scale. Certain components of the excretions may have a eutrophic (fertilizing) effect upon algal growth in the area of the farm, but this, again, is a potential problem for the farm rather than for the environment. Algal fouling of fish cages reduces water exchange, and it must be removed. Even if this is done by pressure hosing, and the dead algae allowed to deposit sediment under the farm, investigations show that it will contribute no more than about 1.5% to 6% of the farm's total organic contribution to the area.

In conclusion, the total organic contribution from excretion, including faeces, non-faecal end products, mucus and epidermal tissue, as well as any eutrophic effect, is a relatively minor proportion of the 'total effect' a fish farm has on its immediate environment. (This conclusion applies to seawater cage farming. In enclosed fresh-waters the situation might be otherwise.) It has been estimated that excretory products account for about 29% of the energy of the feed consumed by the fish. It is, however, unconsumed feed which usually accounts for the major part of a fish farm's organic loading on the environment.

Waste feed

Modern salmon feeds are much more sophisticated than they were only a few years ago. They provide balanced diets as small, energy-rich pellets. Not all fish farmers use commercially formulated feeds, but most do. Pellets are more easily stored than fresh, frozen or acid conserved feeds, and they require less labour in use. For smolt producers, commercial 'dry feeds' are a necessity.

It was calculated, only a few years ago, on the basis on the total amount of commercial salmon feed sold to the industry, and the salmon production figures, that Norwegian salmon farmers had, collectively, been feeding their fish roughly twice the amount of feed necessary. In other words, for every tonne of feed which had been fed to the fish, about half may have ended up on the seabed under the farm. Although this is probably an overestimate, and almost certainly not the case today, all investigations of sedimentation under salmon farms have shown that the principle contributor to bottom sedimentation after the establishment of the farm, has been uneaten feed.

Unlike fish faeces, feed usually sinks rapidly. Swedish studies have indicated that it nearly always sinks faster than the speed of the water current passing through the farm, and that 80–90% of a farm's total organic contribution to its environment deposits sediment directly under the cages. Until recently, salmon farms were most commonly located in sheltered, relatively shallow areas, where tidal water currents provided the water exchange. Under these conditions sedimentation under fish cages due to overfeeding occurs rapidly.

The consequences of continuous addition of organic material to aquatic sediment are well known. The oxygen, required by aerobic bacteria for decomposition, becomes exhausted, the sediment becomes anaerobic, and hydrogen sulphide, produced by anaerobic bacteria, builds up to high concentrations. Hydrogen sulphide is, like ammonia, an extremely toxic gas. Salmon exposed to concentrations as low as 1 mg/l die after a short time. It is, fortunately, highly soluble, and the good water exchange conditions at fish farms probably ensure that most of the hydrogen sulphide escaping from the sediment is diluted and transported safely away. However, after the sulphate in the sediment has been used up in the formation of H_2S, methane gas will be produced. Methane gas, being less soluble, tends to bubble to the surface from anaerobic sediments, and in sediments containing high concentrations of hydrogen sulphide the methane bubbles could act as quick transport for higher concentrations of H_2S than would otherwise reach the upper layers of the water column. Concentrations of hydrogen sulphide of 0.5 mg/l have been measured in fish cages above anaerobic sediment. While not necessarily lethal, this might be enough to stress the fish and reduce performance. Fish deaths, due to wave disturbance of anaerobic sediments and consequent sudden release of hydrogen sulphide, are well documented.

Interestingly, recent studies have indicated that denitrification, the usual mechanism of nitrogen removal from anaerobic sediments, resulting in the formation of gaseous nitrogen, is not a significant nitrogen removal mechanism in the vicinity of a salmon farm. Instead ammonia is produced, and in turn, nitrites and nitrates, contributing to eutrophication. In the long term, however, and at the ecosystem level, bacterial denitrification probably balances the nitrogen input from fish farms.

The large amounts of organic material added to fish farm water also lead to the consumption of large amounts of oxygen, by aerobic decay bacteria at the oxygenated water/sediment interface. 'Biochemical oxygen demand' (BOD) is one of the measures commonly used in evaluating organic loading from human habitations on the environment. It has, therefore, been relatively easy to 'popularize' quantitative data by comparing salmon farms with sewage effluents, in terms of BOD. It can be stated, for example, that a salmon farm producing 60 tonnes of salmon is loading its immediate environment with a quantity of organic matter equivalent to that which the sewage from a habitation of 6000 persons would produce. Comparisons even more dramatic have been published about the industry as a whole.

Such comparisons may be used to shock, or to make the public aware of yet another source of environmental pollution but they are not always put in perspective. Salmon farms, for reasons dictated by the fish themselves, are located in areas of clean, flowing, oxygenated water. The water volumes involved in exchange through and around the farms are truly enormous. For example, the farm mentioned above would have required, in addition to the amount needed for the salmon, about 100–600 kg oxygen daily to satisfy the oxygen demand of the waste organic material. This requirement would be covered if the rate of water flow through the farm always exceeded 9 cm/second, a fairly modest flow.

Waste feed caused by overfeeding is without doubt a problem, but, at least in sea-cage salmon farming, any detrimental effects of organic loading from both fish and feed will be likely to affect mainly the farm itself, rather than the surrounding ecosystem as a whole. In fact, many established salmon farms in Norwegian fjords have caused so much below-cage sediment enrichment that cage rotation has been deemed necessary. Since recovery of the polluted benthos usually takes a very long time in these cool water ecosystems, such a solution is obviously an unsuitable one, likely to lead to relatively large increases in organically overloaded areas. The industry and its feed suppliers are, for these and other reasons, taking the subject of overfeeding very seriously today. Salmon farmers are being made aware of exactly how much feed they should be feeding, and how to monitor this closely so as to avoid wastage.

Antifoulants and disinfectants

Fouling of salmon cages by marine organisms restricts water exchange. This can be a serious problem, especially since high fouling rates coincide with high water temperatures, high oxygen requirement by the fish and minimum oxygen levels in the seawater. High pressure hosing and frequent net changing are common methods employed to combat the problem. It is also possible to impregnate cage nets with antifoulants.

Antifoulant paints or impregnating solutions release biocides, producing toxic concentrations close to the treated surface. A common biocide was copper, the toxic effects of which have already been mentioned, but more recently, antifoulants based on tributyltin (TBT) have been used. This substance is about 1000 times more toxic than copper at the same concentrations. By the mid-1980s, concern was being expressed, based on reports of TBT effects on non-target organisms, such as mussels. As a result, the use of TBT is now banned. Large scale environmental effects, if any, of antifoulants in the sea are undoubtedly difficult to estimate, but, certainly, the bulk of antifouling substances in use today are used on boats, and structures other than fish farm cages.

Disinfectants are often used in salmon hatcheries to wash fish tanks before use, and in the course of the general cleaning operations. Common disinfectants are formalin, chlorine and chloramine, the latter often in conjugation with ammonia, which enhances its stability. These substances are highly toxic, not only to the target microorganisms, but also to the fish. (Salmon fry die fairly quickly if exposed to 'town water' which is chlorinated for reasons of public health). However, the quantities used are small and the flushing volumes large, and hatcheries almost certainly do not pollute the environment in this way.

Medicaments

In recent years the disease problem has become increasingly serious in salmon farming. As has been pointed out earlier, the salmon on a farm are being kept in a very unnatural environment. It is to be expected that aspects of this environment will act as stressors, making the fish more susceptible to disease than they would otherwise be. In addition, as in all animal and plant husbandry, the fact that the stocks are being farmed at high densities makes the outbreak of an infectious disease potentially disastrous.

All illustration of the difference between a farm environment and the natural situation is provided by a common, and serious 'pest' in salmon farms, salmon lice. These parasites are not, of course, lice, but parasitic crustaceans. It is not uncommon for wild fish to be infested with a few fish lice. Farms, however, with their concentrated stocks of captive salmon provide ideal situations for the infective stages of the parasite, allowing infections to reach enormous levels. If left untreated, major lice infections can destroy salmon farm stocks, and even attacks of minor proportions stress the salmon, as well as causing sores open to fungal and bacterial infection.

Salmon lice can be removed by bathing the fish in freshwater, a simple and environmentally harmless remedy. This is not always possible, however, and a common method of treatment is the use of the organophosphate neguvon, or a derivative known as Nuvan. These substances attack the nervous system of the parasite, preventing transmission across synapses. They are extremely toxic, also to humans, and must be used with great care. Whether or not the use of Nuvan actually does any real environmental damage is still a matter of debate. However, it is clear that release of organophosphates into the environment is undesirable. They may affect non-target organisms, and they have a tendency to accumulate in the ecosystem. They are the fish farming equivalent of insecticides. Realizing this, the salmon farming industry is seeking alternatives. A novel, environmental 'friendly' possibility being tried by some farmers, is the use of 'cleaner fish', wrasses, to pick the lice from the salmon.

For some of the major diseases, such as vibriosis, reliable and easily administered vaccines are now available, but in cases where unvaccinated fish are diagnosed as having contracted a disease, the usual treatment is to administer medicated feed, that is feed containing a broad-spectrum antibiotic.

It has already been pointed out that all, or at least most salmon farmers overfeed to some degree. It is also the case that one of the earliest effects of infection by most fish diseases is that the fish lose appetite. As a result, much of the medicated feed, like the normal feed, ends up not in the fish, but in the sediments under the farm. In addition, most of the antibiotics actually consumed are excreted in active form by the fish. Herein lies a potential danger greater than that of normal organic loading.

The use of antibiotics in human and veterinary medicine is usually subject to various restrictions, for the very good reason that long-term exposure to low concentrations of antibiotics is known to enhance development of resistant microbial strains. It is, therefore, regarded as extremely important that in treatment with antibiotics, the drugs are restricted as much as possible to the target, and not allowed to enter the surrounding environment. Fish farming stands, until now, in dramatic contrast to this practice. Medicated feeds containing large quantities of antibiotics have been simply thrown into the sea, significant amounts of them missing the target organism completely.

In recent Norwegian studies, samples of bottom sediments from under four salmon farms all contained the antibiotic oxytetracycline, even at considerable depths under the farms. Concentrations of the drug in the sediments 12 weeks after medication were found to be higher than the minimum inhibitory concentration for bacteria.

The implications of this are alarming. Japanese investigators have detected as many as 111 strains of *Vibrio alguillarum* out of 139 tested, to be resistant to tetracyclines. In other studies, 88% of the bacterial strains isolated from rainbow trout were found to be resistant to at least one drug. Not only Vibrio bacteria, but also *Aeromonas*, the causative agent of the disease furunculosis, as well as several intestinal bacteria have been shown to have the ability to transfer resistance factors to other bacteria.

It is, therefore, clear that the hitherto fairly indiscriminate use of antibiotics in fish farming is having potentially very dangerous side-effects. The best solution to this undesirable situation is the development of vaccines and more environmentally suitable drugs. However, where these do not exist, or are unavailable, salmon farmers have little choice but to continue to use medicated feed.

With the intention of minimizing environmental effects from antibiotics, restrictions on their use have now been introduced in Norway. These include tighter control of access to medicated feed, limitations on fish size and stocking density in fish to be treated, and control of feeding method. Antibiotic residues must also be collected, where possible, and stored prior to destruction. Such legislation will certainly have the effect of reducing the addition of antibiotics to the environment, and it must be welcomed.

As salmon become more domesticated, and knowledge about how to minimize stress factors in the farm environment increases, risk of disease may decrease, but it is hardly likely to disappear.

Salmon farms, as this short account has tried to illustrate, need not be serious polluters of the environment. Attempts to portray them as such are often dramatized and taken out of perspective. The image of salmon as a fish which demands a high quality environment is a correct one, and any pollution deriving from the activities of a salmon farmer is most likely to affect first the farm itself, and thus become 'self-limiting'. Even the last mentioned subject, the release of antibiotics to the environment, although by far the most negative aspect of fish farming today, should not be over-dramatized. With the (hopefully imminent) cessation, or at least reduction of this practice, resistance in natural bacterial populations can be expected to return to pre-exposure levels, since it is normally the presence of an antibiotic, by eliminating non-resistant competition, which allows resistant strains of micro-organisms to multiply to unnaturally high levels.

References and selected bibliography

Aure, J. and Stigebrandt, A. (1990) Quantitative estimates of the eutrophication effects of fish farming fjords. *Aquaculture*, 90, 135–56.

Beveridge, M.C.M. (1984) Cage pen fish farming, carrying capacity models and environmental impact. *FAO Fish Technical Paper 225*.

EIFAC (European Inland Fisheries Advisory Commission, FAO) (1968) Water quality criteria for European freshwater fish. Report on extreme pH values. *EIFAC Technical Paper No. 4*.

EIFAC (1970) Water quality criteria for European freshwater fish. Report on ammonia. *EIFAC Technical Paper No. 11*.

EIFAC (1973a) Water quality criteria for European freshwater fish. Report on dissolved oxygen. *EIFAC Technical Paper No. 19*.

EIFAC (1973b) Water quality criteria for European freshwater fish. Report on chlorine. *EIFAC Technical Paper No. 20*.

EIFAC (1984) Water quality criteria for European freshwater fish. Report on nitrite. *EIFAC Technical Paper No. 46*.

Gowen, R.J. and Bradbury, N.B. (1987) The ecological impact of salmon farming in coastal waters: a review. *Oceangr. Marine Biology*, 25, 563–75.

Iwama, G.K. (1991) Interactions between aquaculture and the environment. *Critical Revue Environmental Control*, 21(2), 177–216.

Liao, P.B. (1971) Water requirements of salmonids, *Progressive Fish Culture*, 33, 210–15.

Mathiesen, C.K. (1990) Fish farming and the environment. Reality and myth – A Danish perspective. *Aqua. Int. Congr. Proc.*, 278–84. Aquaculture International, Vancouver.

Rosenthal, H., Weston, D., Gowen, R. and Black, E. (1988) Report of the *ad hoc* study group on environmental impact of mariculture. *ICES COOP. Res. Rep.*, 154.

Smart, G.R. (1981) Aspects of water quality producing stress in intensive fish culture. In *Stress and Fish* (Ed. by A.D. Pickering), pp. 277–93, Academic Press, London.

Weston, D.P. (1986) *The Environmental Effects of Floating Mariculture in Puget Sound*, University of Washington, Seattle.

Chapter 6
Genetic Effects of Aquaculture on Natural Salmonid Populations

FRED UTTER, KJETIL HINDAR AND NILS RYMAN

Introduction

This book testifies to the current importance of salmonid aquaculture which has recently accelerated to a major activity in a number of temperate areas throughout the world. However, as with other activities that have rapidly blossomed and subsequently thrived, prosperity and optimism commonly mask potential hazards. There is a general tendency to ignore possibly undesirable side-effects until their visibility reaches the point where action is imperative. For instance, while the major benefits derived from the automobile industry have also been accompanied by air pollution and traffic jams, remedies for these detrimental by-products have primarily been sought in circumstances where their presence has become intolerable.

This chapter addresses concerns about genetic dangers to wild populations from aquacultural activities (see Fig. 5.1). We are particularly concerned about the dangers of indiscriminate accidental or intentional releases of cultured organisms into the natural environment. We maintain that the above noted tendency to react only after extensive damage has become apparent is a particularly inappropriate response to these problems. Our concerns are heightened for this specific issue because the genetic effects are irreversible, and by the time they are detected, the damage has already occurred. Therefore, our message is based on a goal of prevention as the only viable strategy to minimize losses of valuable and irreplaceable genetic resources.

We contend that interbreedings among genetically diverged salmonid populations are generally disadvantageous to the natural populations. The direct experimental evidence for this contention is admittedly sparse, basically because – to our knowledge – appropriate experiments to clarify this critical issue have not been carried out. However, we point out (with the support of numerous empirical observations) that the consequences of such releases cannot be predicted; and, based on a body of indirect evidence, we show that intentional or accidental releases of cultured salmonids threaten natural populations (1) by homogenization of genetic population structure and breakdown of locally adapted gene complexes through interbreeding, and (2) by loss of native populations through competition or disease introduction. Consequently, we make specific recommendations as guidelines for effectively dealing with these concerns, including the responsibility of the culturer for monitoring the genetic effects of releases of cultured populations. The reader is referred to Hindar *et al.* (1991) for more detailed and technical discussions of these problems.

Background

In order to justify the conclusions and recommendations of this chapter, we summarize some pertinent background information below concerning genetics and population structure of salmonid fishes.

Population subdivision

Species of salmonid fishes are commonly subdivided in a number of genetically differentiated and more or less reproductively isolated local populations (Allendorf and Utter, 1979; Ryman, 1983; Ryman and Utter, 1987). A remarkably precise homing behaviour (e.g. Quinn and Tallman, 1987) appears to be important for maintaining the genetic integrity of these populations.

New insights from molecular genetic data

Most of the information about genetic population structure of salmonids comes from a very limited proportion of the total genes, and has been obtained within the past 20 years. This information, collected by such molecular procedures as protein electrophoresis or direct examination of DNA, has provided a wealth of insight into breeding structure, migration patterns and ancestral relationships. Surveys of genes detected by electrophoresis have identified a considerable difference among species in the distribution of genetic variation among populations. These differences are apparent in the comparison given in Table 6.1 of nine salmonids and two other species as the estimated percentage of the total genetic variation shared by all populations.

Using humans as an example, we note that an average population contains about 90% of the total genetic variation observed in the species (as measured by electrophoresis). The remaining 10% of the total genetic variability is explained by differences between major races and between populations within races. In comparison, species such as the brown trout and the Atlantic salmon are much more genetically subdivided, that is, a much larger proportion of the total genetic variation is explained by differences between populations (more than 36% in both species). The larger

Table 6.1 Estimated percentage of total genetic variation shared by all populations.

Species	Percentage	Reference
Chum salmon (*Oncorhynchus keta*)	97.0	Okazaki, 1983
Coho salmon (*O. kisutch*)	95.2	Reisenbichler and Phelps, 1987
Chinook salmon (*O. tshawytscha*)	87.7	Utter *et al.*, 1989
Sockeye salmon (*O. nerka*)	90.4	Utter *et al.*, 1984
Rainbow trout (*O. mykiss*)	85.0	Allendorf, 1975; Ryman, 1983
Coastal cutthroat trout (*O. clarki*)	94.2	Campton and Utter, 1987
Atlantic salmon (*Salmo salar*)	63.9	Ståhl, 1987
Brown trout (*S. trutta*)	63.3	Ryman, 1983
Arctic charr (*Salvelinus alpinus*)	75.6	Andersson *et al.*, 1983
Atlantic herring (*Clupea harengus*)	99.2	Ryman *et al.*, 1984
Man (*Homo sapiens*)	90.2	Ryman *et al.*, 1983

degree of genetic differentiation between populations of such species potentially reflects greater local adaptations to environmental conditions, most likely indicating that those species are particularly vulnerable to the effects of excessive gene flow from exogenous fish. Conversely, those species having smaller levels of genetic divergence among populations presumably have greater degrees of gene flow and thus, possibly reduced levels of local adaptation. It is important to emphasize, however, that such potential differences among species with regard to degrees of local adaptation remain speculative in the absence of empirical data.

Genetic differentiation with small levels of divergence

These differences between species with respect to the degree of genetic differentiation between populations undoubtedly reflect geographic factors (e.g. the presence or absence of landlocked populations where gene flow among drainages is precluded), and perhaps innate differences in the tendency to interbreed among conspecific populations. However, as pointed out by several authors (e.g. Wright, 1978) genetic differentiation is by no means negligible even when 95% or more of the total gene diversity is shared among conspecific populations; thus the above data should not be interpreted as a reflection of negligible levels of genetic differentiation in, for example, chum and coho salmon.

Potential for adaptive differentiation within groups distinguished by molecular genetic data

The genetic information provided by electrophoresis and other molecular techniques tells very little about the adaptive characteristics of a population because the variant genes revealed by these methods are usually not strongly affected by the forces of natural selection (Eanes, 1987; Nei, 1987). The distribution of such genes is therefore largely uniformative about, for example, ecological, behavioral and morphological differentiation, and the amplitude of their frequency differences among conspecific populations is expected to be smaller than for genes influenced by directional selection (e.g. Allendorf, 1983). Therefore, it is not surprising that even minimal genetic differences among subpopulations as revealed by protein coding genes are frequently coupled with extensive morphological and ecological differences (e.g. Ryman *et al.*, 1979; Ferguson & Mason, 1981; Utter, 1981; Ryman, 1983; Hindar *et al.*, 1986; Foote *et al.*, 1989; Utter *et al.*, 1989).

Adaptive differences

Differences among local salmonid populations in morphological, behavioural and ecological characteristics are thought to represent adaptation by natural selection to local environmental conditions (e.g. Ricker, 1972; Schaffer & Elson, 1975). The following traits are included in those known to be directly or indirectly influenced by genetic factors (Riddell *et al.*, 1981; Saunders, 1981; Gjedrem, 1983; Naevdal, 1983; Refstie, 1987; McIsaac & Quinn, 1988; Bakke, 1991): body and fin morphology; egg number; egg size; survival to various life stages; growth rate; precocious male maturity; age at smolting; age and size at sexual maturity; migratory behaviour at sea; homing precision; resistance to bacterial diseases and parasites; resistance to low pH.

The impossibility of predicting effects of gene flow

It is presently impossible to predict the genetic consequences of successful interbreeding (i.e. gene flow) between cultured and wild populations. This inability exists because of the near absence of knowledge of mechanisms of genetic adaptation. Furthermore, nothing is known about the evolutionary histories of the wild populations, or of the wild progenitors of cultured populations. We are largely limited to more or less intelligent guesses and generalizations (e.g. 'increased gene flow will reduce the genetic differences between cultured and wild fish') when trying to foresee the effects of gene flow from cultured to wild populations.

The need for caution

This inability to predict a priori the genetic consequences of releasing large numbers of cultured fish into natural environments requires that actions be based on a posteriori evaluations and risk assessments from empirical observations. Caution is mandated as empirical data accumulate.

Genetic changes resulting from aquaculture

Genetic changes within cultured populations

Changes within cultured populations can be either intentional or unintentional (see Hynes *et al.*, 1981). Intentional changes for such characteristics as growth rate, disease resistance, spawning time, age at maturity, etc. are the goal of many aquacultural programmes (e.g. Kirpichnikov, 1981; Gjedrem, 1983; Gall & Busack, 1986). Unintentional changes, resulting primarily from inbreeding and loss of genetic variability, are inevitable in any selective breeding programme; and reduced performance for such quantitative traits as viability, growth, fertility, etc., may be directly related to inbreeding (e.g. Falconer, 1981; Gall, 1987).

Domestication is a special case involving either or both of the above categories. Populations developed for aquaculture must adapt to conditions of growth, maintenance and reproduction that are not encountered in nature, and genetic changes from wild ancestral populations are inevitable. In some cases these changes are desirable and even intentionally sought to produce a fish that is easily manageable under conditions of culture. In other cases domestication is an unwanted by-product of artificial propagation, for example, in hatchery operations aimed at conservation or for releasing fish to support threatened or otherwise weak natural populations. The genetic principles for such hatchery programmes (see Allendorf & Ryman, 1987) focus on minimizing genetic changes through using suitable numbers of breeders, selecting suitable founder populations, and avoiding intentional or unintentional selection. Ideally, a fish culture operation of this kind does not affect the genetic population characteristics because it is designed to imitate nature.

Genetic losses remain a concern, however, even in an ideally operated fish culture facility that prevents domestication and is derived from and enhances only a single wild population. The problem in such supportive breeding lies in the increased survival of the enhanced segment of the population, and thus the increased variance of family size – a critical parameter in the effective size of the overall wild plus enhanced population (Ryman & Laikre, 1991; Ryman, 1991; Ryman *et al.*, in press). For instance, Ryman (1991) notes that an enhanced production of 40% of the total offspring from 20 effective parents taken from an effective wild population of 200 reduces the total effective size to 100 individuals as a consequence of the perturbation of the variance of family size. This trade off in gain of total production at the expense of loss in effective population size can lead to serious depletion of genetic variability, particularly in small populations – those most likely to be enhanced through supportive breeding. This source of genetic loss must be considered in the planning and implementation of salmonid enhancement projects.

Ocean ranching (cf. Isaksson, 1988, for a recent review) is a common culture practice for anadromous salmonids that involves captive culture until the time of seaward migration. At this time the juvenile fish are released to migrate and grow in the natural environments cohabited by wild populations. Successive generations are perpetuated by fish returning as adults to the point of release where they are (in most cases) collected and artificially spawned. The domestication of ocean ranched fish is most likely less intensive than that of totally captive populations because it is limited to the juvenile phases preceding migration, and to the adult phases involving reproduction (Fleming & Gross, 1992).

Potential effects from interactions between cultured and wild populations

Competition and contamination

Excessive releases of cultured fish may result in such overcrowding that a natural population cannot compete, and the long-term natural productivity may drastically decline (see Altukhov & Salmenkova, 1987). Another problem with competition is mixed harvests of wild and more abundant cultured salmonid populations where management for the latter has resulted in severe over-harvest of wild fish (cf. Larkin, 1981). A major problem to salmonid management is to be able to identify in a timely manner when threatened wild populations enter a particular fishery so that appropriate harvest adjustments can be made (Fournier *et al.*, 1984; Lincoln, 1987; Pella & Milner, 1987).

Contamination through introduction of pathogens by exogenous fish has a genetic aspect when the exogenous population has a genetic resistance to the pathogen. A vivid and painful reminder of the seriousness of contamination is the recent decimation of wild Norwegian Atlantic salmon stocks in 35 rivers by the fluke *Gyrodactylus salaris* (Johnsen & Jensen, 1991). The parasite appears to have been introduced by eggs or juveniles from Baltic populations, a region where resistant populations have been reported (Bakke, 1991).

Similarly, furunculosis caused by the bacterium *Aeromonas salmonicida* was introduced to Norway in 1985 from infected smolts from Scotland. Although a differential susceptibility of Scottish and Norwegian populations to this disease has not been tested, genetic differences between families have been observed (Gjedrem *et al.*, 1991). By 1990, the disease had been registered in 30 Norwegian rivers including an epidemic in one river in which the majority of the spawning population (over 700 fish) was killed; by 1991, the apparent incidence of the disease had more than doubled, being registered in 63 Norwegian rivers (Heggberget *et al.*, in press).

Interbreeding

Two opposing points of view have been expressed with regard to the effects of successful breedings between individuals from specific wild populations and any exogenous wild or cultured ones. We favour arguments that species such as salmonids, whose strong homing instincts result in a genetically diverse population structure, yield maladapted progeny in crosses between genetically diverged populations (Shields, 1982). Evidence for such a breakdown of presumably complex genetic adaptations in hybrid progeny (i.e. 'outbreeding depression') has been observed in various species including salmonids (Templeton, 1986; Reisenbichler & McIntyre, 1977; Ryman, 1981; Altukhov & Salmenkova, 1987; Gharrett & Smoker, 1991).

Alternatively, it has been proposed that hybrids between specifically adapted cultured fish stocks and wild populations could be used to add new genetic variation to wild populations, and thereby increase the potential of the wild populations to adapt to conditions of environmental instability and genetic depletion (Moav *et al.*, 1978; Wohlfarth, 1986). Nelson & Soulé (1987) argue against this proposal. They point out that while very limited interbreeding may provide a beneficial stimulus through providing new genes without destroying local adaptation, massive hybridization results in irreversible loss of adaptive variation; numerous instances are cited where naturally adapted gene complexes of salmonids have been irreversibly compromised by hybridization with introduced exogenous stocks. Furthermore, Allendorf & Leary (1988) have concluded that the ultimate outcome of large scale hybridization is loss of biological diversity among populations, and that species originally characterized by morphologically distinct populations would ultimately be replaced by 'mongrel' populations that largely lack the characteristics that existed prior to hybridization.

Empirical observations

Do escaped or intentionally released salmonids from culture operations reproduce among themselves or with established wild populations? A wide variety of outcomes (Table 6.2) is consistent with the conclusion drawn above that there are far too many unknown and unknowable genetic characteristics of both natural and cultured populations to permit accurate predictions about degrees of genetic interactions in a particular natural setting. Truly, each combination of local and exogenous stock is a genetically unique experiment.

At one extreme is displacement of the indigenous stocks from massive introductions of exogenous fish. Such displacement appears to have happened with non-native chum salmon transplanted to a hatchery population on the Naiba River on Sakhalin Island, Russia (Altukhov & Salmenkova, 1987). Returns to the hatchery following these introductions decreased to only 5% of previous averages.

Complete interbreeding of native and exogenous cultured populations has been observed in many instances, for example, in brown trout of Ireland and France (Taggart & Ferguson, 1986; Guyomard, 1989), in rainbow trout in the State of Washington in the USA (Campton & Johnston, 1985), and within native subspecies of cutthroat trout from exogenous subspecies and rainbow trout in the western USA (Allendorf & Leary, 1988). In some cases hybridized and native populations coexist in the same regions (e.g. Allendorf *et al.*, 1980; Allendorf & Leary, 1988) and there is

Table 6.2 Observations bearing on the genetic effects that cultured salmonids may have on indigenous populations of the same or related species.

Trait studied	Species (region)	Observation	Reference
Genetic structure	Atlantic salmon (Norway)	Landlocked salmon not detectably affected by releases of offspring of anadromous salmon.	Vuorinen and Berg 1989
	Brown trout (N. Ireland)	Released hatchery stock reproduces successfully and interbreeds with local stock.	Taggart and Ferguson 1986
	Brown trout (France)	Released hatchery stocks may interbreed with local stocks.	Guyomard 1989
	Brown trout (Sweden)	Local genetic structure broken down, presumably as a result of releases of hatchery stock.	Ryman 1981
	Chinook salmon (USA, Canada)	Genetic homogenization in areas of intense culture; heterozygosity and number of genes the same in comparable wild and hatchery populations.	Utter *et al.* 1989
	Chum salmon (Sakhalin, Russia)	Significant gene frequency changes following mass introduction of eggs from foreign stocks.	Altukhov and Salmenkova 1987
	Rainbow trout (Kootenai River, Montana, USA)	Some native populations persist in spite of introduced stocks; admixture also seen.	Allendorf *et al.* 1980
	Rainbow trout (Yakima River, Washington, USA)	Local genetic structure represents an admixture of introduced and native stocks.	Campton and Johnston 1985
	Rainbow trout (Skagit River, Washington, USA)	Local genetic structure not detectably affected by introduction of hatchery stocks.	Donald Campton pers. comm.
	Rainbow trout (Washington, USA)	Little genetic differentiation probably a result of releases of hatchery stocks.	Reisenbichler and Phelps 1989
	Rainbow trout (Idaho, USA)	Local genetic structure not detectably affected by introduction of hatchery rainbow trout.	Wishard *et al.* 1984
Hybridization	Atlantic salmon (Newfoundland, Canada)	Increased rate of hybridization in areas where brown trout has been introduced.	Verspoor 1988
	Cutthroat trout (western USA)	Extensive hybridization in areas where nonnative sub-species and rainbow trout have been introduced.	Allendorf and Leary 1988

Stock size	Atlantic salmon (Norway)	Eradication of local stocks following introduction of fish from hatcheries infected by *Gyrodactylus salaris*.	Johnsen and Jensen 1991
	Chum salmon (Sakhalin, Russia)	Total stock size declines to 5% of original number after mass introduction of eggs from foreign stocks.	Altukhov and Salmenkova 1987
	Pink salmon (Alaska, USA)	Apparent second generation outbreeding depression in crosses between even and odd year classes.	Gharrett and Smoker 1991
	Coho salmon (Oregon, USA)	Decline in juvenile densities in stocked streams despite high densities of wild plus stocked juveniles in the previous generation.	Nickelson *et al.* 1986
	Rainbow trout (Montana, USA)	Approximately ten-fold increase in population size of wild fish in two streams following cessation of stocking programme.	Vincent 1987
Reproductive success	Rainbow trout (Washington, USA)	Success of hatchery steelhead in producing smolt offspring less than that for wild fish.	Chilcote *et al.* 1986 Campton *et al.* 1991
	Coho salmon (British Columbia, Canada)	Differences in reproductive behaviour of hatchery and wild fish increased variance of reproductive success in competitive situations.	Fleming and Gross 1992
Juvenile survival	Brown trout (New York, USA)	Survival of wild fingerlings higher than that of hatchery produced fingerlings.	Schuck 1948
	Rainbow trout (Washington, USA)	Fry of wild origin have higher survival than either fry of hatchery origin or hybrid fry.	Reisenbichler and McIntyre 1977
Juvenile physical fitness	Atlantic salmon (Russia)	Stamina of pond-raised hatchery fish 1.5–2 times weaker than for wild fish.	Shustov and Shchurov 1988
	Brook trout (New York, USA)	Stamina in offspring of wild parents higher than in offspring of domestic parents.	Green 1964
Territorial behaviour	Atlantic salmon (Sweden)	Weaker territorial behavior in offspring from hatchery stock than from wild stock.	Norman 1987
Concealment behaviour	Brook trout (New York, USA)	Offspring of domestic fish do not seek shelter and are less frightened by human presence than offspring of wild fish.	Vincent 1960

Table 6.2 *Continued*

Trait studied	Species (region)	Observation	Reference
Ocean survival	Atlantic salmon (Canada)	Hatchery-reared salmon have lowered survival when released in rivers other than their own.	Ritter 1975
Recapture rate	Coho salmon (Oregon and Washington, USA)	Negative correlation between recapture rate and distance transferred from natal stream.	Reisenbichler 1988
Straying rate	Pink salmon (British Columbia, Canada)	Lower straying rate in offspring from local stock than in either foreign stock or hybrid offspring.	Bams 1976
Return rate	Atlantic salmon (Canada)	Lower return rates following mixing of stocks.	Bailey 1987
	Chinook salmon (Washington, USA)	Higher return rate of native stock than of either foreign stock or hybrid offspring.	Brannon 1982
Competitive ability	Atlantic salmon (Norway)	Adult males of farmed origin show little fighting ability compared with either wild or sea ranched males.	Bror Jonsson, pers. comm.
Disease resistance	Atlantic salmon (Norway)	Susceptibility to vibriosis much higher in one stock from the Baltic than in 14 stocks from Norway.	Gjedrem and Aulstad 1974
	Atlantic salmon (Norway)	Susceptibility to *Gyrodactylus salaris* much higher in two Norwegian stocks than in one stock from the Baltic.	Bakke 1991
	Coho salmon (Oregon, USA)	Susceptibility to *Ceratomyxa shasta* higher for foreign stock than for either local stock or hybrid offspring.	Hemmingsen *et al.* 1986

considerable concern that many native populations are in imminent danger of loss through hybridization or displacement (e.g. Ryman, 1981). In areas of intense fish culture, genetic homogenization of wild populations appears to be the ultimate result (Ryman, 1981; Allendorf & Leary, 1988; Utter *et al.*, 1989).

At the opposite extreme are reports of no detectable gene flow into indigenous populations in spite of substantial introductions of exogenous stocks. Presumably, the exogenous fish have been unable to reproduce in the new environments. A notable example of apparent lack of gene flow occurs in arid environments in southern Idaho (USA) which support populations of locally adapted rainbow trout called 'redband', and which have been heavily planted with exogenous cultured rainbow trout without any detectable interbreeding (Wishard *et al.*, 1984).

It is evident from the above examples that exogenous fish often interbreed with indigenous populations, and sometimes displace them. This evidence alone justified protective actions for natural populations based on the rationale of such legislation as the US Endangered Species Act of 1973 which grants legal protection to threatened or endangered distinct populations within species (cf. Utter, 1981; Meffe, 1987). Some of the empirical data summarized in Table 6.2 indicate that this right to protection carries direct short-term economic value in addition to the less tangible (but in the long term more critical) implications related to conservation of genetic diversity. A direct genetic basis for reduced productivity resulting from intraspecific hybridization is supported by experimental evidence in steelhead (anadromous rainbow) trout where both reproductive success and juvenile survival in native populations were higher than in either exogenous cultured fish or in wild cross-cultured hybrids (Reisenbichler & McIntyre, 1977; Chilcote *et al.*, 1986; Campton *et al.*, 1991).

What are the genetic implications where empirical evidence for displacement or interbreeding is lacking? Bearing in mind the limitations of negative data, let us assume that there is actually no displacement or interbreeding occurring. Natural populations still are placed in jeopardy from other causes related to aquaculture. Three such examples are mentioned below:

(1) Management of a fishery of steelhead trout in the Skagit River, USA was based on the assumption that natural reproduction was sustained by both uncaught wild and hatchery fish prior to the finding that gene frequencies of naturally reproduced juvenile fish indicated a predominance of wild parentage; consequently, wild fish were being overharvested (D. Campton, University of Florida, Gainesville, pers. comm.). Presently, all wild fish that are caught (identified by the presence of an adipose fin – removed from all hatchery releases – and by the absence of fin erosion) are required to be released in order to reduce the harvest of the wild populations.

(2) A decline of natural populations in the Yaquina River adjacent to a large ocean ranching operation in Oregon (USA) was concluded to be caused by environmental changes resulting from the ranching facility. Among other things, it appears that predators were attracted in large numbers to the points of release and rearing of the cultured fish; these artificially attracted predators also fed excessively on the wild fish. Thus the environmental change induced by the culture operation adversely affected the local natural populations in spite of apparent absence of interbreeding or displacement (Anon., 1987).

(3) The devastation of Norwegian Atlantic salmon by the parasite *Gyrodactylus*

salaris introduced through aquacultural operations has affected naturally reproducing fishes independently of large scale interbreeding or displacement.

In instances such as these, where neither displacement nor interbreeding occurs, there are still actual and potential genetic losses to natural populations that are directly related to cultural activities. In the absence of appropriate remedial actions, the observed reductions of genetic variability through reduced population sizes that have already occurred within populations will inevitably lead to losses of genetic variation between populations through extinctions.

Escape rates of cultured salmonids

Most empirical data concerning the fate of accidental or intentional releases of cultured salmon are from Atlantic salmon in Norway. Escapes from Norwegian fish farms have roughly increased linearly with the expansion of the industry. The production of farmed fish has approximately doubled every two years since 1970 to 158 000 tons in 1990 (Heggberget *et al.*, in press), while the yield of wild salmon has remained stable at about 1500 tons (Gausen & Moen, 1991; Hindar *et al.*, 1991; Hindar, 1992). During the winter of 1988–89, accidents resulted in net-pen escapements of approximately 1500 tons, that is, escapements from culture facilities of the same magnitude as the entire annual harvest of wild fish!

The proportion of escaped fish included in naturally spawning populations of Norwegian rivers has increased with the increasing numbers of escaped fish (Fig. 6.1). It was estimated in 1987 and 1988, respectively, that about 13% and 28% of the spawners were of farmed origin (Gausen, 1988; Moen & Gausen, 1989). This proportion increased to about 35% in 1989, where it has subsequently remained stable (Økland *et al.*, 1991; Lund *et al.*, 1992). It is evident from Fig. 6.1 that the proportion of escaped fish varies considerably among rivers; in areas of intense fish culture, up to 80% of the spawners are escapees from fish farms.

There is direct evidence that these escaped fish are successfully reproducing. For example, in the River Oselven, where 56% of the female spawners in 1988 were escaped fish, 38% of the redds contained eggs from escaped fish (V. Moen, University of Trondheim, pers. comm.). This reduced reproductive success of the exogenous females was not significant because of a small sample size; however, it is consistent with evidence from other salmonids (Table 6.2).

Strategies for minimizing genetic impacts on natural populations

At this point it is apparent (1) that extensive gene flow from cultured to wild stocks is undesirable, (2) that a priori predictions of the outcome of such interactions are impossible, (3) that increasing numbers of escapes are occurring as the number of pen-reared fish increases, and (4) that these escaped fish are also migrating to rivers in increasing numbers and are reproducing.

What actions can be taken to assure that the critical amount of genetic diversity inherent in wild salmonid populations does not undergo continual erosion without crippling a thriving and growing salmon culture industry? We discuss below measures

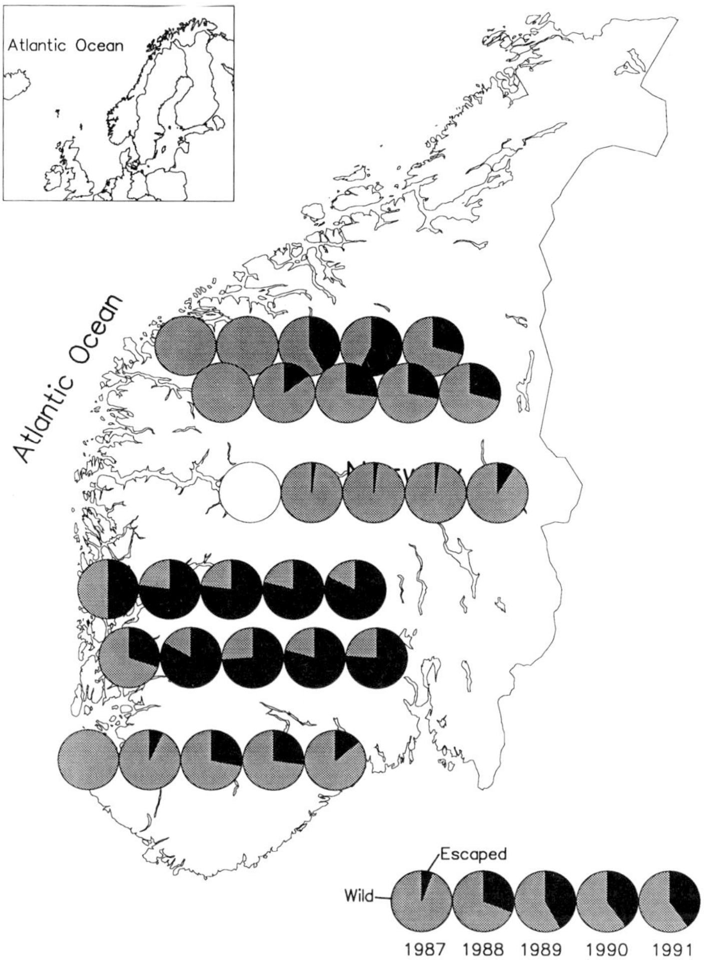

Fig. 6.1 Proportions of wild and escaped net-pen adult salmon on spawning grounds of six rivers in southern Norway, 1987–91. The pie diagrams for 1987 mark the locations of the rivers. The data were taken from Gausen (1988), Moen and Gausen (1989), Økland *et al.* (1991) and Lund *et al.* (1992). The empty circle indicates no data available.

for precluding or minimizing gene flow from cultured to wild populations, including strategies for ocean ranching and gene banks.

Closed culture

Escape-proof containment of closed cultured fish is seen as an ultimate goal that can probably never be achieved from a marine-based operation. However, this goal can be reached through land-based facilities. Recent experiences in Norway indicate that land-based technology is a realistic alternative to marine netpen rearing (Gracey, 1992; Per Heggelund, Seattle, USA, pers. comm.). Initial investment costs of such facilities exceed those of netpen operations but are compensated by reduced operational expenses through, for example, better food conversion.

Sterilization

Sterilization is an option for reducing gene flow from either closed cultured or ocean ranched fish that can presently be induced without difficulty (either through hormonal treatments of juvenile fish or by induction of triploidy shortly after fertilization – see Chourrout (1987) for a recent review). However, potential problems persist even with sterility. Sterile fish still have the potential to spread pathogens. Also, little is known about the behaviour of sterilized salmonids. Of particular concern are triploid males which, while sterile, mature sexually (Bye & Lincoln, 1986) and have the potential to interfere with spawning processes of normal diploid fish; similar concerns apply to some hormonally sterilized males (E. Donaldson, Can. Dept. Fish. and Oceans, pers. comm.). The possibility of sterile but sexually active male salmonids competing in courtship with fertile males brings to mind the successful eradication of blowfly populations intentionally flooded with sterile but sexually active males (Patterson *et al.*, 1970). This problem possibly could be averted with the mandatory use of sterile all-female populations (Bye & Lincoln, 1986), but even here it is unsettling to consider a marine environment populated with increasing numbers of perpetually immature and growing female salmonids. Much more empirical data are required before large-scale releases of sterilized salmonids can be considered as a responsible option for aquaculture.

Localization

Gene flow may be reduced in restricting locations of net-pen operations or points of release for ocean ranching. It appears that escaped fish migrate in greater numbers to streams proximal to the point of escapement (Moen & Gausen, 1989) and that ranched fish return mainly to single drainages if released within a drainage or a bay or estuary affected by a single drainage (Quinn & Fresh, 1984). We encourage further experimentation in artificial imprinting with such chemicals as morpholine (Hasler & Scholz, 1983) with the hope that a unique set of artificial chemical cues for particular cultural operations may ultimately result of increased attraction of released or escaped fish.

Concurrent with implementation of non-genetic procedures for minimizing gene flow, it is important to minimize the genetic differences between escaping fish and recipient wild populations to make any gene flow that might occur as harmless as possible. Use of indigenous populations for aquaculture programmes is strongly encouraged. Although such cultured stocks will inevitably diverge from their wild ancestors through domestication and selective breeding, their escapement and potential gene flow will not introduce exotic genes or pathogens. Furthermore, since exogenous fish may stray more than those from local stocks (as previously mentioned), the use of local stocks might keep the majority of escaped fish from spawning over large geographical areas.

The need for localized management to protect natural populations has been recognized by management agencies. For example, the Washington State Department of Fisheries (USA) set aside 25% of Puget Sound for full harvest of hatchery coho salmon while managing the remainder of Puget Sound on the basis of natural reproduction (Wright, 1981). Segregation of areas within regions is the only option for preserving wild populations within a region where the goal of ocean ranching is to maintain consistently large populations for commercial or sport harvest. An extension of this strategy is setting aside totally protected areas such as national parks. Such actions are probably needed in countries like Norway where escapes are increasing dramatically and no populations except those located sufficiently distant from aquacultural operations are likely to remain reasonably uncontaminated.

Monitoring the genetic effects of aquaculture

Information on the amount of genetic interactions between cultured and wild populations can only come from empirical observations of genetic change of the wild populations. Such observations require easily observed genetic differences between the wild and the cultured populations so that invading genes are readily detected. Such genetic differences between cultured and natural populations commonly exist (e.g. at protein coding loci), but they must be identified and quantified prior to any escape if effective monitoring of gene flow is to occur. It is therefore very important that adequate amounts of this kind of genetic information be available both for cultured populations and for wild populations that may be affected by effective migration from them.

Fish culture stocks have not been established with the intention of creating detectable genetic differences from wild populations. However, such differences frequently can be straightforwardly induced without altering the performance of the cultured stock (Parker *et al.*, 1990). Such 'genetic marking' should be a requirement for cultured stocks that are not otherwise genetically identifiable. We consider this action the most important step to be taken in order to obtain even the vaguest idea of the genetic effects of release of cultured fish on wild populations. Without such a step, these effects will always remain conjectural, and we will never be able to produce anything but more or less meaningless speculations that are based on no empirical evidence.

A second kind of information that is potentially useful in a programme for monitoring genetic changes in populations is from morphological variation measured by fluctuating asymmetry of bilateral characters (e.g. Leary *et al.*, 1985). Increase in levels of asymmetry have been shown to be sensitive indications of loss of genetic variation (Leary & Allendorf, 1989). The monitoring applications of this kind of data are summarized by Allendorf & Ryman (1987); they point out the complementary nature of asymmetry measurements to biochemical genetic information, particularly with regard to early detection of actual morphological effects due to losses of genetic variability in a population. As with biochemical genetic studies, adequate sets of baseline data must be available from potentially affected wild populations for subsequent changes in asymmetry to be measured.

A systematic monitoring of wild populations for gene flow from intentional or unintentional releases of cultured fish can proceed once adequate genetic distinction exists between cultured and wild populations. This process not only permits detection of genetic interactions, but also makes it possible to carry out remedial actions after such interactions have been detected. The detection of a substantial genetic intrusion into wild populations is a very serious matter that requires immediate adjustments of aquacultural practices.

A detailed discussion concerning the appropriate actions to be taken following the detection of gene flow to wild populations lies beyond the scope of this chapter. However, it is pertinent to consider how much gene flow is acceptable. We have previously proposed (Ryman, 1991; Ryman *et al.*, in press) that the maximum permissible level of *effective* migration per generation be based on the degree of genetic differentiation (F_{st}, Wright, 1969; or G_{st}, Nei, 1975) between a particular cultured and wild population. These indices are related to those values previously given as percentages of shared genetic variation in different species; that is, (1 − percentage/100). They are readily estimated from allele frequencies determined from biochemical genetic data and commonly fall between 0.1 and 0.01 for salmonid populations, permitting respective effective migrations of between two to 25 individuals per generation within this range. This guideline based on real and readily measurable biological differences is distinct from and preferable to arbitrary ones such as '10% per generation' (Anon., 1988) that lack biological relevance and provide false security while permitting rapid deterioration of natural populations.

Recommendations

The International Symposium on Fish Gene Pools, Preservation of Genetic Resources in Relation to Wild Fish Stocks, was held in Stockholm in January of 1980. The participants of this meeting formulated a set of recommendations to serve as guidelines for the management of genetic resources in natural populations (Ryman, 1981). These recommendations are summarized in the following action statements:

- Identify genetic resources
- Maintain natural ecosystems
- Avoid selective harvest of natural populations
- Preserve genetic variability within cultured populations
- Release fish into natural environments with great care
- Provide adequate funding for basic and applied research
- Inform those responsible for management of existing knowledge.

Nothing has changed during the intervening decade or so to reduce the vitality and currency of these recommendations. Indeed, the only change is that the urgency for their implementation has increased because of aquacultural growth, continued habitat degradations and, unfortunately, continued losses of genetic resources due to failure to adequately carry out these recommendations. We stress that the primary issue of the present chapter is *not* to point to the need for more research. Rather we stress the need for immediate action based on everything that is already known. We therefore affirm these existing recommendations from the Fish Gene Pools Symposium as the underlying basis for the ones stated below that are more specifically related to the genetic effects of aquaculture on natural fish populations.

(1) All aquacultural activities should be conducted in a manner that minimizes potential gene flow to wild populations.
(2) Aquacultural facilities should be located and designed to minimize the possibility of escaping.
(3) Locations for aquacultural rearing and release should be selected on the basis of

having minimal impact upon natural populations; whenever needed, areas for complete protection should be set aside and located to assure minimum impact from aquaculture operations.

(4) Movements among areas should be restricted to guard against the importation of individuals that may carry pathogens to which local forms lack resistance.

(5) A cataloguing of genetic characteristics of natural populations that would potentially be genetically affected by aquacultural releases (e.g. allele frequencies, spawning times, life history variables, etc.) should accompany any aquacultural activity to serve as baseline data for subsequent monitoring.

(6) Cultured populations subject to potential escape or to release should have heritable characteristics (i.e. genetic marks) that permit monitoring introgression with natural populations or establishment of self-sustaining populations.

(7) A variety of experiments designed to provide empirical data concerning the genetic effects of aquacultural activities on natural populations should be carried out. Particularly important are experimental releases of genetically identifiable fish, which should precede any proposed large-scale rearing or release, and the results of which should dictate the scope of such activities.

(8) Rearing of sterile fish for aquaculture should be encouraged through use of sterile hybrids and applications of such procedures as induced triploidy and hormonal sterilization. However, an understanding of the potential ecological and genetic effects of releasing such individuals should precede any large-scale rearing or release.

References

Allendorf, F. (1975) Genetic variability in a species possessing extensive gene duplication: genetic interpretation of duplicate loci and examination of genetic variation in populations of rainbow trout, PhD thesis, University of Washington, Seattle.

Allendorf, F. (1983) Isolation, geneflow, and genetic differentiation among populations. In *Genetics and Conservation* (Ed. by C.M. Schonewald-Cox, S.M. Chambers, B. MacBryde and W.L. Thomas), pp. 51–65, Benjamin-Cummings, Menlo Park, California.

Allendorf, F. and Leary, R. (1988) Conservation and distribution of genetic variation in a polytypic species, the cutthroat trout. *Conservative Biology*, 2, 170–84.

Allendorf, F. and Ryman, N. (1987) Genetic management of hatchery stocks. In *Population Genetics and Fishery Management* (Ed. by N. Ryman and F. Utter), pp. 141–59, University of Washington Press, Seattle.

Allendorf, F. and Utter, F. (1979) Population genetics. In *Fish Physiology*, vol. 8, (Ed. by W.S. Hoar, D.J. Randall, and J.R. Brett), pp. 407–54, Academic Press, New York.

Allendorf, F., Espeland, D., Scow, D. and Phelps, S. (1980) Coexistence of native and introduced rainbow trout in the Kootenai River drainage. *Proceedings of the Montana Academy of Sciences*, 39, 28–36.

Altukhov, Yu and Salmenkova, E. (1987) Stock transfer relative to natural organization, management, and conservation of fish populations, In *Population Gen-*

etics and Fishery Management (Ed. by N. Ryman and F. Utter), pp. 333–43, University of Washington Press, Seattle.

Andersson, L., Ryman, N. and Ståhl, G. (1983) Protein loci in the Arctic charr, *Salvelinus alpinus* L.: electrophoretic expression and genetic variability patterns. *J. Fish. Biol.*, 23, 75–94.

Anon. (1987) The impact of Oregon Aqua-Foods operation on the wild coho salmon in the Yaquina River. Staff Report, Oregon Dept. Fish. Wildlife, Fish Division, Portland, OR.

Anon. (1988) Fiskeristyrelsens kungorelse med foreskrifter och allmanna rod om utplantering och flyttning av fish. Fiskeristyrelsens forfattningssamling, FIFS 1988: 10. (Legislative note on release and translocation of fish.) Swedish National Board of Fisheries (in Swedish).

Anon. (1989a) Fishing and rearing of salmon, etc. (1987) Central Bureau of Statistics of Norway, Oslo (in Norwegian, English summary).

Anon. (1989b) Annual report No. XXXIII for 1988. Salmon Research Trust of Ireland Inc., Dublin.

Bailey, J. (1987) Canadian sea ranching program (east coast). In *Proceedings of the World Symposium on Selection, Hybridization and Genetic Engineering in Aquaculture*, vol. II (Ed. by K. Tiews), pp. 443–8, Heeneman, Berlin.

Bakke, R. (1991) A review of the inter- and intraspecific variability in salmonid hosts to laboratory infections with *Gyrodactylus salaris*. *Aquaculture*, 98, 303–10.

Bams, R. (1976) Survival and propensity for homing as affected by presence or absence of locally adapted paternal genes in two transplanted populations of pink salmon (*Oncorhynchus gorbuscha*). *J. Fish. Res. Board Can.*, 33, 2716–25.

Brannon, E. (1982) Orientation mechanisms of homing salmonids. In *Proceedings of the Salmon and Trout Migratory Behavior Symposium* (Ed. by E.L. Brannon and E.O. Salo), pp. 219–28, University of Washington, Seattle.

Bye, V. and Lincoln, R. (1986) Commercial methods for the control of sexual maturation in rainbow trout (*Salmo gairdneri* R.). *Aquaculture*, 57, 299–309.

Campton, D. and Johnston, J. (1985) Electrophoretic evidence for a genetic admixture of native and nonnative rainbow trout in the Yakima River, Washington. *Trans. Am. Fish Soc.*, 114, 782–93.

Campton, D. and Utter, F. (1987) Genetic structure of anadromous cutthroat trout (*Salmo clarki clarki*) populations in the Puget Sound Area: evidence for restricted gene flow. *Can. J. Fish. Aquat. Sci.*, 44, 573–82.

Campton, D., Allendorf, F., Behnke, R., Utter, F., Chilcote, M., Leider, S. and Loch, J. (1991) Reproductive success of hatchery and wild steelhead. *Trans. Amer. Fish. Soc.*, 120, 816–27.

Chilcote, M., Leider, S. and Loch, J. (1986) Differential reproductive success of hatchery and wild summer-run steelhead under natural conditions. *Trans. Am. Fish. Soc.*, 115, 726–35.

Chourrout, D. (1987) Genetic manipulations in fish: review of methods. In *Proceedings of the World Symposium on Selection, Hybridization and Genetic Engineering in Aquaculture* (Ed. by K. Tiews), vol. II, pp. 111–26, Heeneman, Berlin.

Eanes, W. (1987) Allozymes and fitness: evolution of a problem. *Trends Evol. Ecol.*, 2, 44–8.

Falconer, D. (1981) *Introduction To Quantitative Genetics*, 2nd edn. Longman, London.

Ferguson, A. and Mason, F. (1981) Allozyme evidence for reproductively isolated sympatric populations of brown trout *Salmo trutta* L. in Lough Melvin, Ireland. *J. Fish Biol.*, 18, 629–742.

Fleming, I. and Gross, M. (1992) Reproductive behavior of hatchery and wild coho salmon (*Oncorhynchus kisutch*): does it differ? *Aquaculture*, 103, 101–21.

Foote, C., Wood, C. and Withler, R. (1989) Biochemical genetic comparison of sockeye salmon and kokanee, the anadromous and nonanadromous forms of *Oncorhynchus nerka*. *Can. J. Fish. Aquat. Sci.*, 46, 149–58.

Fournier, D., Beacham, R., Riddell, B. and Busack, C. (1984) Estimating stock composition in mixed stock fisheries using morphometric, meristic, and electrophoretic characteristics. *Can. J. Fish. Aquat. Sci.*, 41, 400–8.

Gall, G. (1987) Inbreeding. In *Population Genetics and Fishery Management* (Ed. by N. Ryman and F. Utter), pp. 47–87, University of Washington Press, Seattle.

Gall, G. and Busack, C. (Eds) (1986) Genetics in aquaculture II. *Aquaculture*, 57, 1–386.

Gausen, D. (1988) Registrering av oppdrettslaks i vassdrag. In *Fragmøte om Sikringssoner for Laksefisk* (Ed. by B. Lindgren), pp. 58–69, LENKA, Miljøverndepartementet, Oslo, Norway (in Norwegian).

Gausen, D. and Moen, V. (1991) Large-scale escapes of farmed Atlantic salmon (*Salmo salar*) into Norwegian rivers threaten natural populations. *Can. J. Fish. Aquat. Sci.*, 48, 426–8.

Gharrett, A. and Smoker, W. (1991) Two generations of hybrids between even- and odd-year pink salmon (*Oncorhynchus gorbuscha*); a test for outbreeding depression? *Can. J. Fish. Aquat. Sci.*, 48, 1744–9.

Gjedrem, T. (1981) Conservation of fish populations in Norway. In *Fish Gene Pools* (Ed. by N. Ryman), pp. 33–6, *Ecological Bulletins* (Stockholm), 34. (Proceedings of an international symposium arranged by the Commission for Research on Natural Resources of the Swedish Council for Planning and Coordination of Research, 23–25 January 1980.)

Gjedrem, T. (1983) Genetic variation in quantitative traits and selective breeding in fish and shellfish. *Aquaculture*, 33, 51–72.

Gjedrem, T. and Aulstad, D. (1974) Selection experiments with salmon. I. Differences in resistance to vibrio disease of salmon parr (*Salmo salar*). *Aquaculture*, 3, 51–9.

Gjedrem, T., Salte, R. and Gjoen, H.M. (1991) Genetic variation in susceptibility of Atlantic salmon to furunculosis. *Aquaculture*, 97, 1–6.

Gracey, K. (1992) Land-based salmon farming – is there a future? *Fish Farm News*, July, 8–10.

Green, D., Jr. (1964) A comparison of stamina of brook trout from wild and domestic parents. *Trans. Am. Fish. Soc.*, 93, 96–100.

Guyomard, R. (1989) Diversité génétique de truite commune. *Bull. Fr. Peche Piscic.*, 314, 118–35.

Hasler, A. and Scholz, A. (1983) *Olfactory Imprinting and Homing in Salmon*. Springer-Verlag, Berlin.

Heggberget, T., Johnsen, B., Hindar, K., Jonsson, B., Hansen, J., Hvidsten, N. and Jensen, A. Interactions between wild and cultured Atlantic salmon: a review of the Norwegian experience. *Fish. Res.* (in press).

Hemmingsen, A., Holt, R., Ewing, R. and McIntyre, J. (1986) Susceptibility of progeny from crosses among three stocks of coho salmon to infection by *Ceratomyxa shasta*. *Trans. Am. Fish. Soc.*, 115, 492–5.

Hindar, K. (1992) Conservation and sustainable use of Atlantic salmon. In *Conservation of Biodiversity for Sustainable Development* (Ed. by O. Sandlund, K. Hindar and A. Brown), pp. 168–86, Scandinavian University Press, Oslo.

Hindar, K., Ryman, N. and Ståhl, G. (1986) Genetic differentiation among local populations and morphotypes of Arctic charr, *Salvelinus alpinus*. *Biol. J. Linn. Soc.*, 27, 269–85.

Hindar, K., Ryman, N. and Utter, F. (1991) Genetic effects of cultured fish on natural fish populations. *Can. J. Fish. Aquat. Sci.*, 48, 945–57.

Hynes, J., Brown, D., Helle, H., Ryman, N. and Webster, D. (1981) Guidelines for the culture of fish stocks for resource management. *Can. J. Fish. Aquat. Sci.*, 38, 1867–76.

Isaksson, A. (1988) Salmon ranching: a world review. *Aquaculture*, 75, 1–33.

Johnsen, B. and Jensen, A. (1991) The *Gyrodactylus* story in Norway. *Aquaculture*, 98, 289–302.

Kirpichnikov, V. (1981) *Genetic Bases of Fish Selection*. Springer-Verlag, Berlin.

Larkin, P. (1981) A perspective on population genetics and salmon management. *Can. J. Fish. Aquat. Sci.*, 38, 1469–75.

Leary, R. and Allendorf, F. (1989) Fluctuating asymmetry as an indicator of stress: implications for conservation biology. *Trends Ecol. Evol.*, 4, 214–17.

Leary, R., Allendorf, F. and Knudsen, K. (1985) Developmental stability as an indicator of the loss of genetic variation in hatchery trout. *Trans. Am. Fish. Soc.*, 114, 230–5.

Lincoln, R. (Ed.) (1987) Genetic stock identification (GSI) of chinook salmon: status, needs and future. Washington Department of Fisheries, Olympia, Washington.

Lund, R., Oakland, F. and Hansen, L. (1992) Escapes of reared salmon in marine homewater and in riverine fisheries in 1991. NINA, Trondheim, *Oppdragsmelding*, 143, 1–16 (in Norwegian, English summary).

Lura, H. and Saegrov, H. (in prep) First documentation of gene-flow from farmed to wild stocks of Atlantic salmon based on different ratios of optical isomers of astaxanthin.

McIsaac, D. and Quinn, T. (1988) Evidence of a hereditary component in homing behavior of chinook salmon (*Oncorhynchus tshawytscha*). *Can. J. Fish. Aquat. Sci.*, 45, 2201–5.

Meffe, G. (1987) Conserving fish genomes: philosophies and practices. *Environ. Bio. Fish.*, 18, 3–9.

Moav, R., Brody, T. and Hulata, G. (1978) Genetic improvements of wild fish populations. *Science* (Wash., DC), 201, 1090–4.

Moen, V. and Gausen, D. (1989) Rømt oppdrettsfisk i vassdrag 1988, Direktoratet for naturforvaltning, Trondheim, Report, 3, 1989 (in Norwegian).

Naevdal, G. (1983) Genetic factors in connection with age at maturation. *Aquaculture*, 33, 97–106.

Nei, M. (1975) *Molecular Population Genetics*, Elsevier, New York.

Nei, M. (1987) *Molecular Evolutionary Genetics*, Columbia University Press, New York, NY.

Nelson, K. and Soulé, M. (1987) Genetical conservation of exploited fishes. In

Population Genetics and Fishery Management (Ed. by N. Ryman and F. Utter), pp. 355–68, University of Washington Press, Seattle.

Nickelson, T., Solazzi, M. and Johnson, S. (1986) Use of hatchery coho salmon (*Oncorhynchus kisutch*) to rebuild wild populations in Oregon coastal streams. *Can. J. Fish. Aquat. Sci.*, 43, 2443–9.

Norman, L. (1987) Stream aquarium observations of territorial behaviour in young salmon (*Salmo salar* L.) of wild and hatchery origin. Salmon Research Institute, Rep. 1987, 2, Äkvkarleby (in Swedish, English summary).

Okazaki, T. (1983) Genetic structure of chum salmon *Oncorhynchus keta* river populations. *Bull. Jap. Soc. Sci. Fish.*, 49, 189–96.

Økland, F., Lund, R. and Hansen, L. (1991) Escapees of reared salmon in marine fisheries, sport fisheries and brood stock fisheries. NINA, Trondheim, *Forsskningsrapport,* 13, 1–36 (in Norwegian, English summary).

Parker, N., Giorgi, A., Heidinger, R., Jester, D., Jr., Prince, E. and Winans, G. (Eds) (1990) Fish-marketing techniques, proceedings, *American Fisheries Society Symposium 7*, American Fisheries Society, Bethesda, Maryland.

Parsons, J. and Thorgarrd, G. (1985) Production of androgenetic diploid rainbow trout. *J. Hered.*, 76, 177–81.

Patterson, R., Weidhass, D., Ford, H. and Lofgren, C. (1970) Suppression and elimination of an island population of *Culex pipiens quinquefasciatus* with sterile males. *Science* (Wash., DC), 168, 1368–70.

Pella, J. and Milner, G. (1987) Use of genetic marks in stock composition analysis. In *Population Genetics and Fishery Management* (Ed. by N. Ryman and F. Utter), pp. 247–76, University of Washington Press, Seattle.

Quinn, T. and Fresh, K. (1984) Homing and straying in chinook salmon (*Oncorhynchus tshawytscha*) from Cowlitz River Hatchery, WA. *Can. J. Fish. Aquat. Sci.*, 41, 1078–82.

Quinn, T.P. and Tallman, R.F. (1987) Seasonal environmental predictability and homing in riverine fishes. *Environmental Biology of Fishes*, 18, 155–9.

Refstie, T. (1987) Selective breeding and intraspecific hybridization of cold water finfish. In *Proceedings of the World Symposium on Selection, Hybridization and Genetic Engineering in Aquaculture*, vol. I (Ed. by K. Tiews), pp. 293–302, Heeneman, Berlin.

Reinsenbichler, R. (1988) Relation between distance transferred from natal stream and recovery rate for hatchery coho salmon. *N. Am. J. Fish. Manage.*, 8, 172–4.

Reisenbichler, R. and McIntyre, J. (1977) Genetic differences in growth and survival of juvenile hatchery and wild steelhead trout. *J. Fish. Res. Board Can.*, 34, 123–8.

Reisenbichler, R. and Phelps, S. (1987) Genetic variation in chinook, *Oncorhynchus tshawytscha*, and coho, *O. kisutch*, salmon from the north coast of Washington. *Fish., Bull.* (USA), 85, 681–701.

Reisenbichler, R. and Phelps, S. (1989) Genetic variation in steelhead (*Salmo gairdneri*) from the north coast of Washington. *Can. J. Fish. Aquat. Sci.*, 46, 66–73.

Ricker, W. (1972) Hereditary and environmental factors affecting certain salmonid populations. In *The Stock Concept of Pacific Salmon* (Ed. by R.C. Simon and P.A. Larkin), pp. 19–160, H.R. Macmillan Lectures in Fisheries, University of British Columbia, Vancouver.

Riddell, B., Leggett, W. and Saunders, R. (1981) Evidence of adaptive polygenic variation between two populations of Atlantic salmon (*Salmo salar*) native to tributaries of the S.W. Miranichi River, N.B. *Can. J. Fish. Aquat. Sci.*, 38, 321–33.

Ritter, J. (1975) Lower ocean survival rates for hatchery-reared Atlantic salmon (*Salmo salar*) stocks released in rivers other than their native streams. ICES Ana. Cat. Comm. C.M. 1975/M: 26, Copenhagen, Denmark.

Ryman, N. (Ed.) (1981) *Fish Gene Pools. Ecological Bulletins* (Stockholm), 34. (Proceedings of an international symposium arranged by the Commission for Research on Natural Resources of the Swedish Council for Planning and Coordination of Research, 23–25 January 1980.)

Ryman, N. (1983) Patterns of distribution of biochemical genetic variation in salmonids: differences between species. *Aquaculture*, 33, 1–21.

Ryman, N. (1991) Conservation genetics considerations in fishery management, *J. Fish Biol.*, 39 (Suppl. A), 211–24.

Ryman, N. and Laikre, L. (1991) Effects of supportive breeding on the genetically effective population size. *Cons. Biol.*, 5, 325–9.

Ryman, N. and Utter, F. (Eds) (1987) *Population Genetics and Fishery Management.* University of Washington Press, Seattle.

Ryman, N., Allendorf, F. and Ståhl, G. (1979) Reproductive isolation with little genetic divergence in sympatric populations of brown trout. *Genetics*, 92, 247–62.

Ryman, N., Chakraborty, R. and Nei, M. (1983) Differences in the relative distribution of human gene diversity between electrophoretic and red and white cell antigen loci. *Hum. Hered.*, 33, 93–102.

Ryman, N., Utter, F. and Hindar, K. (in press) Introgression, supportive breeding and genetic conservation. In *Population Management for Survival and Recovery* (Ed. by J.D. Ballou, T. Foose and M. Gilpin), University of Chicago Press, Chicago.

Ryman, N., Lagercrantz, U., Andersson, L., Chakraborty, R. and Rosenberg, R. (1984) Lack of correspondence between genetic and morphological variability patterns in Atlantic herring (*Clupea harengus*). *Heredity*, 53, 687–704.

Saunders, R. (1981) Atlantic salmon (*Salmo salar*) stocks and management implication in the Canadian Atlantic Provinces and New England, USA. *Can. J. Fish. Aquat. Sci.*, 38, 1612–25.

Schaffer, W. and Elson, P. (1975) The adaptive significance of variations in life history among local populations of Atlantic salmon in North America. *Ecology*, 56, 577–90.

Schuck, H. (1948) Survival of hatchery trout in streams and possible methods of improving the quality of hatchery trout. *Prog. Fish-Cult.*, 10, 3–14.

Shields, W. (1982) *Philopatry, Inbreeding, and the Evolution of Sex.* SUNY Press, Albany, New York.

Shustov, Yu and Shchurov, I. (1988) Quantitative estimation of stamina of wild and hatchery-reared Atlantic salmon (*Salmo salar* L.). *Aquaculture*, 81, 81–7.

Simon, R. (1972) Gene frequency and the stock problem. In *The Stock Concept of Pacific Salmon* (Ed. by R.C. Simon and P.A. Larkin), pp. 161–9, H.R. Macmillan Lectures in Fisheries, University of British Columbia, Vancouver.

Ståhl, G. (1987) Genetic population structure of Atlantic salmon. In *Population Genetics and Fishery Management* (Ed. by N. Ryman and F. Utter), pp. 121–40,

University of Washington Press, Seattle.

Stoss, J. (1983) Fish gamete preservation and spermatozoan physiology. In *Fish Physiology*, vol. 9B (Ed. by W.S. Hoar, D.J. Randall and E.M. Donaldson), pp. 305–50, Academic Press, New York.

Taggart, J. and Ferguson, A. (1986) Electrophoretic evaluation of a supplemental stocking programme for brown trout, *Salmo trutta* L. *Aquaculture Fish Manage*, 17, 155–62.

Templeton, A. (1986) Coadaptation and outbreeding depression. In *Conservation Biology: the Science of Scarcity and Diversity* (Ed. by M. Soulé), pp. 105–16, Sinauer Associates, Sunderland, Massachusetts.

Utter, F. (1981) Biological criteria for definition of species and distinct intraspecific populations of salmonids under the U.S. Endangered Species Act of 1973. *Can. J. Fish. Aquat. Sci.*, 38, 1626–35.

Utter, F., Aebersold, P., Helle, J. and Winans, G. (1984) Genetic characterization of populations in the southeastern range of sockeye salmon. In *Proceedings of the Olympic Wild Fish Conference* (Ed. by J.M. Walton and D.B. Houston), pp. 17–32, Peninsula College, Fisheries Technology Program, Port Angeles, Washington.

Utter, F., Milner, F., Ståhl, G. and Teel, D. (1989) Genetic population structure of chinook salmon in the Pacific Northwest, *Fish. Bull.* (USA), 87, 239–64.

Verspoor, E. (1988) Widespread hybridization between native Atlantic salmon, *Salmo salar*, and introduced brown trout, *S. trutta*, in eastern Newfoundland. *J. Fish. Biol.*, 32, 327–34.

Vincent, R. (1960) Some influences of domestication upon three stocks of brook trout (*Salvelinus fontinalis* Mitchill). *Trans. Am. Fish. Soc.*, 89, 35–52.

Vincent, R. (1987) Effects of stocking catchable-size hatchery rainbow trout on two wild trout species in the Madison River and O'Dell Creek, Montana. *N. Am. J. Fish. Manage.*, 7, 91–105.

Vuorinen, J. and Berg, O. (1989) Genetic divergence of anadromous and non-anadromous Atlantic salmon (*Salmo salar*) in the River Namsen, Norway. *Can. J. Fish. Aquat. Sci.*, 46, 406–9.

Wishard, L.N., Seeb, J.E., Utter, F.M. and Stefan, D. (1984) A genetic investigation of suspected reband trout populations. *Copeia*: 120–32.

Wohlfarth, G. (1986) Decline in natural fisheries – a genetic analysis and suggestion for recovery. *Can. J. Fish. Aquat. Sci.*, 43, 1298–306.

Wright, S. (1969) *The Theory of Gene Frequencies. Evolution and the Genetics of Populations*, vol. II, University of Chicago Press, Chicago.

Wright, S. (1978) Evolution and the genetics of populations. Vol. 4. *Variability Within and Among Natural Populations*. University of Chicago Press, Chicago.

Wright, S. (1981) Contemporary Pacific salmon management. *N. Am. J. Fish. Manage.*, 1, 29–40.

Chapter 7
Salmon Disease

JAN A. OLAFSEN AND RONALD J. ROBERTS

THE MICROBIAL ECOLOGY OF FISH AQUACULTURE BY JAN A. OLAFSEN

Introduction

Fish farming has proved a successful method to ensure a stable supply of high quality fish for food and stock enhancement, but not without problems. Diseases are prevalent, and there is concern about the effect from spill of food, faeces, pathogenic microorganisms and antibiotics to the environment. Some of these problems are not unexpected since all animals are more liable to contract disease when they live in dense populations (Isenberg, 1988; Finlay & Falkow, 1989). Fish are no exception to this. In the aquatic environment fish and microorganisms, including pathogenic ones, share the ecosystem. Bacteria may travel easily between habitats, and from one animal to another. In principle microorganisms in the aquatic ecosystem inhabit a continuous medium, and their spreading is limited only by physical barriers and availability of suitable growth factors. Whether internal or external they are separated from their hosts by a mucosal surface. These simple ecological facts are imperative to understand disease problems related to fish aquaculture.

All aquatic organisms are exposed to a varied microflora that harbours multiple pathogens. Many of these organisms are filter feeders, and some live in areas with a diverse and dense microflora. They accumulate bacteria from the environment, and may carry bacteria on their surfaces and in their tissues and body fluids (Olafsen *et al.*, 1993). Over millions of years complex interrelations have evolved between aquatic organisms, their indigenous microflora and pathogens. Yet we are left with the puzzle that the interplay between poikilothermic organisms and their commensal and pathogenic microflora is governed by cellular and molecular mechanisms that are to a large extent temperature dependent (Hildemann and Cooper, 1963; Rijkers, 1982; Lamers, 1985). Apparently some of their defences are not effectively tuned to the ambient temperatures at which they live, indicating there are still gaps in our current understanding of their interactions with bacteria. Also, even though fish larvae ingest bacteria from the time they are hatched, their immune system apparently only becomes effective after a few months of age (Manning *et al.*, 1982a; Chantanachookhin *et al.*, 1991). It is thus obvious that we still do not understand the factors governing these relationships in the marine environment.

The intensive rearing of fish requires that food is added to water at high concentration. This results in an excellent medium for growth of heterotrophic and opportunistic bacteria. Ingredients from food spillage will affect selection and growth of the microorganisms in the nets or cages. When we feed the fish in the farm, we also feed the microflora. Thus feed composed of substances not abundant in the aquatic environment could result in the proliferation of bacteria that normally have a restricted opportunity to grow. Many fish diseases are caused by opportunistic pathogens, some (e.g. vibrios) with growth characteristics that could benefit from such nutrients. Food substances will thus have a selective effect on the microflora of the fish farm, as well as also on the intestinal mucosal microflora. In the aquatic environment pathogenicity frequently involves a quantitative aspect in the relationship between host and microorganism. Thus diseases caused by the proliferation of opportunistic pathogens can be related to feeding or environmental factors that also affect the commensal microflora.

Stresses caused by handling of fish or overstocking reduce the overall resistance against infection by opportunistic pathogens. Current rearing methods require that salmon are extensively handled and transferred from fresh to sea-water immediately following smoltification, a physiologically challenging process to the fish. Use of antibiotics during transport may greatly alter the intestinal microflora (Austin & Al-Zahrani, 1988; Hansen *et al.*, 1992), which play an important part in the primary defences against pathogenic microorganisms. Thus transfer of smolts to seawater may be followed by severe attacks from opportunistic or pathogenic bacteria.

To develop a suitable environment for the successful rearing of marine organisms it is imperative to understand how food composition and rearing techniques affect the microbial ecology of the fish farm. The factors which affect the proliferation of microorganisms in this environment, like temperature and substrate availability, also affect the disease resistance of the fish and spreading of the pathogens. Mechanisms governing these interactions need to be studied to ascertain whether pathogens are normal members of the free-living or commensal aquatic microflora, or merely surviving in this ecosystem. Better understanding of the factors that cause stress to the animals or impair their immune mechanisms are imperative in order to establish effective techniques for the successful rearing of fish.

Microbes in the aquatic environment

Some aquatic bacteria are highly specialized to a life at extreme conditions, whereas others are extremely adaptable. Marine bacteria may be pleomorphic and may change their shape according to the prevalent conditions, i.e. they shift between microforms transported freely through the water, and more actively metabolic and morphological bulky adherent forms (Morita, 1985; Kjelleberg *et al.*, 1987). Aquatic habitats vary in chemical composition and physical factors such as temperature, pressure, oxygen tension and currents. Even in environments with low substrate availability a multitude of bacteria may persist by down-regulating metabolic activities and increasing their generation times. Resident forms may be 'dwarfs' or cells that multiply with no net increase in biomass. Growth and proliferation of such resting bacteria may be induced by changes in physical factors, or increased availability of growth factors. On the other hand we may find relatively rich growth conditions in the interfaces against coastlines, surfaces or temperature shifts in this ecosystem. Sudden changes in the availability of substrate result from plankton, spawning products or dead organisms. At the seawater surface phytoplankton blooms cause dramatic changes in biomass, and bacteria play a dominant role in the conversion of this biomass. Dead organisms and faeces are converted by microorganisms during sedimentation, and specialist bacteria may obtain energy from simple organic compounds in the sediments. In this variable aquatic environment many microorganisms have adapted by evolving chemotaxis towards nutrients, and some may even utilize available substrate effectively both at high and at low concentration (Kjelleberg *et al.*, 1987; Roszak & Colwell, 1987; Gottschal, 1990).

The marine environment is predominantly cold, with 90% of the seawater at a temperature lower than 5°C (ZoBell, 1963). Most coastal habitats used for salmon farming are also cold with a variable content of organic material, and inhabited by microorganisms well suited to take advantage of the ecological changes introduced by a fish farm. Many aquatic bacteria have evolved sophisticated mechanisms for adhering to favourable surfaces to ensure a stable supply of nutriments and avoid being washed away. In aquatic environments the adherent microflora of particulate organic matter, plankton and solid surfaces may vary from a few per cent to more than 90% of the total bacterial biomass; depending on habitats and nutrient availability (Sieburth, 1979). For fish the epithelium with mucous layer forms the barrier against the external environment with a multitude of potentially harmful microorganisms. A continuous secretion and shedding of mucus produced by goblet cells in skin, gills and mucosa of the gastrointestinal tract may prevent microbial colonization. It has been argued that mucus on the body surfaces of healthy fish normally does not contain detectable amounts of bacteria (Crouse-Eisnor *et al.*, 1985), whereas large numbers may be found in mucus during advanced stages of infections (Kanno *et al.*, 1990). As most bacterial infections begin on mucosal surfaces (Klein, 1990), this may not be unequivocally beneficial, and the mucosal surfaces of fish may serve as substrate for the adhesion of bacteria. Settlement and colonization by invertebrate larvae is induced on surfaces covered with bacteria (Fitt *et al.*, 1989), and in a fish farm this will eventually lead to the proliferation and growth of secondary invaders or fouling organisms with their own microflora. This may account for a varied microbial community in the fish farm. However, fish pathogenic vibrios could also be resident in the invertebrates at low seawater temperatures (Olafsen *et al.*, 1993).

We still have relatively scarce knowledge about the free-living and adherent microflora of the aquatic environment. One reason for this has been the fact that many typical microorganisms of this environment do not grow well on laboratory media (Sieburth, 1979; Morita, 1985; McKay, 1992). There are usually discrepancies between cell numbers isolated from direct and viable counts. In the aquatic environment direct counting methods indicate that there are approximately 10^6 bacteria per ml in lake or seawater. Culturable bacteria may constitute as little as only 0.01–10% of the viable bacterial population from the marine environment. Viable counts may range from 10^1 to 10^4 bacteria per ml, and are normally higher in summer and in eutrophic water, with more than 10^6 bacteria per ml in some coastal areas. Also in cold Antarctic water the percentage of unidentified strains was high (30%). This situation is probably similar in Arctic waters, which have a high biomass production of migrating fish. Thus a multitude of 'new' psychrotolerant bacteria may potentially invade fish farms in cold environments.

The indigenous microflora of aquatic organisms

The relationship between microorganisms and their hosts vary from simple interdependence of nutrients to symbiosis or parasitism. In animals, establishment of the indigenous microflora is regulated by physical or chemical factors and non-specific defence reactions. In higher animals local mucosal and secretory immunity also play a role in controlling microbial colonization. Growth of pathogens is thus prevented, whereas colonization by commensals may result in immune tolerance.

Whereas a protective role of the commensal microflora of warm-blooded animals against pathogens appears established (Savage, 1989), no such role has been unequivocally accepted for the microflora of aquatic animals. Neither is it known at what growth stage a microflora may be established. Most aquatic organisms live in intimate relationships with opportunistic bacteria from their early life stages. Marine aquaculture diseases that affect fish are frequently caused by opportunistic pathogens, like vibrios. Some pathogenic vibrios may be related in biochemical or cell surface characteristics to members of the commensal flora (Strøm & Olafsen, 1990). Thus an understanding of the role of the indigenous microflora of aquatic organisms, and the factors that govern the balance between commensalism and pathogenicity, is essential to improve feed and conditions for the intensive mass rearing of healthy fish.

Most farmed aquatic animals are poikilotherms, comprising species such as fish with temperature dependent immunity (Hildemann & Cooper, 1963; Rijkers, 1982), and primitive invertebrates that lack the protection of inducible immunity and immune-memory. Thus temperature variations can have a marked effect on the adherent microflora, and may be an important factor in the relationships between microorganisms and their hosts. Organisms of the aquatic environment differ in optimum temperature for growth and survival, as also do their commensal microorganisms. The microflora is generally well adapted to the temperatures sustained by their host. Whereas most obligate pathogens of mammals are not adapated to lower temperatures and their growth in the environment is relatively slow, both the commensal as well as the pathogenic microflora of poikilothermic organisms are well adapted to external water temperatures. During and following epizootic diseases of fish considerable numbers of bacteria may be released to the environment, and may survive well (Hoff, 1989a). It was originally conceived that the 'classical' fish disease vibriosis, caused by *Vibrio anguillarum* or *Vibrio ordalii*, was a 'summer-problem', possibly induced by stress at high temperatures. However, when fish farming beyond the Arctic circle became established, vibriosis caused by psychrotrophic *Vibrio*-strains became a 'winter problem' (Olafsen *et al.*, 1981). Since then 'cold water vibriosis' caused by *Vibrio salmonicida* has been the most serious disease problem in this area (Egidius *et al.*, 1986). Thus in general we can expect that opportunistic pathogens like vibrios may take advantage of any external condition which is adverse to the host.

The microflora of marine invertebrates and plankton may be dominated by vibrios at certain times of the year (Huq *et al.*, 1983). Bottom-dwelling marine bivalves accumulate large numbers of microorganisms from the seawater by filter-feeding, and thus harbour an exceptionally rich microflora with *Vibrio* species predominating. The *Vibrio*-group comprises microorganisms with a dynamic role in the marine and brackish environment. They are isolated from seawater, sediment, plankton, invertebrates and fish, and are abundant members of the microflora of aquatic organisms (Colwell, 1984). We know relatively little about how vibrios interact with aquatic organisms, but apparently they are the most commonly isolated members of the 'normal' microflora. Other microorganisms isolated from bivalves include the Gram-negative *Aeromonas*, *Alcaligenes*, *Flavobacterium*, *Pseudomonas* and *Moraxella*-like spp., and the Gram-positive *Bacillus*, *Corynebacterium* and *Micrococcus* (Lauckner, 1983). Marine vibrios are frequently found in association with invertebrate surfaces and also in haemolymph and internal organs of healthy bivalves (Vasta, 1990; Olafsen *et al.*, 1993). A specific relationship exists between some marine vibrios and their invertebrate hosts, and thus vibrios persist in shellfish, haemolymph and soft tissues following depuration in UV-treated seawater (Greenberg, 1982; Eyles & Davery, 1984) whereas *Escherichia coli* counts are rapidly reduced under similar circumstances. This interaction may be specific since shellfish may have specific agglutinins (lectins) that bind vibrios (Olafsen *et al.*, 1992), including the *Vibrio cholera* and *Vibrio vulnificus* which are pathogenic to man (Tamplin & Fisher, 1989; Tamplin & Capers, 1992). Such lectins may also be found in pallial cavity mucus of oysters (Fisher, 1992). *Pseudomonas* and *Vibrio* are natural constituents of the bacterial flora of the molluscan digestive tract, and bacteria may persist in tissues and body fluids of healthy marine bivalves without causing disease (Vasta, 1990; Olafsen *et al.*, 1993). Invertebrates apparently possess effective defences to regulate the proliferation of these bacteria, and thus the relationship between bacteria and hosts is 'fine-tuned' (Olafsen, 1988). On the other hand *Vibrio* and *Pseudomonas* species have been responsible for most diseases affecting marine bivalves (Lauckner, 1983).

Aquaculture has resulted in intensive systems for incubation of eggs and hatching of fish larvae. In incubators with a relatively large biomass, newly hatched larvae are exposed to organic debris and high bacterial densities, and viable counts may typically increase from 10^3 to 10^6 per ml during hatching. The chorion of fish eggs appears suitable for the adhesion of bacteria, and within a couple of days fish eggs may be heavily colonized with bacteria (Hansen & Olafsen, 1989). Stalking and budding bacteria appear to be abundant (Austin, 1982). In particular *Cytophaga*, but also bacteria belonging to the *Flavobacterium* and *Pseudomonas* group may dominate the microflora on salmon eggs incubated in fresh water (Barker *et al.*, 1989). On unfertilized cod eggs *Caulobacter* sp. with long holdfasts and *Seliberia*-like bacteria have been observed, and in addition *Pseudomonas*, *Alteromonas*, *Aeromonas* and *Flavobacterium* dominated the adherent microflora on cod and halibut eggs (Hansen & Olafsen, 1989). Thus the genera *Flavobacterium*, *Pseudomonas* and *Cytophaga* appeared to dominate the microflora on both freshwater and marine fish eggs. Despite their widespread occurrence in the marine environment, vibrios have not been abundant colonizers of the chorion of marine fish eggs. Fish eggs may be colonized by bacteria prior to spawning, indicated by the presence of *Seliberia* and *Caulobacter* spp. on eggs dissected from the ovary. Whether or not this represents an autochthonous flora of eggs remains to be investigated. The filamentous bacterium *Leucothrix mucor* has been observed particularly on eggs close to hatching, and could result in a 'hairy appearance' with filaments reaching lengths of more than 1.5 cm (Hansen & Olafsen, 1989). It is thus possible that in hatcheries overgrowth of this bacterium could be mistaken for a fungal infection.

L. mucor is not pathogenic, but heavy overgrowth could possibly cause asphyxia or effect the buoyancy of the eggs. Massive bacterial growth on the chorion could affect hatching or chorion strength, but a diverse non-pathogenic adherent flora would probably be an effective barrier against colony formation by pathogens. Adherent organisms on fish eggs may also erode the chorion by exoenzymatic activity, resulting in exposure of new attachment sites on the egg chorion for microorganisms. Some pathogens may dissolve the chorion and zona radiata of the egg shell, such as the psychrotrophic *Flexibacter ovolyticus* that may constitute a major part of halibut egg epiflora, and result in high larval mortalities after hatching (Hansen *et al.*, 1992).

Poor quality and mass mortalities of eggs have been serious problems in fish farming. Bacterial overgrowth could affect the commensal relationship between the normal indigenous microflora and opportunistic pathogens on mucosal surfaces of eggs and larvae. The microflora on fish eggs and in larval incubators may be important for establishment of the indigenous microflora of newly-hatched larvae, and subsequently the health of farmed fish. Use of antibiotics may result in alterations in the microflora that could be unfavourable (Austin & Al-Zahrani, 1988; Hansen *et al.*, 1992), and thus there is an urgent need to control the microflora in incubators by other means.

On the body surfaces of freshwater fish, bacteria of the genera *Acinetobacter*, *Aeromonas*, *Cytophaga*, *Flavobacterium*, *Moraxella* and *Pseudomonas* have been demonstrated. In the marine environment we find many of the same groups, but typically with a much higher incidence of *Vibrio* spp. The intestinal microflora of various fresh and saltwater fishes was found to contain only a limited range of bacteria in some cases, whereas others reported a wider diversity of the gut microflora (Austin & Al-Zahrani, 1988; Cahill, 1990). The majority of studies have been concerned with adult fish, and little information exists on the microflora of fish larvae. It has been inferred that the intestinal microflora of fish reflects the bacterial content of ingested food and of the environment, and thus the existence of a truly indigenous gut microflora in fish has been disputed (Horsley, 1973). However, several reports describe bacteria firmly attached to the gut mucosa of marine fish, and it is now generally accepted that fish contain an indigenous gastrointestinal microflora consisting of both aerobic, facultative anaerobic and obligate anaerobic bacteria (Cahill, 1990; Sakata, 1990; Strøm & Olafsen, 1990). Members of this microflora appear to be resident, but its gross composition may change following starvation (Conway *et al.*, 1986) or change in food and habitat (Strøm & Olafsen, 1990). Evidence is accumulating that lactic acid bacteria are among the resident flora of both fresh and seawater fish. Similarly to other animals a protective role has been implied for this flora (Conway *et al.*, 1986; Strøm *et al.*, submitted), even though this can still not be unequivocally accepted (Conway, 1989).

Also other bacteria, including non-pathogenic vibrios, have the potential of repressing the growth of fish pathogenic vibrios, and thus act as a probiotic flora in fish (Olsson *et al.*, 1992). It was found that the intestinal microflora of six-month-old wild cod juveniles consists mainly of *Vibrio*, *Lactobacillus* and *Bacillus*. Of a total viable count of 10^3 to 10^5 bacteria per cm^2 vibrios accounted for more than 50%. Of these gut-associated vibrios 30% crossreacted with monoclonal antibodies against the fish pathogen *V. salmonicida*, and they were also related to this pathogen in biochemical traits (Strøm and Olafsen, 1990). These typical gut-vibrios disappeared from the intestinal microflora after feeding a commercial diet to the juvenile cod, and also following an outbreak of coldwater vibriosis. It thus appears that changes in food regime affect the equilibrium on the gut mucosa between commensal vibrios and related opportunists occupying the same microenvironments.

Defence reactions and the microflora of fish

For all organisms fundamental reactions of defence involves discrimination between self and non-self. In the aquatic environment the outer compartment, consisting of seawater with food, pathogens and predators forms a continuum only separated from the internal body fluids by the epithelial surfaces covered by mucus. A continuous secretion and shedding of mucus produced by goblet cells in skin, gills and mucosa of the gastrointestinal tract may prevent microbial colonization, and various antimicrobial factors in mucus such as immunoglobulins, complement, C-reactive proteins, lysozyme and lectins may furnish protection (Fletcher, 1982). Mucus secretion may increase following stress or infection (Blackstock & Pickering, 1982), or increase in the number of bacteria in the seawater (Ottesen and Olafsen, submitted). However, as most bacterial infections begin on mucosal surfaces (Klein, 1990), increased production of mucus may not be unequivocally beneficial unless it can be easily shed.

Thus a great number of bacteria will be ingested with the natural food available to fish, and by drinking seawater. Ingested bacteria may affect the intestinal microflora, furnish nutrients or enzymes, or affect immune reactions by exposing antigenic determinants. In adult fish it now appears that local mucosal and secretory immunity is important in protection against bacterial infections (Hart *et al.*, 1988). Teleosts possess intraepithelial lymphoid tissue that is less organized than in mammals, and macrophages, lymphoid cells and secretory immunoglobulin-forming cells are infiltrated within the intestinal epithelium (Tomonaga *et al.*, 1985; Rombout & van den Berg, 1989). Cells capable of sequestering intact proteins or antigens are found in the digestive tract, and in particular in the posterior segment of the hindgut (Vernier, 1990; Olafsen, 1984; Olafsen and Hansen, 1992). Fish apparently lack specialized intraepithelial (microfold) M-cells for the uptake of antigens, and it appears that enterocytes serve a similar purpose of antigen sampling and presentation to lymphoid cells (Davina *et al.*, 1982; Rombout *et al.*, 1989; Rombout and van den Berg, 1989; Rombout *et al.*, 1989). Absorptive cells may present antigens to the immune system, and result in production of secretory immunoglobulins (Georgopoulou *et al.*, 1986; Rombout *et al.*, 1989; Rombout and van den Berg, 1989). Secretion of interleukin-like factors also suggests that the epithelial cells may have immunomodulatory functions (Sigel *et al.*, 1986). However, local secretory mucosal immunoglobulins similar to the IgA class found in mammals have not been detected in fish, and it has been suggested that mucosal and serum antibodies are identical (Harrell *et al.*, 1976).

Since also marine fish larvae ingest substantial numbers of bacteria before active feeding commences (Olafsen, 1984), this may affect the establishment of a primary intestinal microflora or antigen priming of the larvae. Some fish larvae having a straight gut, like herring (*Clupea harengus*), accumulate large quantities of bacteria in the posterior part of the gut. The bacteria may be propulsed to the rear by the help of ciliated epithelial cells found in the alimentary canal of most fish larvae (Hansen *et al.*, 1992). This process may also serve to sustain yolksac larvae with exogenous nutriments.

Relatively little is known about ontogeny of immunity in fish. It has been inferred that the immunological capacity of fish larvae is not fully developed until they are several months old (Chantanachookhin *et al.*, 1991). Until then fish larvae may rely on non-specific immunity, active macrophages or immunity acquired from the mother (Mor & Avtalion, 1988). The lymphoid functions are asynchronously developed in fish. At four days post-hatching, and before lymphocytes mature, functional macrophages are present in skin, gills and kidney, and thereafter cellular immunity develops rapidly (Manning *et al.*, 1982a). Ig-positive cells can be detected by day 16, and from then the thymus resembles that of adult fish (Secombes *et al.*, 1983). Antibody response and memory can be induced in young (about two months old) fish, whereas the ability to respond to some antigens that are considered 'T-cell dependent' may develop later (Lamers, 1985). Too early injection (before three months) may result in tolerance to soluble protein antigens, whereas response to bacteria was observed three weeks post hatching (Manning *et al.*, 1982b; Mughal & Manning, 1985). Fish of 2–3 weeks have been successfully vaccinated by injection (Khalifa & Post, 1976), and some protection was obtained following immersion of young fish. The 'immunological competence' appears to be correlated with evolvement of a critical number of immune cells rather than with age (Tatner & Horne, 1983).

In addition to uptake in the hindgut, immature yolksac larvae of cod also sequester intact bacterial antigens in columnar epithelial cells of the fore-gut. These cells have some preference in the uptake of antigens (Olafsen & Hansen, 1992). This is of interest, since vaccination or stimulation of non-specific defence mechanisms of fish larvae has to be by the oral route. However, it must be noted that the production of specific antibodies has not been observed in fish at this age, and it is believed that the immature larvae rely mostly on non-specific defence mechanisms (Fletcher, 1982), perhaps resembling the responses of invertebrates.

Oral administration of vaccine against various fish pathogenic bacteria usually yields less protection than injections, but high levels of protection were obtained when antigens reach the posterior part of the intestine unaltered (Johnson & Amend, 1983). The oral route may be important to prevent colonization by fish pathogenic bacteria, and it has been demonstrated that oral immunization against *V. anguillarum* prevented adhesion of the bacterium to the body surfaces, but did not result in decreased mortality following injection of the bacterium (Kawai *et al.*, 1981). Local mucosal and secretory immunity may also be important in regulation of the intestinal microflora. It is widely observed that vibrios are common members of the commensal microflora of marine fish, and wild-living fish may have antibodies to pathogenic *Vibrio anguillarum* strains (Olafsen *et al.*, 1981). *Vibrio* species may dominate the intestinal microflora of marine fish species. This microflora will be 'observed' by the immune system of the fish, and it can be speculated that sequestering of intact antigens of commensal bacteria by fish larvae could result in tolerance. Loss of commensal 'gut-vibrios' during domestication could ease invasion of related pathogens on the gut mucosa. Such opportunistic proliferation could be facilitated through acquired tolerance to shared antigens.

The opportunistic ecology of some fish pathogens

Obligately pathogenic bacteria are unable to survive in the environment, and subsequently spread by contact between their hosts. The rates of infectious disease caused by these microorganisms can be decreased by good husbandry and the proper use of vaccines and antibiotics. However, in aquaculture many diseases are caused by facultative or opportunistic pathogens. Vibrios are members of this class of opportunistic pathogens that may also inhabit healthy fish (Rødsæther *et al.*, 1977). These bacteria may become pathogenic when the fish is stressed or weakened, or be induced by factors in the environment. However, once an infection by opportunistic bacteria has been established, it becomes epizootic, and can then only be stopped by chemotherapy.

A relatively wide variation has been observed among *V. anguillarum* strains isolated from different fish species and different locations. It is thus customary to divide the fish pathogenic vibrios into sub-groups with different biochemical features and serological characteristics (Larsen *et al.*, 1988). Fish that are infected or have survived infections may release viable bacteria through pores, macrophages, exposed lesions or faeces to the environment. Vibrios survive well in the marine environment. *V. anguillarum* is biochemically well adapted to proliferation in marine sediments, and may survive for months as free-living cells in seawater and for more than a year attached to particles (Olafsen *et al.*, 1981; Enger *et al.*, 1989; Hoff, 1989b). Pathogenic strains of *V. anguillarum* have been isolated from apparently healthy fish, that may carry pathogenic strains for more than a year in freshwater (Rødsæther *et al.*, 1977). However, the most pathogenic serotypes are not frequently found in water or sediments (Larsen *et al.*, 1988).

Fish are relatively resistant to disease, but stress due to handling, transport, overstocking or malnutrition may compromise the host and affect the immune responses. Also pollutants such as copper may affect disease resistance (Rødsæther *et al.*, 1977). The routes of infection for *V. anguillarum* or *V. salmonicida* are not known, but it has been demonstrated that these bacteria colonize the mucosal layer of larvae, and may result in extensive damage to the microvilli (Hansen and Olafsen, manuscript in preparation). Growth and proliferation of opportunistic bacteria in the fish farm or on mucosal surfaces may be affected by the chemical composition of the food. Fish food may contain components that directly affect the microflora, or indirectly its adherence to the mucosal surfaces of the fish. Such components may be fibre, starch, enzyme inhibitors, saponins, tannins, or lectins. Plant material contains agglutinins (lectins) that also affect the attachment and subsequent colonization by bacteria of the intestinal mucosal surfaces (Grant, 1989) by interacting with their receptors. Since farmed fish are kept essentially on the same diet throughout their life, the effect of such food compounds could be detrimental. We know relatively little about such effects of the fish diet, and thus a sensible food regime should be varied and close to the natural food available to fish.

The use of antibiotics in fish farming may affect the intestinal microflora of fish (Austin & Al-Zahrani, 1988; Hansen *et al.*, 1992). Antibiotics have routinely been used in the mass rearing of fish larvae, but may adversely affect the composition or biochemical versatility of the intestinal microflora (Hansen *et al.*, 1992). It has long been known that the marine invertebrates contain numerous antimicrobial compounds that may be extracted (Li, 1969), and possibly could be used in aquaculture. Similarly the use of immuno-stimulants could enhance the non-specific defence reactions of fish. Such stimulants may be compounds that are extracted from or mimic the cell-walls of microorganisms, such as lipopolysaccharides of bacteria (Fletcher & White, 1987) or β-1,3 glucans of fungi (Robertsen *et al.*, 1991). However, the effect of the use of such compounds in aquaculture will have to be investigated.

The possibility thus exists that in incubation of larvae the extensive use of antibiotics could be substituted by a diet caring also for the maintenance of a suitable microflora, and perhaps including probiotic microorganisms in a 'tailored' microbial environment. The use of a diet containing probiotic microorganisms could also be beneficial at critical events resulting from stress or handling, or at low temperature. We have observed that a diet including lactic-acid bacteria isolated from fish proved to be beneficial during smoltification and transport of fish (Strøm and Olafsen, unpublished results). However, before such food regimes become part of the 'husbandry of aquaculture', a better understanding of aquatic organisms and their relations to the commensal and pathogenic microflora is needed.

DISEASE IN RELATION TO SALMON AQUACULTURE
By Ronald J. Roberts

Introduction

The exploitation of salmon by means of stock enhancement, and for farming purposes is of long standing. The full impetus to research on their diseases, however, only came with the twin imperatives of losses associated with intensification and the availability of research funds, associated with the growth and vulnerability of the industry. The following section provides and overview of diseases affecting salmon reared in captivity, with particular emphasis on Atlantic salmon, including identification of pathogens, conditions facilitating infection, and treatments of specific diseases.

Bacterial diseases

Furunculosis

One microbial disease of salmon has a pedigree almost as long as that of microbiology itself. Furunculosis, described in brown trout in 1894 by Emmerich and Weibel, was first found in Germany. Its subsequent 'spread' is just as likely to have been a result of the growth in capability to diagnose it as to any actual progression of infection. There are, however, some grounds for suspecting that it was originally a pathogen of rainbow trout, which are considerably more refractory to it than Atlantic salmon or brown trout. Be that as it may, the first serious experience of epizootic furunculosis, caused by the Gram-negative bacillus *Aeromonas salmonicida*, was in wild Atlantic salmon in river fisheries on the East Coast of Scotland over the period 1927 to 1935. So severe were the losses that a parliamentary commission of inquiry was launched by the British government to enquire into the condition. The scientific research carried out for the commission by Professor Mackie and his colleagues between 1930 and 1935 was an exemplary model for all subsequent strategic research on fish diseases. (Mackie *et al.*, 1930–33). Mackie succeeded in providing not only a wide ranging body of information on *Aeromonas salmonicida* and its disease, furunculosis, but also the first detailed description of a second, almost equally serious disease, now known as bacterial kidney disease (BKD).

As the farmed salmon industry has developed, a number of other bacterial diseases have manifested themselves. Even now, with the industry mature and in many ways technically sophisticated, it is still fair to consider furunculosis and BKD as the two most serious of all diseases of Atlantic salmon.

Furunculosis is a problem at both the freshwater and the marine stages of Atlantic salmon production. In freshwater, it is principally associated with farms which draw from rivers or lakes containing migratory fish, and often outbreaks occur in line with the rise in water temperatures of late spring and early summer and the first runs of adult fish into the watershed.

In the sea, again, it is principally a summer disease, and the bacterial septicaemia which induces the clinical signs of haemorrhage and myonecrosis, is particularly difficult to treat because affected fish go off their food. Antibiotics such as oxytetracycline or sulphonamides incorporated in the food are traditionally used as treatment, but development of transmissible drug resistance by many strains of the pathogen has led to the need for an ever wider range of therapeutants.

The obvious and ideal answer to the control of furunculosis in Atlantic salmon would be the development of a vaccine, preferably an oral one, which could be applied during the freshwater stage of growth and which would protect against subsequent exposure to the virulent agent. There are, however, a number of technical difficulties in attempting production of such a vaccine. Not least of these are the apparent ability of the pathogen to produce metabolites which are immunosuppressive in salmon and the problems associated with infections occurring around the time of smoltification, when the young salmon has a very high circulating cortisol level and is virtually incapable of mounting any type of immune response.

Furunculosis has been a most serious problem in salmon farming in Scotland and Ireland for many years. In Norway, however, although some outbreaks of disease associated with atypical non-pigmented strains, originally described as *Pasteurella* infections, were reported in the 1960s, it was only in the early 1980s that significant outbreaks in farmed fishes occurred in Norway. These have gradually extended their range until the condition is now as serious there as in the UK.

Currently a number of biochemically engineered vaccines are under development and newer antibiotics with potentially greater potency are being tested for maketing. It seems likely, however, that these will serve to maintain no more than the present level of control of this most evasive and capable pathogen. Total control of it is, probably, no more than a long-term aspiration.

Bacterial kidney disease (BKD)

At the time of the Furunculosis Committee's work in the 1930s, they also defined a disease of wild salmon which they called Dee disease because of its high incidence in the river Dee in North-east Scotland.

This disease was also subsequently shown to be a serious cause of losses of wild Pacific salmon on the West Coast of North America where it was called kidney disease. Neither name was particularly useful, indeed the newer name bacterial kidney disease (BKD) is only marginally better, but BKD is now the accepted name. The condition is a chronic granulomatous disease, not unlike tuberculosis in its pathology, but generally associated primarily with the kidney and spleen, where it causes destruction of both haemopoietic and renal or splenic functional elements. It is also capable of causing granulomatosis of other organs, notably the heart and skeletal muscle. In severe cases, the large cavitating lesions in the muscle are the most significant. Muscular lesions were not originally a feature of the disease in wild Atlantic salmon, but common in Pacific salmon. It has been suggested that the emergence of the cavitating form in farmed salmonids in Europe in the 1970s was a result of egg importations from the Pacific.

One of the most significant features of the causative agent, the dipththeroid bacterium *Renibacterium salmoninarum* is its capacity for vertical transmission because of its ability to survive within the yolk of salmon eggs. Here it is safe from egg disinfection, and able to induce disease in the next generation. It is this vertical transmission, coupled with the density of the lesions, preventing antibiotic penetration, which make BKD a virtually impossible condition to treat.

It is now endemic in many salmon-producing areas, and control depends almost exclusively on ensuring eggs and smolts are from non-infected stocks and that fish in the sea are kept on a good nutritional plane during the winter and spring, when clinical disease seems at its most active.

Vibriosis

Infection of eels and other marine fishes with *Vibrio anguillarum*, a close relative of the cholera bacterium, is probably the oldest known bacterial fish disease. It was well described by Bonaveri in 1718 in eels from the Italian coast. Its importance for Atlantic salmon, however, was not recognized until the establishment of intensive salmon production in Norway. There in the 1970s and early 1980s, major losses occurred associated with infection with both *Vibrio anguillarum* and a new and highly pathogenic, distinctive, species called *Vibrio salmonicida* (Egidius *et al.*, 1986). The latter was particularly associated with farms around the island of Hitra and was the principal cause of the winter condition known as Hitra disease (coldwater vibriosis), although nutritional and husbandry deficiencies are also considered to have been a major component of that condition.

The advent of vaccination, using killed bacterins given by injection or bath treatment to young fish, has greatly improved the situation with regard to vibriosis. Antibiotic treatment is still needed but use of vaccines along with sensible stocking densities and good husbandry regimes has made vibriosis a problem of much less significance than it was in the early days of Atlantic salmon culture.

Several other bacterial genera are associated with disease in farmed Atlantic salmon (Roberts, 1989). These include the Gram-negative Enterobactero bacterium *Yersinia ruckeri* which although primarily known as the agent of red-mouth disease in American rainbow trout, is also on occasion associated with entero-septicaemia in salmonids. Another group which is spasmodically responsible for losses, especially of young fish, is the *Myxobacteria*, aquatic bacteria which are highly motile and principally associated with the surface mucus. On occasion their number can reach pathogenic levels when water flows or water quality are deficient.

Parasitic diseases

Parasitic diseases in farmed fishes are a constant and often limiting factor to production. Many of the protozoan parasites such as *Costia* sp. and *Trichodina* sp. in fry can wipe out whole populations should water quality fail, or other environmental factors stress the fish. In older salmon, parasitism is important in two ways. First, infestation of marine stages with the crustacean parasites known as sea lice is one of the most serious and production-threatening diseases of all, and second, infection with certain relatively benign parasites such as cestodes can imperil the market perception of particular stocks on aesthetic grounds.

There are two species of sea lice of importance to Atlantic salmon. They are normally parasites of wild fish, and indeed are esteemed by anglers. Because they are obligatory marine species, their presence on a captured sport-fish denotes a fresh-run fish which entered freshwater less than 48 hours earlier. The larval, infective stages, however, exist in large numbers offshore, in the very areas where most salmon farms are located. As a captive stock, exposed to possible infection over a long period, caged Atlantic salmon can build up high levels of infection, with both *Lepeophtheirus salmonis* and *Caligus* spp and unless control measures are taken, several hundred parasites may develop on each fish. They feed on cutaneous and subcutaneous tissues and tissue fluids and cause extensive ulceration, particularly on the head (Sommerville *et al.*, 1989).

Sea louse infection can lead to catastrophic losses unless levels of infection can be controlled and this can only be done by treatment with chemical ectoparasiticides although experiments with wrasse as 'cleaner' fish are under way. Currently the preferred treatments are the organohalide insecticides. There is considerable environmental concern about the extensive use of such compounds in sensitive coastal waters. Despite the very rapid breakdown of the licensed compounds in water and lack of any evidence of detrimental ecological effects even on sites where usage has extended over a decade or more, this has led to current research being focused on the possibilities of use of different types of therapeutant or even the development of vaccines against the mouth parts or digestive enzymes of the parasites.

The other important effect of parasitism in salmon is in relation to aesthetics. Neither the processor nor consumer, is prepared to accept the presence of tapeworms such as diphyllobothrids in the peritoneal cavity. Usually these are of no public health significance, merely aesthetically unattractive. The larval anisakid worms, occasionally found in the abdominal cavity of wild salmonids and associated with human illness are not found in farmed fishes.

Two emerging parasitic problems in farmed Atlantic salmon are currently confined to a single country, but nevertheless cause major problems there and represent a source of great concern elsewhere. *Thecamoeba* sp. presumed to be free-living ameoba under normal conditions, have, in Tasmania, become a serious problem to be industry. They normally remain on the surface of the gills but in circumstances of high summer temperature and especially if there are large numbers of *Myxobacteria* on the fish, they can become invasive, inducing massive gill lamellar proliferation and death. In Norway, recently, *Gyrodactylus salaris*, a parasite alleged to have been introduced with transfer of farmed stock from the Baltic has been implicated in the disappearance of wild Atlantic salmon juveniles from many Norwegian rivers. It has not yet been associated with disease in farmed stock (Johnsen & Jensen, 1986).

Virus diseases

Atlantic salmon show only a limited number of virus diseases in culture. Their distribution is varied, with only infectious pancreatic necrosis (IPN) and salmon papilloma occurring universally.

Infectious pancreatic necrosis occurs with a virtually world-wide distribution. It is a birnavirus infection of the pancreas and intestine and the disease was first described as an enteritis of hatchery fry in the Atlantic provinces of Canada by M'gonigle (1940). IPN virus was the first fish virus to be isolated, by Wolf and his colleagues in 1960, and a wide range of serotypes have now been isolated from salmon and trout throughout the world as well as from other fish species. Although it is ubiquitous, there are considerable differences in pathogenicity between strains, which make control difficult.

The characteristic disease is a density related acute necrotizing infection principally of the pancreas in newly hatched fry. Losses may vary but recovered fish remain carriers and also more susceptible to a wide range of other infections, particularly bacterial (Roberts & Horne, 1978). The disease is treated with much greater respect in Atlantic salmon rearing areas than in rainbow trout culture, because of this effect on resistance to other infections.

Salmon papilloma or pox, is a benign neoplastic condition occurring on wild or farmed salmon in the first or second year of life. The white hyperplastic plaques of wart-like epithelial growth are unsightly and persist for several months. They ultimately slough, usually with little after-effect. The virus has not been isolated but has been observed by electron microscopy (Carlisle & Roberts, 1977).

Infectious haematopioetic necrosis (IHNV) is a composite species name for at least three rhabdovirus isolates (strains?) from Pacific salmonids. The disease is discussed extensively elsewhere in relation to Pacific salmon. In the few populations of Atlantic salmon where it has occurred, due to fish transfers, the condition has been similar to that found in Pacific fish.

Viral haemorrhagic septicaemia (VHS) is another rhabdovirus disease, this time of mainland European origin. It is primarily a condition of rainbow trout but on occasion has been shown to be pathogenic for Atlantic salmon also. VHS is a notifiable disease which is of particular importance to the European salmon industry since if it became widespread, it would greatly limit exports, particularly to the United States.

Pancreas disease is a condition of putative viral origin, which has caused major losses in Scottish and Irish salmon culture for many years. The principal feature is a severe acute dissolution of pancreatic acini, with little or no cellular response, following transfer of smolts to seawater. In many fish, the pancreatic acini grow back and the fish progress normally, but in others the loss is long-term or permanent. They lose weight, become vulnerable to secondary conditions and many die.

The disease has now been described in most parts of the world where salmon are cultured and is one of the most serious causes of loss. Experimental studies have shown that it can be transmitted using kidney tissue filtrates from affected fish, but there is also a close correlation with nutrition and in particular the vitamin E status of affected fish.

Infectious Salmon Anaemia, ISA, has reached serious proportions in Norway in recent years and has now been found to have a viral origin, although the research has yet to be published.

Fungal diseases

The only fungal disease of Atlantic salmon of any significance is *Saprolegnia* infection of adult broodstock in fresh water. The fungus appears to be ubiquitous. The disease problem associated with its invasion of the skin is linked to the presence of high circulating levels of androsteroids and thus infection occurs predominantly in adult mature males.

Saprolegnia

This infection is also associated with the disease of wild Atlantic salmon known as ulcerative dermal necrosis (UDN). The primary cause of this disease is unknown but it is confined to wild fishes and has never been found in farmed salmon.

Other fungus problems

Occasional problems are associated with ingestion of feed contaminated with soil fungi such as *Phoma herbarum* or other septate phycomycetes which can result in massive granulomatosis of internal organs, particularly the kidneys. A much more serious problem, limited to spasmodic losses only because of the vigilance of feed compounders, is aflatoxicosis. This results from the feeding of oil seeds, which have been contaminated by the toxin produced by *Aspergillus fumigatus*. The toxin is neoplastic for all vertebrates but salmonids appear to be particularly susceptible. The result is the production of hepatoma, a malignant cancer of the hepatic epithelium which can cause heavy losses.

Nutritional and other non-infectious diseases

Salmon diets, although largely empirically derived, at least for the fattening stages, are now generally adequate and reliable. It is only rarely that a particular batch is deficient or contaminated. There is still concern about correct levels of vitamin E to incorporate in salmon diets, particularly at marine stages, and whether there is a direct connection with pancreas disease.

A problem which regularly asserts itself in Atlantic salmon culture, spasmodic, but serious when it occurs, is zinc deficiency in freshwater parr. The pathogenesis is generally associated not with absolute zinc deficiency in the diet, but rather with inhibition of absorption of available zinc by relatively high levels of calcium, phytate or ash in the diet. The result is a central or sublenticular cataract, associated with darkened colour and generally also with poor growth.

Toxic algal blooms have been a very severe scourge of the Atlantic salmon farming industry in Ireland, Scotland and particularly Norway. The cause is usually a massive bloom, in calm summer weather, of one of the naked dinoflagellates of the plankton, such as *Gyrodinium aureolum*. They can under certain circumstances produce a cell-associated toxin which acts on gills or digestive tract (Roberts *et al.*, 1983) but in many cases it is suspected that losses are associated purely with the mechanical effect of such numerous and abrasive micro-organisms in the water.

Marine farms affected may lose all their fish and the only possible remedy would appear to be towing cages out to sea or sinking them below the level of the surface bloom, both of which present serious logistical problems.

Conclusion

Atlantic salmon culture has developed very considerably over the last 25 years. At the same time, a major improvement has taken place in our understanding not only of the clinical diseases to which salmon are vulnerable, but also of the basic pathological and immune processes that take place in such fish. This has greatly assisted in the ability to control salmon diseases but there are nevertheless major disease problems, extant almost from the start of the industry, which still represent an ever-present threat to the viability of farms.

Furunculosis is the most serious of these conditions, but BKD, sea lice and algal blooms all appear to be equally intractable. High quality husbandry and proper stocking densities are major factors in avoiding the heaviest losses from such conditions but they remain serious concerns in relation to the welfare of both the fish and the Atlantic salmon farming industry.

References

Austin, B. (1982) Taxonomy of bacteria isolated from a coastal, marine fish-rearing unit. *J. Appl. Bacteriol.*, 53, 253–68.

Austin, B. and Al-Zahrani, A.M.J. (1988) The effect of antimicrobial compounds on the gastrointestinal microflora of rainbow trout, *Salmo gairdneri* Richardson. *J. Fish Biol.*, 33, 1–14.

Barker, G.A., Smith, S.N. and Bromage, N.R. (1989) The bacterial flora of rainbow trout *Salmo gairdneri* Richardson, and brown trout, *Salmo trutta* L., eggs and its relationship to developmental success. *J. Fish Dis.*, 12, 281–93.

Blackstock, N. and Pickering, A.D. (1982) Changes in the concentration and histochemistry of epidermal mucous cells during the alevin and fry stages of the brown trout, *Salmo trutta. J. Zool. Lond.*, 197, 463–71.

Bonaveri, G. (1718) quoted by Cannestrini, G. (1893) *La Mallatia dominante delle anguiller.* Atti 1st Veneto 7:809–14.

Brown, W. (1862) *The Natural History of the Salmon*. Thos. Murray and Son, Glasgow.

Cahill, M.M. (1990) Bacterial flora of fishes: a review. *Microb. Ecol.*, 19, 21–41.

Carlisle, J. and Roberts, R.J. (1977) An epidermal papilloma in the Atlantic salmon (*S. salar*). *J. Wildlife Dis.*, 13, 235–9.

Chantanachookhin, C., Seikai, T. and Tanaka, M. (1991) Comparative study of the ontogeny of the lymphoid organs in three species of marine fish. *Aquaculture*, 99, 143–55.

Colwell, R.R. (Ed.) (1984) *Vibrios in the Environment*, John Wiley, New York.

Conway, P.L. (1989) Lactobacilli: fact and fiction. In *The Regulatory and Protective Role of the Normal Microflora* (Ed. by R. Grubb, T. Midtvedt and E. Norin), pp. 263–81, Macmillan Basingstoke.

Conway, P.L., Maki, J., Mitchell, R. and Kjelleberg, S. (1986) Starvation of marine flounder, squid and laboratory mice and its effect on the intestinal microbiota. *FEMS Microbiol. Ecol.*, 38, 187–95.

Crouse-Eisnor, R.A., Cone, D.K. and Odense, P.H. (1985) Studies on relations of bacteria with skin surface of *Carassius auratus* L. and *Poecilia reticulata*. *J. Fish Biol.*, 27, 395–402.

Davina, J.H.M., Parmentier, H.K. and Timmermanns, L.P.M. (1982) Effect of oral administration of *Vibrio* bacterin on the inestine of cyprinid fish. *Dev. Comp. Immunol.*, Supplement 2, 157–66.

Egidius, E., Wiik, R., Andersen, K., Hoff, K.A. and Hjeltnes, B. (1986) *Vibrio salmonicida* sp. nov., a new fish pathogen. *Int. J. System. Microbiol.*, 36, 518–20.

Enger, Ø., Husevåg, B. and Goksøyr, J. (1989) Presence of the fish pathogen *Vibrio salmonicida* in fish farm sediments. *Appl. Environ. Microbiol.*, 55, 2815–18.

Eyles, M.J. and Davery, G.R. (1984) Microbiology of commercial depuration of the Sydney rock oyster *Crassostrea commercialis*. *J. Food Prot.*, 47, 703–6.

Finlay, B.B. and Falkow, S. (1989) Common themes in microbial pathogenicity. *Microb. Rev.*, 53, 210–30.

Fisher, W.S. (1992) Occurrence of agglutinins in the pallial cavity mucus of oysters. *J. Exp. Mar. Biol. Ecol.*, 162, 1–13.

Fitt, W.K., Labare, M.P., Fuqua, W.C., Walch, M., Coon, S.L., Bonar, D.B., Colwell, R.R. and Weiner, R.M. (1989) Factors influencing bacterial production of inducers of settlement behaviour of larvae of the oyster *Crassostrea gigas*. *Microb. Ecol.*, 17, 287–98.

Fletcher, T.C. (1982) Non-specific defense mechanisms in fish. *Dev. Comp. Immunol.*, Supplement 2, 123–32.

Fletcher, T.C. and White, A. (1987) Metabolic and immunological effects of endotoxin in the plaice, *Pleuronectes platessa* L. *J. Fish Biol.*, 31 (Suppl. A), 81–5.

Georgopoulou, U., Sire, M.F. and Vernier, J.M. (1986) Immunological demonstration of intestinal absorption and digestion of protein macromolecules in the trout (*Salmo gairdneri*). *Cell Tissue Res.*, 245, 387–95.

Gottschal, J.C. (1990) Phenotypic response to environmental changes. *FEMS Microb. Ecol.*, 74, 93–102.

Grant, G. (1989) Anti-nutritional effects of soybean. A review. *Prog. Food Nutr. Sci.*, 13, 317–48.

Greenberg, E.P., Dubois, M. and Palhof, H. (1982) The survival of marine vibrios in

Mercenaria merceneria, the hardshell clam. *J. Food Saf.*, 4, 113–23.

Hansen, G.H. and Olafsen, J.A. (1989) Bacterial colonization of cod (*Gadus morhua* L.) and halibut (*Hippoglossus hippoglossus*) eggs in marine aquaculture. *Appl. Environ. Microbiol.*, 55, 1435–46.

Hansen, G.H., Strøm, E. and Olafsen, J.A. (1992) Effect of different holding regimens on the intestinal microflora of herring (*Clupea harengus*) larvae. *Appl. Environ. Microbiol.*, 58, 461–70.

Hansen, G.H., Bergh, Ø., Michaelsen, J. and Knapskog, D. (1992) *Flexibacter ovolyticus* sp. nov., a pathogen of eggs and larvae of Atlantic halibut, *Hippoglossus hippoglossus* L. *Int. J. Syst. Bact.*, 42, 451–8.

Harrell, L.W., Etlinger, H.M. and Hodgins, H.O. (1976) Humoral facors important in resistance of salmonid fish to bacterial disease. II. Anti-*Vibrio anguillarum* activity in mucus and observations on complement. *Aquaculture*, 7, 363–70.

Hart, S., Wrathmell, A.B., Harris, J.E. and Grayson, T.H. (1988) Gut immunology in fish: A review. *Dev. Comp. Immunol.*, 12, 453–80.

Hildemann, W.H. and Cooper, E.L. (1963) Immunogenesis of homograft reactions in fishes and amphibians. *Fed. Proc.*, 22, 1145–51.

Hoff, K.A. (1989a) Occurrence of the fish pathogen *Vibrio salmonicida* in faeces from Atlantic salmon (*Salmo salar*). *J. Fish Dis.*, 12, 595–7.

Hoff, K.A. (1989b) Survival of *Vibrio anguillarum* and *Vibrio salmonicida* at different salinities. *Appl. Environ. Microbiol.*, 55, 1775–86.

Horsley, R.W. (1973) The bacterial flora of the Atlantic salmon (*Salmo salar* L.) in relation to its environment. *J. Appl. Bact.*, 36, 377–86.

Huq, A., Small, E.B., West, P.A., Huq, M.I., Rahman, R. and Colwell, R.R. (1983) Ecological relationships between *Vibrio cholerae* and planktonic crustacean copepods. *Appl. Environ. Microbiol.*, 45, 275–83.

Isenberg, H.D. (1988) Pathogenicity and virulence: another view. *Clin Microbiol. Rev.*, 1, 40–53.

Johnsen, B.O. and Jensen, A.J. (1986) Infestations of Atlantic salmon, *Salmo salar.* L. by *Gyrodactylus salaris* in Norwegian waters, *J. Fish Biol.*, 29, 233–41.

Johnson, K.A. and Amend, D.F. (1983) Efficacy of *Vibrio anguillarum* and *Yersinia ruckeri* bacterins applied by oral and anal intubation of salmonids. *J. Fish Dis.*, 6, 473–6.

Kanno, T., Nakai, T. and Muroga, K. (1990) Scanning electron microscopy on the skin surface of ayu *Plecoglossus altivelis* infected with *Vibrio anguillarum*. *Dis. Aquat. Org.*, 8, 72–5.

Kawai, K., Kusuda, R. and Itami, T. (1981) Mechanisms of protection in ayu orally vaccinated for vibriosis. *Fish Pathology*, 15, 257–62.

Khalifa, K.A. and Post, G. (1976) Immune-response of advanced rainbow trout fry to *Aeromonas liquefaciens*. *Prog. Fish. Cult.*, 38, 66–8.

Kjelleberg, S., Hermansson, M., Mården, P. and Jones, G.W. (1987) The transient phase between growth and non-growth of heterotrophic bacteria, with emphasis on the marine environment. *Annu. Rev. Microbiol.*, 41, 25–49.

Klein, J. (1990) *Immunology* Blackwell Scientific Publications Cambridge, Massachusetts.

Lamers, C.H.J. (1985) The reaction of the immune system of fish to vaccination, Agricultural University, Wageningen, The Netherlands.

Larsen, J.L., Rasmussen, H.B. and Dalsgård, I. (1988) Study of *Vibrio anguillarum* strains from different sources with emphasis on ecological and pathobiological properties. *Appl. Environ. Microbiol.*, 54, 2264–7.

Lauckner, G. (1983) Diseases of mollusca: bivalvia. In *Diseases of Marine Animals* (Ed. by O. Kinne), pp. 477–962, Biologische Anstalt Helgöland, Hamburg.

Li, C.P. (1969) Antimicrobial activity of certain marine fauna. *Proc. Soc. Exp. Biol. Med.*, 104, 366–8.

McKay, A.M. (1992) Viable but non-culturable forms of potentially pathogenic bacteria in water – review. *Letters in Applied Microbiology*, 14, 129–35.

Mackie, T.J., Arkwright, J.A., Pryce-Tannatt, T.E., Mottram, J.C., Johnstone, W.D. and Menzies, W.J.M. (1930; 1935) *Furunculosis Committee Reports.*, HMSO, Edinburgh.

Manning, M.J., Grace, M.F. and Secombes, C.J. (1982a) Developmental aspects of immunity and tolerance in fish. In *Microbial Diseases of Fish* (Ed. by R.J. Roberts), pp. 31–46, Academic Press, London.

Manning, M.J., Grace, M.F. and Secombes, C.J. (1982b) Ontogenetic aspects of tolerance and immunity in carp and rainbow trout: studies on the role of the thymus. *Dev. Comp. Immunol.*, Suppl. 2, 75–82.

M'gonigle, R.H. (1940) Acute catarrhal enteritis of salmonid fingerlings. *Trans. Am. Fish Soc.*, 70, 297–303.

Mor, A. and Avtalion, R.R. (1988) Evidence of transfer of immunity from mother to eggs in tilapias. *Bamidgeh*, 40, 22–8.

Morita, R.Y. (1985) Starvation and miniaturization of heterotrophics with special emphasis on maintenance of the starved viable state. In *Bacteria in their Natural Environment*, (Ed. by M. Fletcher and G.D. Floodgate), pp 111–30, Academic Press, New York and London.

Mughal, M.S. and Manning, M.J. (1985) Antibody responses of young carp, *Cyprinus carpio*, and grey mullet, *Chelon labrosus*, immunized with soluble antigen by various routes. In *Fish Immunology* (Ed. by M.J. Manning and M.F. Tatner), pp. 313–25, Academic Press, London.

Olafsen, J.A. (1984) Ingestion of bacteria by cod (*Gadus morhua* L.) larvae. In *The Propagation of Cod*, Gadus morhua *L.* (Ed. by E. Dahl, D.S. Danielsen, E. Moksness and P. Solemdal), pp. 627–43, Institute of Marine Research, Bergen.

Olafsen, J.A. (1988) Role of lectins in invertebrate humoral defense. *Am. Fish. Soc. Special Publ.*, 18, 189–205.

Olafsen, J.A. and Hansen, G.H. (1992) Intact antigen uptake by intestinal epithelial cells of marine fish larvae. *J. Fish Biol.*, 40, 141–56.

Olafsen, J.A., Christie, M. and Raa, J. (1981) Biochemical ecology of psychrotrophic strains of *Vibrio anguillarum* isolated from outbreaks of vibriosis at low temperature. *System. Appl. Microbiol.*, 2, 339–48.

Olafsen, J.A., Fletcher, T.C. and Grant, P.T. (1992) Agglutinin activity in Pacific oyster (*Crassostrea gigas*) hemolymph following in vivo *Vibrio anguillarum* challenge. *Dev. Comp. Immunol.*, 16, 123–38.

Olafsen, J.A., Mikkelsen, H.V., Giæver, H. and Hansen, G.H. (1993) Indigenous bacteria in hemolymph and soft tissues of marine bivalves at low temperature. *Appl. Environ. Microbiol*, 59, 1848–54.

Olsson, J.C., Westerdahl, A. Conway, P.L. and Kjelleberg, S. (1992) Intestinal colonization potential of turbot (*Scophthalmus maximus*) and the dab (*Limanda limanda*) – associated bacteria with inhibitory effects against *Vibrio anguillarum.*

Appl. Environ. Microbiol., 58, 551–6.
Ottesen, O., Amin, A. and Olafsen, J.A. Responses of mucous-producing cells of halibut (*Hippoglossus hippoglossus*) larvae to bacteria in the seawater, submitted.
Rijkers, G.T. (1982) Kinetics of humoral and cellular immune reactions in fish. *Dev. Comp. Immunol.*, Suppl. 2, 93–100.
Roberts, R.J. (1989) *Fish Pathology*, Baillière Tindall, London.
Roberts, R.J. and Horne, M.T. (1978) Bacterial meningitis in farmed rainbow trout *Salmo gairdneri* Richardson affected with chronic pancreatic necrosis. *J. Fish Dis.*, 1, 157–64.
Roberts, R., Bullock, A.M., Turner, M. and Jones, K.J. and Tett, P. (1983) Mortalities of *Salmo Salar* exposed to *Gyrodinium aureolum*. *J. Mar. Biol. Assoc.*, UK., 63, 741–3.
Robertsen, B., Rørstad, G., Engstad, R. and Raa, J. (1991) Enhancement of the non-specific disease in Atlantic salmon (*Salmo Salar* L.) by a glucan from *Saccharomyces cerevisiae* cell walls. *J. Fish Dis.*, 13, 391–400.
Rødsæther, M.C., Olafsen, J., Raa, J., Myhre, K. and Steen, J.B. (1977) Copper as an initiating factor of vibriosis (*Vibrio anguillarum*) in eel (*Anguilla anguilla*). *J. Fish Biol.*, 10, 17–21.
Rombout, J.H.W.M., Bot, H.E. and Taverne-Thiele, J.J. (1989) Immunological importance of the second gut segment of the carp. II. Characterization of mucosal leucocytes. *J. Fish Biol.*, 35, 167–78.
Rombout, J.H.W.M. and van den Berg, A.A. (1989) Immunological importance of the second gut segment of the carp. I. Uptake and processing of antigens by epithelial cells and macrophages. *J. Fish Biol.*, 35, 13–22.
Rombout, J.H.W.M., van den Berg, A.A., van den Berg, C.T.G.A., Witte, P. and Egberts, E. (1989) Immunological importance of the second gut segment of the carp. III. Systemic and/or mucosal immune responses after immunization with soluble or particulate antigen. *J. Fish. Biol.*, 35, 179–86.
Roszak, D.B. and Colwell, R.R. (1987) Survival strategies of bacteria in the natural environment. *Microbiol. Rev.*, 51, 365–79.
Sakata, T. (1990) Microflora in the digestive tract of fish and shellfish. In *Microbiology in Poecilotherms* (Ed. by R. Lésel), pp. 171–6, Elsevier, Amsterdam.
Savage, D.C. (1989) The normal human microflora – composition. In *The Regulatury and Protective Role of the Normal Microflora* (Ed. by R. Grubb, T. Midtvedt and E. Norin), pp. 3–18, Stockton Press, New York.
Secombes, C.J., Van Groningen, J.J.M., Van Muiswinkel, J.B. and Egberts, E. (1983) Ontogeny of the immune system in carp (*Cyprinus carpio* L.). The appearance of antigenic determinants on lymphoid cells detected by mouse anti-carp thymocyte monoclonal antibodies. *Dev. Comp. Immunol.*, 7, 455–64.
Sieburth, J.M. (1979) *Sea Microbes*, Oxford University Press, New York.
Sigel, M.M., Hamby, B.A. and Huggins Jr., E.M. (1986) Phylogenetic studies on lymphokines. Fish lymphocytes respond to human IL-1 and epithelial cells produce an IL-1 like factor. *Vet. Immunol. Immunobiol.*, 12, 47–58.
Sommerville, C., Jones, M.W., and Brown, J. (1990) The histopathology associated with the juvenile stages of *Lepeophtheirus salmonis* on the Atlantic salmon. *Salmo salar* L. *J. Fish Dis.*, 13, 303–10.
Strøm, E. and Olafsen, J.A. (1990) The indigenous microflora of wild-captured juvenile cod in net-pen rearing. In *Microbiology in Poecilotherms* (Ed. by R.

Lésel), pp. 181–5, Elsevier, Amsterdam.

Strøm, E., Olafsen, J.A. and Raa, J. Gut-associated *Carnobacteria* with inhibitory properties against pathogenic *Vibrios* in fish. (Manuscript in preparation).

Tamplin, M.L. and Capers, G.M. (1992) Persistence of *Vibrio vulnificus* in tissue of Gulf coast oysters, *Crassostrea virginica*, exposed to seawater disinfected with UV light. *Appl. Environ. Microbiol.*, 58, 1506–10.

Tamplin, M.L. and Fisher, W.S. (1989) Occurrence and characteristics of agglutination of *Vibrio cholerae* by serum of the eastern oyster, *Crassostrea virginica. Appl. Environ. Microbiol.* 55, 2882–7.

Tatner, M.F. and Horne, M.T. (1983) Susceptibility and immunity to *Vibrio anguillarum* in post-hatching rainbow trout fry, *Salmo gairdneri* Richardson 1836. *Dev. Comp. Immunol.*, 7, 465–72.

Tomonaga, S., Kobayashi, K., Hagiwara, K., Sasaki, K. and Sezaki, K. (1985) Studies on immunoglobulin and immunoglobulin-forming cells in *Heterodontus japonicus*, a cartilagous fish. *Dev. Comp. Immunol.*, 9, 617–26.

Vasta, G.R. (1990) Invertebrate lectins, C-reactive proteins and serum amyloid – structural relationships and evolution. In *Defense Molecules* (Ed. by J.J. Marchalonis and C.L. Reinisch), pp. 183–99, Wiley-Liss, New York.

Vernier, J.M. (1990) Intestine ultrastructure in relation to lipid and protein absorption in teleost fish. In *Animal Nutrition and Transport Processes, 1. Nutrition in Wild and Domestic animals* (Ed. by J. Mellinger), pp. 166–75. S. Karger, Basle.

Wolf, K., Snieszko, S.F., Dunbar, C.E. and Pyke, E. (1960) Virus of infectious pancreatic necrosis in trout. *Proc. Soc. Exp. Biol. Med.*, 104, 105–8.

ZoBell, C.E. (1963) Domain of the marine microbiologist. In *Symposium on Marine Microbiology* (Ed. by C.H. Oppenheimer), pp. 3–24, Charles W. Thomas, Springfield.

Chapter 8
Breeding and Selection

BJARNE GJERDE

Introduction

In farm animals and plants, selective breeding has played an important role in domestication, increasing yields and survival rates and improving product quality. The rate of change has been particularly rapid in the last two or three decades and today one cannot really imagine animal husbandry and plant production without selection programmes.

Quantitative genetics is the theoretical basis on which animal and plant breeding programmes are founded. Although established around 1920, this theory has only recently been applied to fish culture and is not widely practised. This is rather surprising because aquaculture has reached a level of economic importance in many countries. For some aquatic species, the explanation for this can be found in the lack of education of researchers, advisory persons and fish farmers in quantitative genetics and planning of breeding programmes. However, for most aquatic species, the greatest obstacle to developing efficient breeding and selection programmes is lack of complete control of artificial reproduction and controlled mating, mainly due to the difficulties or impossibilities of hatching and feeding larvae and fry. This has been the case, for instance, in the farming of eel, yellowtail and halibut.

Compared with farm animals, most aquaculture species have the advantage of high reproductive capacity. This, together with the species' method of external fertilization, permit the design of more efficient breeding programmes than are available for farm animals. High fertility also makes the expense of maintaining broodstock very low.

The main objective of this chapter is to focus on methods and strategies for improvement of genetic traits which would be of economic importance in the farming of salmonids.

Main objective

The main objective for animal breeders is to change the economically important characteristics or traits of the animals in the desired direction. This change involves the following steps:

- Definition of the traits to be improved;
- Evaluation of the overall genetic capacity (breeding value) of potential breeding animals;
- Selecting the highest ranking individuals as parents for the next generation.

Procedures for how to best attain the objective are the topic for this chapter. At first it is necessary to introduce the reader briefly to some key words and expressions used in animal breeding terminology.

Variation around the mean

When considering some of the most commonly farmed fish species, it is not difficult to see that there are quite large differences among them in many characteristics or traits. Neither is it difficult to find differences within species, for example, between the different strains or breeds.

For a number of traits, there are also quite large individual differences within a group or population. The existence of difference or variation among individuals within a population is a prerequisite for genetic improvement of a trait.

An example of observed differences between individuals is given in Fig. 8.1, showing the frequency distribution of body weight for 1625 Atlantic salmon reared in the same floating net cage in the sea for about two years. Average body weight for this group of salmon was 5.8 kg. The smallest fish was 0.8 kg and the largest, 10.8 kg. The differences in body weight among individuals are obviously large, and therefore, we say that this trait shows large phenotypic variation.

Most of the fish in Fig. 8.1 are distributed around the mean value, while the number of fish per class decreases towards the extremes. Such a character is said to be *normally distributed*. There are two main reasons that a trait like body weight shows a normal distribution:

(1) Individual variation for a trait is partially determined by the many genes the individuals possess. The number of individuals with all genes extremely 'bad' or 'good' are few, while the number of individuals possessing an 'average'

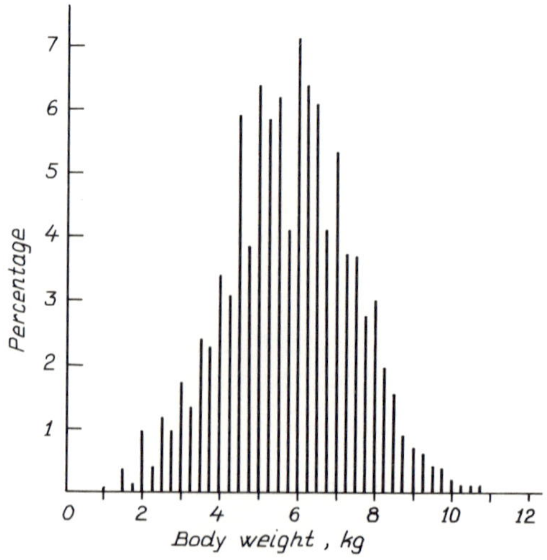

Fig. 8.1 The body weight frequency distribution of 1625 Atlantic salmon.

genotype, that is, the particular assemblage of genes possessed by the individual, is higher.

(2) A trait is to a large extent determined by a number of environmental factors. These factors might influence the trait in a positive or a negative way. The number of individuals influenced by extremely negative or positive factors is low, while the number of individuals affected by average environmental influences increases.

Thus, each of these two sources of variation, genetic and environmental, tend to give a normal distribution. It is therefore not surprising that a characteristic influenced by both factors shows a normal distribution.

However, for some traits, the observations fall into two or several distinct classes. Such traits are called *either/or* or *categorical* traits. Examples of such traits are survival (two classes, i.e. dead or alive); age at sexual maturity (two or more classes, i.e., salmon maturing as precocious males, or during their first, second or third year in the sea or later); and traits being subjectively scored into a few classes due to convenience or difficulties in measuring them. Meat colour, abdominal fat and belly thickness are examples of traits for which a scoring procedure may be used. The above categorical traits are assumed to have a non-observable underlying normal distribution.

The traits so far discussed are called *quantitative traits*. This group of traits is characterized by (1) having a normal or underlying normal distribution, (2) being influenced by a large number of genes and thus having a quite complicated mode of inheritance, and (3) being influenced by environmental factors. Many of these characteristics are of great economic importance.

Another group of traits are called *qualitative traits*. These traits are characterized by (1) being determined by few genes with a simple mode of inheritance, (2) being influenced minimally by environmental factors, and as a result, (3) the observations fall into a few distinct classes. Examples of qualitative traits are eye colour, blood groups and skin colour, for example, the albino and the normal-coloured trout determined by a single locus with two alleles or genes. Normally, qualitative traits are of little or no economic value.

Components of phenotypic value and variance

The value observed when a character is measured or scored on an individual is called the *phenotypic value* of that individual. As an example, the distribution in Fig. 8.1 is constructed based on phenotypic values for body weight.

The phenotypic value can be partitioned into two components; one attributable to the influence of *genotype*, that is, the particular assemblage of genes possessed by the individual, and one attributable to the influence of *environment*, that is, all non-genetic circumstances that influence the phenotypic value. Or, symbolically.

$$P = G + E$$

where P is the phenotypic value, G the genotypic value and E the environmental deviations.

Thus we may think of the genotype conferring a certain value on the individual and the environment causing a deviation from this value. The mean environmental deviation in the population as a whole is then taken to be zero, so that the mean phenotypic value is equal to the mean genotypic value.

The genetics of a character centres around the study of its variation. The amount of variation is measured and expressed as the *variance*, that is, simply the mean of the squared values when the values are expressed as deviations from the population mean. The variance of phenotypic values is the phenotypic variance (V_P), also termed the total variance. The variance of genotypic values is termed the genetic variance (V_G) and the variance caused by environmental deviations, the environmental variance (V_E). Or symbolically,

$$V_P = V_G + V_E = V_A + V_D + V_I + V_{ES} + V_{ER}$$

In the above equation, V_G is further partitioned into three separate components: V_A is the *additive genetic variance* or the variance due to the average (additive) value of the genes, or the variance of breeding values (see later). V_D is the *dominance genetic variance* or the variance due to the value of intra-locus interaction among genes. V_I is the *interaction genetic variance* or the variance due to the value of inter-locus interaction among genes. The sum of V_D and V_I is termed the *non-additive genetic variance*.

The environmental or non-genetic component is portioned into two separate components: V_{ES} is the variance due to the value of systematic or recognizable environmental causes. Examples of systematic causes that are at least partly under experimental control in farming of fish are nutritional factors; water temperature; age of the fish; tank, cage and pond effects, and sex effects. V_{ER} is the variance due to the

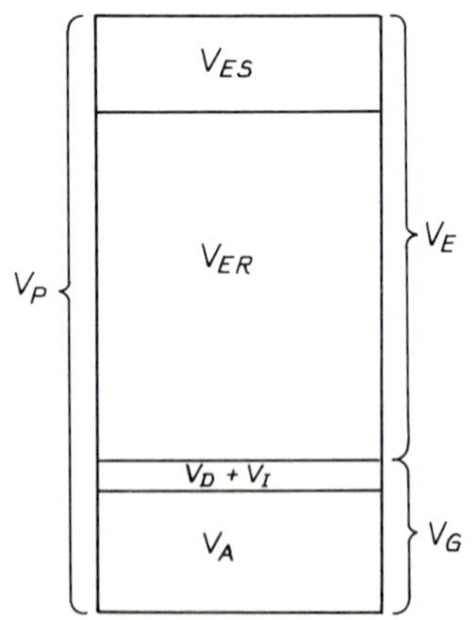

Fig. 8.2 Portioning of the phenotypic variance (V_P) into its genetic ($V_G = V_A + V_D + V_I$) and environmental ($V_E = V_{ER} + V_{ES}$) components.
Notes:
V_A = Additive genetic variance
V_D = Dominance genetic variance
V_I = Interaction genetic variance
V_{ES} = Systematic environmental variance
V_{ER} = Random environmental variance

value of unknown or random environmental causes which therefore cannot be eliminated by experimental design.

The portioning of the total phenotypic variance into its different components is illustrated in Fig. 8.2.

The portioning of the total variance into its components allows us to estimate the relative importance of the various determinants of the phenotype, in particular the role of heredity versus environment. The ratio V_G/V_P expresses the extent to which individuals' phenotypes are determined by their genotypes. This is called the *heritability* in the broad sense. The ratio V_A/V_P expresses the extent to which phenotypes are determined by the additive value of their genes. This is called the heritability in the narrow sense, or simply the heritability. The heritability ($h^2 = V_A/V_P$) determined the degree of resemblance between relatives and is of the greatest importance in additive genetic improvement of the animals.

The non-additive genetic part of the total variance ($V_D + V_I$) is of vital importance in determining whether crossbreeding should be implemented in breeding programmes or not.

Heritabilities are ratios with values between zero and one. If the heritability is high (in the narrow sense), this means that the observed variation for a trait is to a larger extent determined by additive genetic effects, while a low heritability indicates that the variation is determined to a larger extent by environmental or non-additive genetic causes.

Breeding objective

Breeding objective should be exactly defined. However, consumers' preferences may change over time and so may the breeding objective. In addition, the outcome of selection decisions today can only be harvested in the future. Therefore, the decision for selecting traits for the objective should be a long-term consideration. For these reasons it is important that those involved in a coordinated breeding programme agree upon the objective in order to avoid working in different directions.

For a trait to be included in the breeding objective, the following prerequisites must hold:

- The trait must be of economic importance;
- It must be possible to measure or judge (score) the trait;
- The heritability of the trait must be greater than zero.

Traits to be considered in the breeding objective for salmonids are discussed below. Approximate values of means, standard deviations and heritabilities of various traits in Atlantic salmon and rainbow trout are shown in Table 8.1.

Food conversion efficiency

Food costs constitute a large part of the total production costs and are usually the most important economic trait in fish farming. It is, however, very difficult to measure the amount of food consumed by individuals or groups of fish. Food conversion efficiency can therefore not yet be included in a selection programme for fish.

Table 8.1 Means, phenotypic standard deviations and heritabilities of various traits in Atlantic salmon and rainbow trout.

Trait	$\bar{x}$	SD	h^2
Atlantic salmon			
Body weight of fingerlings (g)[1]	33.0	17.0	0.06
Body length of fingerlings (cm)[1]	13.7	2.4	0.20
Early sexual maturity (%)[2]	21.4	–	0.15
Traits recorded at slaughter[4]			
gutted body weight (kg)	6.22	1.81	0.36
body length (cm)	81.2	6.4	0.32
condition factor	1.20	0.14	0.37
dressing percentage	93.9	0.8	0.20
meat colour (score 1–4)	3.1	0.6	0.09
belly thickness (score 1–4)	3.3	0.6	0.13
abdominal fat (%)	4.8	0.8	0.27
fat content in the meat (%)	16.0	2.5	0.21
Rainbow trout			
Body weight of fingerlings (g)[3]	10.0	7.6	0.08
Body length of fingerlings (cm)[3]	8.9	2.3	0.12
Early sexual maturity (%)[2]	12.2	–	0.05
Traits recorded at slaughter[5]			
gutted body weight (kg)	2.98	0.65	0.21
body length (cm)	59.0	4.4	0.18
condition factor	1.58	0.15	0.19
dressing percentage	87.3	1.9	0.36
meat colour (score 1–4)	3.3	0.6	0.27
belly thickness (mm)	12.3	2.4	0.37
abdominal fat (%)	10.9	2.4	0.30
fat content in the meat (%)	14.8	2.6	0.47

Sources:
[1] Refstie and Steine (1978).
[2] Gjerde (1986).
[3] Refstie (1980).
[4] Rye and Refstie (submitted); Rye and Gjerde (submitted).
[5] Gjerde (1989) and Gjerde and Schaeffer (1989).

Further development of systems for direct and/or indirect measurement of food consumed should be given high priority.

Growth rate

Growth rate is one of the most important traits in fish farming. High growth rate speeds production turnover and a fast growing fish will reach a higher body weight before the onset of sexual maturation. Frequently, larger fish also fetch a higher price per unit of body weight than smaller ones. Moreover, preliminary results from rainbow trout indicate a medium to high genetic correlation between growth rate and food conversion so selection for high growth rate results in improvement of food conversion (Refstie, pers. comm.).

Age at sexual maturity

Sexual maturation in salmonids results in decreased growth rate, reduced meat quality and increased mortality. The general breeding objective for this trait should therefore be a fish that reaches market size before sexual maturation. Precocious males should be eliminated in all species, while the objective for age at sexual maturity in the sea may vary from one species to another. The objective and importance for this trait may also change over time as a result of genetic improvement of growth rate.

The possibility for genetic improvement of this trait is to a large extent dependent on the frequency of maturing fish at a given age. Family selection is therefore recommended for this trait, particularly at low frequencies of the trait (<10–15%).

Survival/disease resistance

High survival rate is of considerable economic importance, particularly for ongrowing and adult fish and less so for eggs and younger individuals since these are less economically valuable. Survival is, however, a very complex trait. The causes for death may be different from one farm to another, and from one year to another. Survival is therefore a difficult trait to work with and heritability is commonly low.

Measuring death rate to specific diseases may give more scope for improvement, but such measurement is very expensive if exposure of the fish to the disease agent is needed (Gjedrem *et al.*, 1991). In the future, direct measurements of resistance to specific diseases may be obtainable by applying immunological methods (Chevassus and Dorson, 1990; Fjalestad, 1990).

Body quality traits

Salmonids are looked upon as a luxury food. To retain a high, even, and well-known quality is therefore of great economic importance.

Until the present, quality has been a question of species, body size, meat colour and external appearance. Other traits that might be of future economic importance are those related to the meatiness of the carcass (e.g. belly thickness), the amount and distribution of fat in the meat, and the amount of abdominal fat. In Atlantic salmon and rainbow trout, these traits have shown fairly high heritabilities indicating that it is possible to obtain genetic progress if desired (Table 8.1).

Fecundity

Salmonids have a high reproductive rate (more than 1000 eggs/kg body weight). This makes the expense of keeping brood stock low. Therefore, it is not important to consider egg number as an economically important trait in a breeding programme. It is, however, important to record fecundity traits and particularly egg quality traits to study whether or not they are changing over time.

Recapture frequency

In salmon-ranching systems (reviewed by Issakson, 1988), recapture frequencies are of great economic importance. This is, however, a complex trait where survival is an important component. In Atlantic salmon there are strong indications that recapture frequency is a heritable trait (Carlin, 1969), and a large sea-ranching experiment is now in progress in Iceland for further investigations of the possibilities for genetic improvement of this trait (Isaksson, 1989).

Breeding strategies

The main objectives in a breeding programme are to move the mean value of the trait in the desired direction for a normally distributed trait, or to increase the frequency of the desired class(es) for a trait with discrete classes. Changes in the population mean or in class frequencies from one generation to the next are termed genetic gain. For a normally distributed trait, this is illustrated in Fig. 8.3.

Genetic gain can be obtained by applying different breeding strategies. For a long-term breeding objective, the only suitable strategy for a breeding nucleus will be some type of purebreeding approach for additive genetic improvement. The breeding strategies that may be applied to produce commercial fry are less restricted than when breeding within the nucleus. Any kind of crossbreeding, ploidy manipulations, sex manipulation, etc., may be applied if the productivity of the commercial fry can be further improved. Any breeding strategy for commercial fry production that imposes limitation on the progress in additive genetic performance in the nucleus should be avoided, (for example, the use of highly inbred lines in a cross-breeding programme – see below).

All breeding programmes should start with collection, comparison and selection of the best genetic material available. The value of testing strains and selecting the best for farming may be equivalent to several generations of within-strain selection.

Figure 8.4 gives an example of differences in growth rate measured as body weight at slaughter found between strains of Atlantic salmon and between fullsib families (derived from the same parents) within a strain. The great differences found

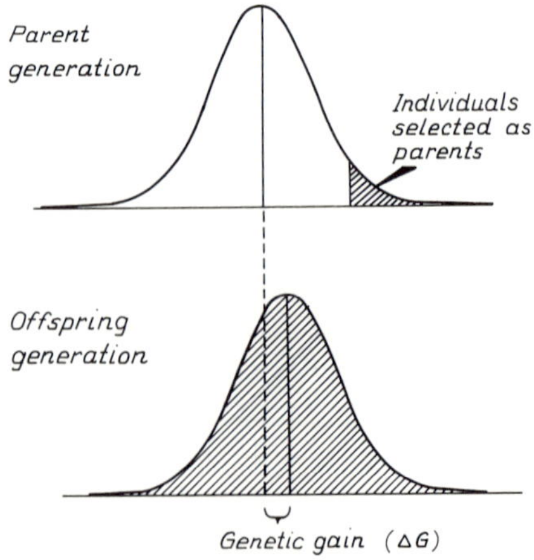

Fig. 8.3 Illustration of genetic gain obtained by one generation of selection.

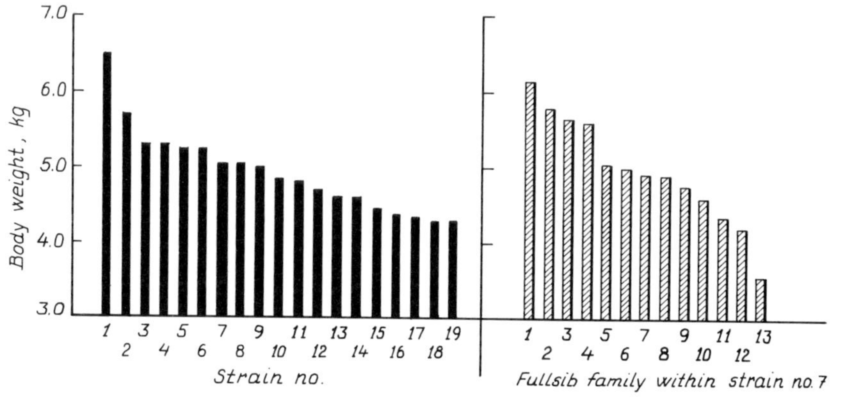

Fig. 8.4 Ranking of strains of Atlantic salmon and fullsib families within a strain for body weight after two years in cages in the sea.

between strains and between fullsib families within a strain emphasize the importance of starting a breeding programme with the best available genetic material. The differences between fullsib families within a strain also illustrate the possibilities for further improvement through selection.

Inbreeding

Inbreeding is the mating of individuals that are more closely related than the average of the population. Generally, inbreeding results in the phenomenon known as inbreeding depression, that is, reduced performance, particularly for traits connected with reproductive capacity and viability. It is therefore important to keep the rate of inbreeding at a low level in a breeding programme. Examples of observed inbreeding depression for survival and growth rate in rainbow trout is in Table 8.2.

Knowing the effective population size (N_e) for any breeding structure, the rate of

Table 8.2 Effect in rainbow trout of three levels of inbreeding on survival of eyed eggs, alevins and fry and on body weight of fingerlings and adult fish (Gjerde *et al.*, 1983).

	Survival (%)						Body weight			
	Eyed eggs		Alevins		Fry		Fingerlings (g)		Adults (kg)	
Inbreeding coefficient[1]	Control	ID[2]	Control	ID[2]	Control	ID[2]	Control	ID[3]	Control	ID[3]
0.25	95.1	8.9	99.0	2.8	81.4	9.1	12.0	9.8	2.50	11.3
0.375	94.6	15.3	98.4	8.3	72.7	5.6	47.8	22.5	3.15	19.9
0.50	96.8	5.8	96.3	4.8	72.6	18.6	12.2	−4.0	2.96	30.5

Notes:
[1] Obtained by one (F = 0.25), two (F = 0.375) and three (F = 0.50) successive generations of fullsib mating.
[2] Inbreeding depression (ID = Control − Inbred)
[3] Inbreeding depression (calculated by the alternative method) $\left(\text{ID} = \frac{\text{Control} - \text{Inbred}}{\text{Control}} 100\%\right)$

inbreeding per generation (ΔF) can be obtained as $\Delta F = 1/2N_e$ where N_e is a function of the number of sires and dams used as parents for each new generation. Assuming no genetic relationship among sires and dams, for 50 sires and 50 dams, $N_e = 100$ and $\Delta F = 0.0005$ or 0.5% per generation. This is likely an acceptable rate of inbreeding for most traits.

The rate of inbreeding depends strongly on the number of the less numerous sex. If the number of sires is reduced to 30, the number of dams must be increased to 150 in order to keep $N_e = 100$. And if the number of sires is 20, N_e can never exceed 80, regardless of the number of dams used.

Within a population under selection, it is impossible to prevent the level of inbreeding increasing over generations. If the level of inbreeding becomes unacceptably high, some animals from an unrelated population for example, a different year-class, should be used as parents to reduce the inbreeding problem.

In practical breeding work, deliberate inbreeding is only of interest when inbred lines are produced for crossing in order to exploit non-additive genetic variance.

Crossbreeding

Crossbreeding is mating between species, breeds, strains or inbred lines. The main objective in crossbreeding programmes is to exploit non-additive genetic variance (heterosis or hybrid vigour). Crossbreeding alone does not produce any additive genetic improvement over time, and should therefore be looked upon as a supplement to a programme for additive genetic improvement.

Little heterosis is found by crossing Norwegian strains of Atlantic salmon (Table 8.3) and inbred lines of rainbow trout (Gjerde, 1988). Crossing inbred lines can only be justified by large heterotic effects. The resources and time needed to produce, maintain and replace inbred lines will be better utilized by increasing the efforts to improve the additive genetic performance of a purebred nucleus.

Purebreeding

The breeding method or strategy for additive genetic improvement within a population is known as purebreeding, and is the method to choose for continuous genetic improvement over a long period of time. By this method, we avoid the mating of close relatives, and we select individuals that possess a majority of positive (desirable) genes as parents for the next generation. Individuals that possess a majority of positive genes normally show good production results. These 'good genes' and properties are transferred to their offspring. Individuals that possess a majority of positive genes are said to have a high *breeding value*.

The breeding value of an individual cannot be measured directly. Neither can it be measured with 100% accuracy. The true breeding value will therefore remain unknown and to a greater and lesser extent be masked by systematic and stochastic environmental effects and also by effects caused by interactions among the genes.

The breeding value can only be estimated by recording the product of the genes, that is, the phenotypic values of the trait(s). Such records may be obtained from the individual itself or from relatives as full and halfsibs, progeny or parents. Records on relatives can be used because the individual and its relatives share common genes. Information from close relatives is in general more valuable than information from

Table 8.3 Average body weight at slaughter (kg) and survival in the sea rearing period (%) of crosses among five Norwegian strains of Atlantic salmon. (Figures in parentheses are average performance of purebred strains.) (Gjerde & Refstie, 1984)

Strain		Body weight			Survival		
		Average of reciprocal crosses	Average of purebred strains	Heterosis[1]	Average of reciprocal crosses	Average of purebred strains	Heterosis[2]
Rana (4.48, 57.2)	× Nidelva (3.86, 31.9)	3.90	4.17	−6.5	50.9	44.6	6.3
"	× Surna (2.97, 57.9)	3.64	3.73	−2.4	51.3	57.6	−6.3
"	× Driva (3.98, 52.7)	3.85	4.23	−9.0	52.1	55.0	−2.9
"	× Eidfjord (3.97, 35.4)	4.34	4.10	5.9	63.7	46.3	17.4
Nidelva	× Surna	3.45	3.42	0.9	47.0	44.9	2.1
"	× Driva	4.10	3.92	4.6	40.5	42.3	−1.8
"	× Eidfjord	4.01	3.92	2.3	38.3	33.7	4.6
Surna	× Driva	3.33	3.48	−4.3	36.7	55.3	−18.6
"	× Eidfjord	3.56	3.47	2.6	53.0	46.7	6.3
Driva	× Eidfjord	3.66	3.98	−8.0	49.1	44.1	5.0
Average		3.78	3.84	−1.6	48.3	47.1	1.2

Notes:

[1] $\frac{\text{Crosses} - \text{Purebred}}{\text{Purebred}} 100\%$

[2] Crosses − Purebred

distant relatives. Records on fullsibs are thus more valuable than records on halfsibs because the individuals share a larger proportion of common genes with their fullsibs than with their halfsibs.

Records on progeny are of particular interest as the breeding value of an individual is strictly defined as the value of an individual judged by the mean value of its progeny.

Selection methods

We distinguish between six different methods of selection of breeding animals, depending on from which individuals or relatives we obtain the information that will be considered in the breeding value estimation. The objective for all methods is to maximize the probability of correct ranking of the potential breeding animals. This is equivalent to maximizing the correlation between the true and estimated breeding value. This correlation (r_{TI}) is frequently termed the accuracy-parameter of the breeding values, and is important because it is directly proportional to the expected response to selection (see 'Response to selection').

What method to choose is dependent on several factors, among which the heritability of the trait(s), the nature of the trait(s) and the reproductive capacity of the species are the most important. A description of each method is given in the following. The relative merits of some of the methods are illustrated in Fig. 8.5.

Pedigree selection

By this method of selection, breeding animals are chosen based on the performance or breeding values of their parents, grandparents or more remote ancestors. However, an individual receives a random sample of half of its chromosomes or genes from

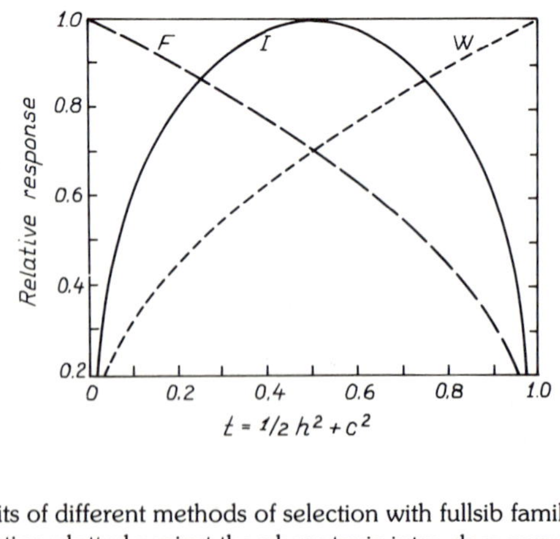

Fig. 8.5 Relative merits of different methods of selection with fullsib families – responses relative to that for combined selection plotted against the phenotypic intra-class correlation (Falconer, 1981).
Notes:
I = Individual selection
F = Family selection
W = Within-family selection

each parent, thus opening a vast number of new chromosome or gene combinations among the offspring. This segregation of genes of each new generation may result in substantial deviation in the breeding values among offspring.

The accuracy of this selection method will therefore not be high. As a result, pedigree selection is infrequently used as the sole method of selection in modern animal breeding.

The accuracy of a breeding value based on both parents' phenotype is equal to $r_{TI} = \sqrt{1/2\,h^2}$, and thus $r_{TI} = 0.35$ for $h^2 = 0.25$, where h^2 is the heritability of the trait in question.

Individual selection

Selection based on an individual's own performance or phenotype is called individual selection. This is a well-known and widely-used method of selection in animal breeding, and for most aquaculture species, the only method practised.

A prerequisite for using individual selection is that the trait(s) can be measured on the breeding individual itself while still alive. The method is thus difficult to practise for carcass quality traits and disease resistance and for age at sexual maturity when the frequency of maturing or immature fish is low (<10–15%). The accuracy of the method is equal to the square root of the heritability of the trait, and thus $r_{TI} = 0.5$ for $h_2 = 0.25$.

In salmonids, individual selection is only of interest for growth rate as the heritability for this trait is fairly high ($h^2 = 0.20–0.40$), and for age at sexual maturity at intermediate frequencies of the trait.

When applying individual selection, it is of vital importance that the environmental influences are kept the same for all individuals that are to be compared at any stage of the life cycle. In this manner, the probability of correct ranking of the individuals can be reasonably high. Differences between individuals or groups of individuals for environmental factors like water temperature and salinity, fish density, light condition, and type of food and feeding regimes, may reduce the accuracy of the selection substantially and thus reduce the possibility for genetic improvement.

To obtain the most equal environmental conditions possible, all individuals that are to be compared should be hatched on the same day, or within a period of a few days, and thereafter reared under identical environmental conditions. This restriction arises because individuals hatched on a different day but recorded at the same age may not actually to comparable due to known or unknown systematic effects which acted upon them during their non-common period.

If we want to select individuals reared in different environments – in different tanks, or in different cages on the same or different farms – then a particular precaution is needed. Due to environmental differences, absolute trait values are not directly comparable. Instead, the mean value for each environmental group should be calculated and the individuals ranked and selected based on their deviations from their group mean. If the genetic level can be assumed to be equal for all groups, we should select an equal number of individuals from each environment or group.

If the environmental groups are significantly different in their mean values, then selection based on deviations will result in more individuals being selected from the group having the highest mean value. This bias is a result of the dependency between the mean and the variance for many traits and frequently observable for body weight. Comparable deviations can be obtained by first multiplying all observations in an environmental group with a constant that is equal to the ratio between the mean value for the group chosen as the base and the mean value for the actual group.

When individual selection is the only selection method practised, it is not usual to tag individuals and thus keep track of the genetic relationship among them. Neither is it possible from a practical point of view to keep many full and halfsib groups separate until time of selection. Therefore, to avoid mating of close relatives (inbreeding), the number of breeding animals should be kept fairly high – 50 or more for each sex.

Family selection

Whole families are selected or rejected as units according to the mean phenotypic value of the family. Individual values are thus not acted on except insofar as they determine the family mean. The families may be of fullsibs or halfsibs, families of more remote relationships being of little practical significance.

The efficiency of family selection rests on the fact that the environmental deviations of the individuals tend to cancel each other out in the mean value of the family. So the phenotypic mean of the family comes close to being a measure of its genotypic mean, and the advantage gained is greater when environmental deviations constitute a large part of the phenotypic variance. Thus, the chief circumstances under which family selection is to be preferred is when the trait selected for has a low heritability.

On the other hand, environmental variation common to members of a family impairs the efficiency of family selection. If this component is large, it will tend to swamp the genetic differences between families and family selection will be correspondingly ineffective.

To reduce the common environmental component to a minimum, the environment for all groups should be standardized as far as possible in the period the groups are kept separate. In addition, individuals from all groups should be tagged as early as possible and thereafter reared together in the same tank, pond or cage.

Another important factor affecting the efficiency of family selection is the number of individuals in the families, or the family size. The larger the family, the closer is the correspondence between mean phenotypic value and the mean genotypic value. The high reproductive capacity in fish make family selection important for these species.

Summarizing up to this point, the conditions that favour family selection are low heritability, little variation due to the common environment and large families.

Another great advantage to family selection is that, based on phenotypic observations on fullsib and/or halfsibs, we can estimate breeding values for traits that cannot be measured on the individuals that are to be used as parents. Carcass quality traits and disease resistance can therefore be included in the breeding objective by applying family selection. Family selection is also far more effective than individual selection for either/or traits such as age at sexual maturity, particularly at low frequencies of the trait.

In order to keep the rate of inbreeding low and the intensity of selection high, the number of family groups bred and measured should not be smaller than 100. In the period prior to marking in which the family groups are to be kept separate, family

Table 8.4 Correlation between true and estimated breeding value (r_{TI}) for a normally distributed trait recorded on n fullsibs and $n \times (m - 1)$ halfsibs and when the environmental variance common to fullsibs (c^2) is 0% and 10% of the total variance. (Figures in parantheses are the correlation when the trait is also recorded on the breeding candidate itself.)

h^2	No. of fullsib groups* (m)	$c^2 = 0.00$, $n = 10$	$c^2 = 0.00$, $n = 50$	$c^2 = 0.10$, $n = 10$	$c^2 = 0.10$, $n = 50$
0.10	1	0.42 (0.48)	0.60 (0.64)	0.33 (0.42)	0.39 (0.47)
	2	0.44 (0.50)	0.61 (0.64)	0.35 (0.44)	0.41 (0.49)
	3	0.45 (0.51)	0.61 (0.64)	0.37 (0.45)	0.43 (0.50)
	4	0.47 (0.52)	0.62 (0.65)	0.39 (0.46)	0.44 (0.51)
	5	0.48 (0.53)	0.62 (0.65)	0.40 (0.47)	0.45 (0.52)
	10	0.50 (0.55)	0.62 (0.65)	0.44 (0.50)	0.48 (0.54)
0.25	1	0.54 (0.65)	0.66 (0.73)	0.45 (0.60)	0.51 (0.64)
	2	0.56 (0.66)	0.66 (0.73)	0.47 (0.61)	0.52 (0.64)
	3	0.56 (0.66)	0.67 (0.73)	0.49 (0.62)	0.54 (0.64)
	4	0.57 (0.66)	0.67 (0.73)	0.50 (0.62)	0.54 (0.65)
	5	0.57 (0.67)	0.67 (0.73)	0.51 (0.63)	0.55 (0.66)
	10	0.58 (0.67)	0.67 (0.73)	0.53 (0.64)	0.56 (0.66)

Note:
* Assuming a nested mating design where each male is mated to m different females.

selection is thus costly of space. If breeding space is limited in this period, the intensity of selection that can be achieved under family selection may be quite small.

The accuracy of family selection is dependent on several parameters: the heritability of the trait (h^2), the family size (n) and family type (full and/or halfsibs), and the variation due to common environment (c^2). Accuracies (r_{TI}) for some parameter combinations are in Table 8.4.

With a family structure of fullsibs only, the upper limit of the accuracy is

$$r_{TI} = 0.71,$$

that is, the square root of the additive genetic relationship among fullsibs, assuming the environmental variance common to fullsibs is equal to zero.

Within-family selection

The criterion of selection is the deviation of each individual from the mean value of the family to which it belongs. This type of selection is the reverse of family selection, the family means being given zero weight. The chief condition under which this method would have an advantage over the others is when there is a large component of environmental variance common to members of a family (Fig. 8.5). Selection within families would eliminate this large non-genetic component from the variation operated on by selection.

An important practical advantage of selection within families, especially in laboratory experiments, is that it economizes on breeding space, for the same reason that family selection is, on the other hand, costly of space.

For selection within fullsib families the accuracy is $r_{TI} = \sqrt{1/2\,h^2}$ and thus $r_{TI} = 0.35$ for $h^2 = 0.25$. This is equal to the accuracy of a breeding value based on both parents' phenotype (see 'Pedigree selection').

Progeny testing

This method of selection is widely applied in breeding programmes of less prolific species such as dairy and beef cattle, sheep and goats. In prolific species like fish, it is of much less advantage since family selection can be applied.

The criterion of selection, as the name implies, is the mean value of an individual's progeny. The progeny families are usually halfsibs. The selection is made among the parents on the basis of their progeny means. Progeny testing suffers from the serious drawback of a protracted generation interval, because the selection of the parents cannot be carried out until the offspring have been measured. Frequently, the generation interval will be doubled. This limitation makes progeny testing difficult or impossible in salmonids because most of the species either spawn only once or suffer high mortality during or after spawning. Deep-freezing of semen is now possible in salmonids (Stoss & Refstie, 1983), but is not sufficiently reliable to be recommended in practice.

Progeny testing is the only selection method for which the accuracy can be 1 (100%). This high accuracy can be obtained with large progeny groups and regardless of the size of heritability.

Combined selection

This method combines in an optimal way, all available sources of information that can add to our knowledge of the breeding value of an animal: it would include information recorded on the animal itself, information about fullsibs and/or halfsibs and progeny as well as pedigree information. It represents the general solution for obtaining the maximum rate of genetic gain, and the other simpler methods are special cases of this method. It is therefore in principle always the best method.

The simpler methods usually combined are family and within-family selection. The within-family deviations are taken into account in addition to the mean phenotypic value of the family. Accuracies (r_{TI}) for some parameter combinations are in Table 8.4.

Genotype by environment interaction

Genotype by environment interaction implies that strains, progeny groups or individuals rank differently when kept under different environmental conditions. Significant genotype × environment interaction in a breeding programme may reduce the response to selection. If the interaction is substantial, a separate breeding population may be needed for each particular type of environment.

Both for Atlantic salmon and rainbow trout, a significant genotype by farm interaction is reported for growth rate when strains and sib groups were reared along the coast of Norway. However, the interaction accounted for a relatively small proportion of the total phenotypic variance. It was therefore concluded that in

Norway only one breeding population of Atlantic salmon and one for rainbow trout is needed (Gunnes & Gjedrem, 1978, 1981).

More serious genotype by environment interactions are, however, reported when sib groups of rainbow trout were reared at quite different temperatures (McKay *et al.*, 1984) and production systems (Sylvén *et al.*, 1991).

In any breeding programme, all testing should take place under commercial production conditions. The possibilities for mistakes are minimized if fish selection is based on data recorded on individuals reared under the same environmental conditions as those of their offspring.

Response to selection

The expected genetic gain (ΔG) or response to selection per year is dependent on four parameters. The formula directly applicable for individual selection is:

$$\Delta G = \frac{(i)(h^2)(\sigma_P)}{L}$$

where

i = The standardized selection differential, also called the intensity of selection. Because of the high fertility in salmonids, a very high intensity of selection can be applied.

h^2 = The heritability of the trait.

σ_P = The phenotypic standard deviation, that is, the square root of the phenotypic variance (V_P). This parameter is a property of the trait and the population, and it sets the units in which the response is expressed, that is, so many kg, cm, percentage units, etc.

L = The generation interval and defined as the average age of the parents at the birth of their selected offspring. This parameter will vary from one species to another and according to management practice. For Atlantic salmon produced from one-year old smolts, the generation interval will be four years when the broodstock is selected after two years in the sea. For rainbow trout, it can be as short as two years, but frequently is three years. It is important to keep the generation interval short to expedite the selection progress.

A more general formula, applicable to all methods of selection is

$$\Delta G = \frac{(i)(r_{TI})(\sigma_A)}{L}$$

where i and L is as described above and

r_{TI} = the accuracy of selection, that is, the correlation between the true and estimated breeding value.

σ_A = the additive genetic standard deviation; that is, the square root of the additive genetic variance (V_A).

The expected response to selection is therefore directly proportional to the size of the accuracy of selection. The efficiency of different methods of selection can therefore be measured as the ratio between their accuracy of selection assuming i, σ_A and L the same for all methods.

Index selection

Selection indices are the most efficient method to combine information about an individual and its relative as well as information about several traits. All information is combined into an index of merit where the traits are weighted according to their relative economic value.

All traits of economic importance should be selected simultaneously. The genetic gain in any one trait by selection based on an index for n equally important and independent (uncorrelated) traits is only $1/\sqrt{n}$ times as much as if selection were directed at that trait alone. However, selection based on an index for n equally important and independent traits is $\sqrt{n}$ times as efficient as selection for the same traits, one at a time.

The Norwegian selection programme for Atlantic salmon

The programme is organized around three main units (see Fig. 8.6):

(1) Two *breeding centres* (The Norwegian Fish Farmers' Breeding Centre A/S and Institute of Aquaculture Research (AKVAFORSK), each having both a freshwater and a marine unit);
(2) Several *marine test stations*, and
(3) Several *marine multiplier stations*.

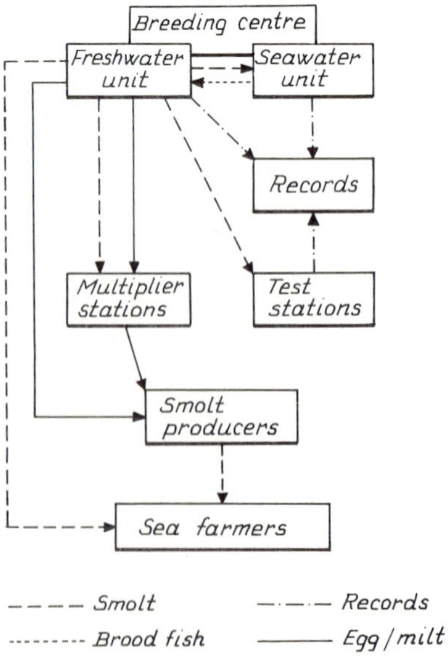

Fig. 8.6 Organization of the Norwegian selection programme for Atlantic salmon.

A breeding council of five members was set up by the Norwegian Fish Farmers' Association, the Norwegian Fish Farmers' Sales Association and the Norwegian Smolt Producers' Association to run the breeding programme (fish farmers constituting the majority in the council).

Breeding centres

The two breeding centres run parallel selection programmes. The centres cooperate, however, to protect the valuable genetic material by now and then sharing it.

The selection method practised is a combination of family and within-family selection. A nested mating design is used where each male is mated to 2–6 females. The two centres have the capacity to test 670 fullsib families each year. About 350 families of Atlantic salmon are produced while the rest of the capacity is used for rainbow trout.

The families are kept in separate units (trays and thereafter tanks) until the fish are big enough (15–20 gm) to be tagged. This occurs about 10 months after the first feeding. The tagging system presently used (a combination of fin-clipping and freeze-branding) has some limitation as it is possible to tag only 120 different families at each breeding centre. Therefore, a preliminary family selection based on growth rate and survival prior to tagging has to be done.

For each family, a sample of 150 fish is tagged and placed together for further rearing in floating net cages at the breeding centre's marine unit. These fish are the potential brood fish for the next generation.

Test stations

For each family, another sample of 100 fish is tagged and placed together for further rearing in floating net cages at the marine test stations placed along the coast from Rogaland county in the south to Finnmark county in the north. Each breeding centre has 3–5 test stations.

Multiplier stations

The industry in Norway has grown to a level where it is impossible to produce enough eggs from fish selected at the two breeding centres. Therefore, multiplier stations are established along the coast. The multiplier stations get smolts from the breeding centres and grow them on to adults and then apply individual selection for body weight before delivering eggs from them to the smolt producers.

The programme is very effective in transferring the genetic gain to the industry because of its hierarchical design and the species' high reproductive capacity. In 1989, about 40 000 litres of Atlantic salmon eggs were produced in the programme.

Data recording

All recordings in freshwater take place at the breeding centres. Among the traits recorded are several survival traits, growth rate, percentage of precocious males, all on a fullsib family basis.

In the autumn of the year after they were transferred to net cages in the sea, the percentage of early sexually maturing fish in each family is recorded at all test stations and at the breeding centres' marine units. Maturing fish are removed from the cages and thus excluded as breeders.

At the test stations, all fish are slaughtered at normal marketing size, i.e. after 16–24 months in seawater. At slaughter, body weight and sex of each fish are recorded.

Selection of broodstock

The selection takes place at the breeding centres' marine unit in June two years after the fish were transferred to seawater. Family selection is practised for survival and growth rate in freshwater and for early sexual maturity, and for body weight recorded on the test stations at slaughter. Males are selected from the 10–15 and females from the 20–30 highest ranking families. In addition, individual selection for body weight is practised within these families. In total, about 150 males and 500 females are selected at each breeding centre. Individual body weight and sex are recorded on all fish and selected fish are individually tagged with a metal tag in the opercular flap.

The selected broodfish are kept in brackish water from September on through the entire mating season.

Data recorded at the breeding centres' marine units are added to the previous data and breeding values for all preliminarily selected broodfish are calculated. At a final selection, the 50–60 highest ranking males and the 200–220 highest ranking females are selected as parents for the next generation. (They are only kept for a single breeding season.)

The known relationships between all individuals in the system permits the calculation of the level of inbreeding for all possible matings. Matings which result in inbreeding coefficient above a stated limit are thereby avoided. If inbreeding results in too many mating restrictions within a year class, the system gives possibilities for mating between unrelated fish from different year classes (distinct lineages). Eggs from selected fish not used for making new progeny groups are sold to commercial smolt producers.

Breeding values for families and for selected individuals are estimated using single-trait-best-linear-unbiased-prediction (BLUP, Henderson, 1973) with the additive genetic relationship among sires and dams included. The traits are weighted together into an overall genetic merit (index) on which the animals are selected.

Disease and health control

A very strict disease and health control is established to secure the delivery of disease-free material to the industry. The breeding programme is safeguarded by disease control precautions. Special attention is paid to minimize the risk for transferring disease agents to the breeding nucleus through water sources, fish, eggs and persons, and from the nucleus to the industry as the improved material is disseminated. This procedure involves an extensive test and diagnosis of each parent fish for several disease agents including infectious pancreatic necrosis (IPN), bacterial kidney disease (BKD) and furunculosis (*Aeromonas salmonicida*). The same tests/diagnoses are performed on a sample of fingerlings from each fullsib group 2–3 months after first feeding. Groups with a positive test/diagnosis are discarded from the programme.

At the multiplier stations, similar tests/diagnoses are performed on a sample of the broodfish.

Possibilities

The system allows for the estimation of genetic parameters for new traits or recalculations of genetic parameters for traits already included in the breeding programme. It also permits new traits to be easily included in the selection programme. Body quality traits, for example, can easily be recorded at the test stations. Criteria for body quality traits are, however, still under discussion.

Selection for better disease resistance is difficult because of the difficulties encountered when measuring this trait. The freeze-branding cannot be read on dead fish and cause of death of individual fish is not recorded. To develop a better tagging system with a reasonable cost is therefore important. Identification of dead fish, together with cause for mortality will give better selection criteria than percentage survival for each family.

Economic aspects

In general, investment in selection programmes gives very high returns, expressed in terms of interest on invested capital. Thus, in sheep, pig and beef cattle production, full investment appraisals have shown net benefit/cost ratios ranging from 5/1 to 50/1 (Barlow, 1982; Mitchell *et al.*, 1982). High ratios are to be expected in salmonids, particularly due to the species' high fecundity and to the high phenotypic and genetic variance observed for economically important traits.

References

Barlow, R. (1982) Benefit-cost analyses of genetic improvement programmes for sheep, beef cattle and pigs in Ireland, PhD thesis, University of Dublin.

Bentsen, H.B. (1990) Application of breeding and selection theory on farmed fish. *Proc. 4th World Congress on Genetics Applied to Livestock Prodn*. Edinburgh, 23–27 July 1990, XVI: 149–58.

Carlin, B. (1969) Salmon tagging experiments. Laxforskningsinstitutets Meddelande 2–4, Swedish Salmon Research Institute, Technical Report, pp. 8–13.

Chevassus, B. (1983) Hybridization in fish. *Aquaculture*, 33, 245–62.

Chevassus, B. and Dorson, M. (1990) Genetics of resistance to disease in fishes. *Aquaculture*, 85, 83–107.

Falconer, D.S. (1981) *Introduction to Quantitative Genetics*, 2nd edn, Longman Group, London and New York.

Fjalestad, K. (1990) Possibilities to include disease resistance in the Norwegian breeding program for salmonids, *Proc. 4th World Congress Applied to Livestock Prodn*, XVI, 461–4.

Gall, G.A. (1990) Basis for evaluating breeding plans. *Aquaculture*, 85, 125–42.

Gjerde, B. (1986) Growth and reproduction in fish and shellfish. *Aquaculture*, 57, 37–55.

Gjerde, B. (1988) Complete diallel cross between six inbred groups of rainbow trout,

Salmo gairdneri. *Aquaculture*, 75, 71–87.

Gjerde, B. (1989) Body traits in rainbow trout. I. Phenotypic means and standard deviations and sex effects. *Aquaculture*, 80, 7–24.

Gjerde, B. and Refstie, T. (1984) Complete diallel cross between five strains of Atlantic salmon. *Livest. Prod. Sci.*, 11, 207–26.

Gjerde, B. and Schaeffer, L.R. (1989) Body traits in rainbow trout. II. Estimates of heritabilities and of phenotypic and genetic correlations. *Aquaculture*, 80, 25–44.

Gjerde, B., Gunnes, K. and Gjedrem, T. (1983) Effect of inbreeding on survival and growth in rainbow trout. *Aquaculture*, 34, 327–32.

Gjedrem, T. (1983) Genetic variation in quantitative traits and selective breeding in fish and shellfish. *Aquaculture*, 33, 51–72.

Gjedrem, T., Salte, R. and Gjøen, H.M. (1991) Genetic variation in susceptibility of Atlantic salmon to furunculosis. *Aquaculture*, 98, 349–54.

Gunnes, K. and Gjedrem, T. (1978) Selection experiment with salmon. IV. Growth of Atlantic salmon during two years in the sea. *Aquaculture*, 15, 19–33.

Gunnes, K. and Gjedrem, T. (1981) A genetic analysis of body weight and length in rainbow trout reared in seawater for 18 months. *Aquaculture*, 24, 161–74.

Henderson, C.R. (1973) Sire evaluation and genetic trend. In *Proceedings of Animal Breeding Genetics Symposium*, Blacksburg, Virginia. American Society of Animal Science, Champaign, Illinois, pp. 10–41.

Isaksson, A. (1988) Salmon ranching: a world review. *Aquaculture*, 75, 1–33.

Isaksson, A. (1989) Stocking and ranching – a general review. In *Aquaculture – a biotechnology in progress* (Ed. by N. De Pauw, E. Jaspero, H. Ackefors, N. Wilkins), European Aquaculture Society, Bredene, Belgium.

Kinghorn, B.P. (1983) A review of quantitative genetics in fish breeding. *Aquaculture*, 31, 283–304.

McKay, L.R., Friars, G.W. and Ihssen, P.E. (1984) Genotype × temperature interactions for growth of rainbow trout. *Aquaculture*, 41, 131–40.

Mitchell, G., Smith, C., Makower, M. and Bird, P.J.W.N. (1982) An economic appraisal of pig improvement in Great Britain, genetic and production aspects. *Anim. Prod.*, 35, 215–24.

Refstie, T. (1980) Genetic and environmental sources of variation in body weight and length of rainbow trout fingerlings. *Aquaculture*, 85, 351–7.

Refstie, T. (1990) Application of breeding schemes. *Aquaculture*, 85, 163–9.

Refstie, R. and Steine, T. (1978) Selection experiment with salmon. III. Genetic and environmental sources of variation in length and weight of Atlantic salmon in the freshwater phase. *Aquaculture*, 14, 221–34.

Rye, M. and Gjerde, B. Phenotypic and genetic parameters of body composition traits and flesh colour in Atlantic salmon. *Aquaculture*, submitted.

Rye, M. and Refstie, T. Phenotypic and genetic paramaters of body size traits in Atlantic salmon, *Aquaculture*, submitted.

Stoss, J. and Refstie, T. (1983) Short-term storage and cryopreservation of milt from Atlantic salmon and rainbow trout. *Aquaculture*, 30, 229–36.

Sylvén, S., Rye, M. and Simianer, H. (1991) Interaction of genotype with production system for slaughter weight in rainbow trout (*Oncorhynchus mykiss*). *Livestock Prod. Sci.*, 28, 253–63.

Chapter 9
The Regulation of Salmon Aquaculture: an International Overview

RICHARD SCHWINDT WITH TROND BJØRNDAL

Introduction

This chapter deals with government regulation of the salmon aquaculture industry. The intention is not to provide an encyclopaedic listing of regulations bearing upon the industry in any particular jurisdiction. Rather, the objective is to show how current government policies in a number of countries 'fit' into general explanations for government intervention in markets.

The exercise is worthwhile for several reasons. It will provide non-specialists with an understanding of the economic rationales for extant regulatory regimes and might well provide the basis for predicting the way in which regulations will develop. It will also provide a basis for evaluating public policy toward this industry. This latter point deserves elaboration. In many jurisdictions, public policy toward fisheries has not been particularly effective. This is especially true of the management of the wild stock salmon fishery which, at least in Pacific Canada, has been in a state of 'crisis' for more than a decade. Admittedly, policymakers have been hampered by seemingly intractable property rights, disputes between nations, between commercial and recreational fishers, between gear types within the fishery, between fishery interests and other extractive industries, and between current and aboriginal rights holders.

Salmon aquaculture, because it is so very new, presents policymakers with a relatively clean slate. Property rights are not terribly muddled and policymakers are not pressured by an imminent collapse of the stock. It is possible that this new industry, which has elements of both a fishery and an agricultural activity, will benefit from a rational, effective regulatory environment.

Why regulate?

Wise regulation in market economies is predicated upon failures of the unregulated market to achieve desired levels of performance. These performance goals include economic efficiency, economic stability, equity and broad political and social objectives. The regulation of salmon aquaculture is influenced by the pursuit of each of these goals.

Economic efficiency

Economic efficiency means the production of the right things in the right quantities using the least amount of resources. Markets can fail to achieve economic efficiency for a host of reasons including imperfect information, monopoly power, ambiguous property rights, and externalities. In the case of salmon aquaculture, regulation is largely a response to perceived potential negative externalities. Therefore the concept deserves some explanation.

Externalities, also called spillovers, occur when someone is affected either positively or negatively by a transaction while not being a party to the transaction. The concept is best explained by way of example. In the production of electricity, coal-fired generating stations discharge pollutants which result in acid rain. The electric company charges prices to its customers which reflect its costs, but it ignores the costs imposed on distant communities which suffer the consequences of the acid rain. As a result, the price of electricity is too low and too much electricity is produced and consumed for economic efficiency.

The most obvious method of correcting an externality is to bring the third party into the transaction. However, this can be difficult (often impossible) because of two factors which characterize externality situations. First, property rights are generally unclear. In the example, it is not obvious that the electric company has the 'right' to emit effluents, nor that distant communities have the 'right' to acid-free rain. Without clear property rights it is difficult to integrate the third party into the transaction; there is no basis for determining who should be buying and who should be selling. Should the afflicted community pay the coal burner to curb emissions, or should the coal burner pay the community for damages inflicted?

The second impediment to creating a market which includes the third party is the outright cost of setting up the transaction. In cases of negative externalities, where effects are diffuse and nebulous, the private costs for a third party victim to document, quantify and collect for damages often exceed the potential gains. Simply, it does not pay the victim to become part of the transaction. In the example, even if it were established that communities had the right to clean air, how could this right be enforced through market transactions? The victims would have to establish who was responsible for specific volumes of pollution, meet with them, determine compensation, and enforce the contract. Clearly, the transaction costs would be very high, probably prohibitively so.

These observations lead to a general conclusion. The vaguer the property right and the more diffuse the affected third party group, the less likely that the market can resolve the externality problem. These two factors, vague property rights and a diffuse third party, are of course not mutually exclusive. Explicit property rights are more likely to exist when the effect of a transaction has a direct. significant, observable impact on a well-defined group.

The failure of the market to remedy an externality does not ensure that government will intervene with corrective regulations. Governments, like markets, respond to supply and demand conditions. In general, when the government is more likely to intervene, the more apparent is the spillover and the more cohesive is the affected third party. An example will clarify this. A proposal to open a pig farm in an urban residential area has little chance of success. The spillover is obvious – odour. The affected group is well defined and numerous – the neighbours. The same farm located in a rural area will generally be acceptable. The spillover is not so clear (the odour is dispersed), and the third party will not be unified (some will be farmers creating similar spillovers).

The clarity of a spillover depends upon its certainty and its significance. An unambiguous spillover with limited significance (e.g. a small dam which floods a single river valley) might be ranked by policymakers with an uncertain spillover which, if it occurred, would have calamitous results (e.g. a reactor meltdown). The clarity of a spillover can be thought of as running along a continuum from clear (significant and probable), to vague (insignificant and improbable).

The strength of a third party group depends upon the number of its members, its organization and articulateness. A very large but unorganized and inarticulate third party might achieve the same attention from policymakers as a small but well organized and vocal group. The strength of the third party can also be thought of as running along a continuum, in this case from strong (large, well-organized and articulate) to diffuse (small, unorganized and inarticulate).

Externalities can be placed along these continuums as reflected in Fig. 9.1. Spillovers with a clear, serious impact on a very cohesive third party would fall in the area close to the origin (point A). At the other extreme, ambiguous spillover effects impinging on a small, diffuse third party would lie at the other extreme (point B). The figure can also be used to get a sense of when regulation will be imposed. The shaded area suggests the combination of characteristics which will bring spillovers under a regulatory regime. Clear externalities affecting a strong third party are very likely to trigger regulation while vague impacts affecting diffuse third parties are not. However, vague spillovers affecting a very strong third party and clear, serious spillovers affecting diffuse third parties might also generate a regulatory response. For example, a large, well organized, articulate group may be able to lobby for regulation in response to an unsubstantiated threat. Similarly, a very clear threat to a diffuse group may move benevolent policymakers to impose regulations.

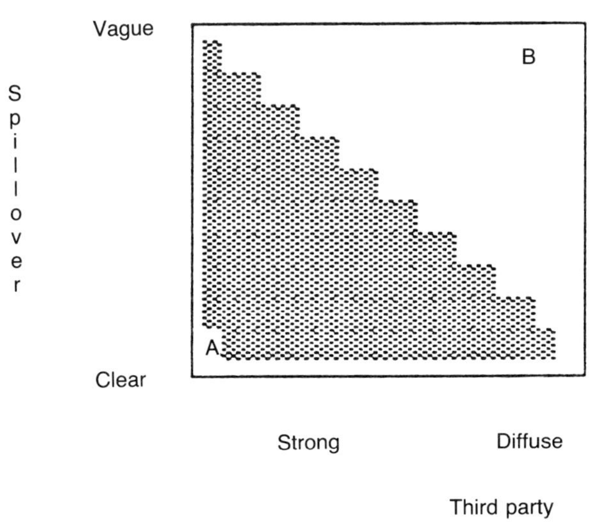

Fig. 9.1 Spillover effect of third parties on salmon farming regulations.

Before leaving the topic of externalities, it is important to note that there is a certain class of third party effects which does not result in the misallocation of resources. When markets respond to changing demand and supply conditions, they create winners and losers in the process. For example, the introduction of the automobile had negative impacts on manufacturers of horse-drawn carriages and positive effects on the petroleum industry. Such effects are known as pecuniary externalities and are the unavoidable result of market change. They are *not* evidence of market failure and therefore normally do not justify government regulation. However, third parties suffering negative pecuniary externalities or enjoying positive ones can be expected to petition government to protect their positions.

Social and political goals

The rationale for government intervention in the pursuit of social and political goals is based upon the failure of markets to automatically achieve socially and politically acceptable results. Unfettered markets will necessarily result in unequal distributions of income. Markets are responsive to changes in supply and demand conditions, which is good, but that responsiveness results in economic instability which is often viewed as undesirable. While markets generally are very effective in the allocation of resources, they do not guarantee achievement of other broader goals such as political, economic or cultural sovereignty.

Regulation: correcting externalities

The discussion of externalities associated with salmon aquaculture requires some familiarity with the rearing process, the effluents generated by the process, and the potential deleterious effects these effluents can have. The potential impacts of a farm on the marine environment are discussed in Chapter 5 and set out pictorially in Fig. 5.1.

The cultivation of large numbers of salmon in a confined pen can generate a substantial buildup of both wasted feed and faeces on the sea floor immediately below the pens. This buildup of organic material has several detrimental effects. It can generate gases (e.g. hydrogen sulphide) which are toxic to the penned fish (in Norway a farm's entire crop was destroyed in this manner) and can contribute to the growth of the phytoplankton which cause lethal (to penned salmon) toxic blooms.

There is also the potential for discharge of pesticides into the seawater: antifoulants are used on nets to control the growth of fouling organisms and the wood used in pen construction is usually treated to slow deterioration. In both cases, these pesticides can leach into the sea.

Salmon farms can also add considerably to solid waste management problems. Farms anticipate normal mortalities in their stock. The volume of these mortalities approximates 3–4% of the weight of the finished product. Methods of disposing of this often repugnant fish waste have included public and private landfill, dumping at sea, and incineration. In many instances the use of public waste management facilities puts pressure on smaller, more remote communities ill-equipped to handle the additional volumes of waste. Private landfills have created problems of surface water pollution, while incineration generates air pollution. Disposal at sea raises wider environmental concerns. Also the farms generate normal household and commercial garbage and sewage.

The establishment of fish farms also results in land-use conflicts. Given that other physical requirements are satisfied (e.g. wave flushing action), farm operators prefer sheltered sites close to infrastructure. This can lead to conflict with marine transportation, other commercial activities and local residents. The negative spillovers include the actual obstruction of marine transportation, the reduction of coastal sites for recreationalists, the obstruction of access to waterfront property and visual degradation of the shoreline. Even when farms are located far from population centres there can be land-use conflicts. Farms reduce the number and quality of pleasure and fishing vessel remote moorage sites and generally degrade wilderness qualities.

It is also in the nature of fish farming that fish escape. In areas where there are wild salmon, this raises questions about the effects of intermingling domestic with wild stocks. There are serious concerns over the potential spread of disease from the farmed to the wild salmon, and over the consequences of interbreeding. Predictably, in those regions where the wild salmon fishery is economically important, these concerns have been strongly voiced.

The risk of disease among penned salmon has led to the widespread use of therapeutic additives (e.g. antibiotics, fungicides) in salmon feed. This practice, which is certainly not unique to salmon aquaculture, raises two concerns. Waste feed

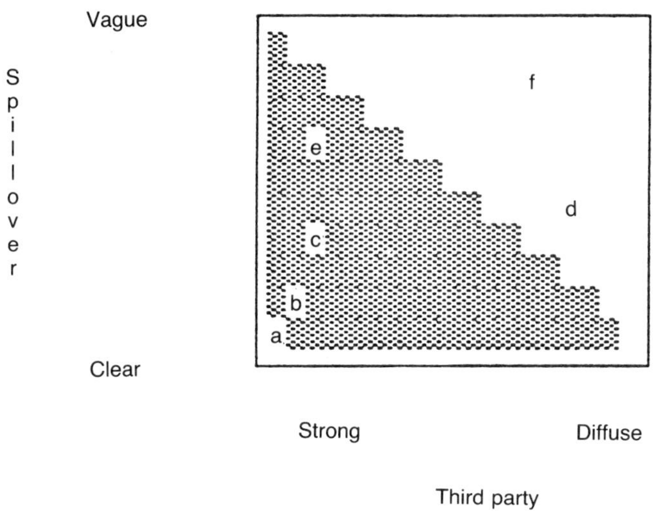

Notes: *Spillover* / *Third party*
a Navigation obstruction — Marine traffic
b Infection/toxic waste — Other aquaculture
c Coastal amenities — Coastal residents
d Marine recreation — Recreationalists
e Wild salmon — Salmon fishery
f Predators — Wildlife conservationists

Fig. 9.2 Spillover effect of salmon farming on specific third parties.

introduces these additives into the marine environment with unknown effects, while additive residues may remain in the slaughtered fish with unknown effects on human consumers.

Finally, salmon farms attract natural predators which include seals, killer whales and eagles. Farms employ predator controls to reduce losses. Wildlife protectionists have argued that these controls threaten protected species.

These are spillovers which have been identified with salmon aquaculture. In order to organize the discussion of regulation, these spillovers have been classified as 'clear' or 'vague' and the affected third parties have been classed as 'strong' or 'diffuse'. The major areas of concern, organized by their spillover and third party characteristics, are set out in Fig. 9.2. This categorization admittedly is subjective, not exhaustive, and certainly not applicable to all jurisdictions.

Clear spillover/strong third party

Clear spillovers affecting cohesive third parties are most likely to be the object of regulation. Three such spillovers are discussed here.

Navigation obstruction/marine traffic

Salmon aquaculture raises novel property rights issues in part because it involves the allocation of coastal resources which previously were not dedicated to an exclusive use. With the exception of land-based operations, the grow-out phase is done in semi-permanent sea pens which exclude any other use at the site. When the proposed location of a planned farm conflicts with established, well-represented users there are generally regulations in effect to curb the establishment of the farm.

Probably the most clearly established property rights over coastal waters are those for navigation channels and waters adjacent to ports and harbours. *Every* jurisdiction reviewed requires that navigation authorities formally provide a clearance for the location of all proposed salmon farms. (The usual procedure is that all farm applications go through a referral process which includes navigation authorities. For example, in British Columbia, applications must be approved by the Coast Guard and Transport Canada; in Chile the Department of the Navy monitors locations. In Scotland, the Department of Transportation, in consultation with yachting associations, the Northern Lighthouse Board, coastguards and harbour authorities decides on navigational issues. In Ireland the Department of Communications has this responsibility.)

Disease/other aquaculture

Salmon farm operations generate effluents which can have adverse effects on other aquaculture activities. Domestic sewage discharge can contaminate shellfish, and salmon diseases can be spread from farm to farm. Disease control is of paramount concern to farm operators. Penned fish, particularly when densely stocked, are extremely susceptible to infection by lethal viral and bacterial pathogens for which there is no known therapy. Once infected, contagion can be rapid and farms have been known to lose entire stocks.

Given this very real and severe threat to adjacent aquaculture activities, most jurisdictions, at the behest of established farms, set minimum distances between farms and accommodate the concerns of other aquaculture operators during the licence application process. Interestingly, in those jurisdictions where licensing procedures were lax, specifically British Columbia and Chile, the industry asked for regulations to control the proximity of farms because densities were perceived to have become dangerous.

Coastal amenities/adjacent coastal residents

The establishment of salmon farms can have direct, detrimental effects on adjacent residents. These effects include loss of access to tidelands, loss of privacy, loss of moorage, degradation of water quality, noise pollution, odour and degradation of visual aesthetics. Spillovers are both straight externalities and pecuniary externalities in that some of these impacts undoubtedly have adverse effects on property values.

Predictably, regulation to protect coastal amenities will be more stringent the more numerous, vocal and politically involved are the affected residents. When farms are established in more densely populated areas, they are usually subjected to more stringent local controls. Every jurisdiction reviewed has some form of referral process for fish farm licence applications, and it generally incorporates referral to local planning authorities. The stringency of these local regulations varies widely.

In Chile the Department of the Navy handles the application referral process and consults with municipal authorities over possible land use conflicts. Very few applications for marine concessions are denied. This is probably attributable to a *laissez-faire* government attitude and the fact that most Chilean farms are located in relatively remote, sparsely populated areas.

At the other extreme is the rigorous Washington State approval system. Development of the coastal areas of Washington State is governed by the Shoreline Management Act (SMA) and the State Environmental Policy Act (SEPA). SMA sets out guidelines for the development of master plans to be created by local governments with extensive public hearings. Once approved by state and federal authorities, power is shifted back to local governments who grant development permits in conformity with the plan.

> Opposition to aquaculture proposals in Washington has been most effective at the local level, where most permits are issued. Obviously, elected officials ignore public opposition at their peril. Some local governments have taken the step of calling moratoriums on net-pen development.
>
> (British Columbia, Ombudsman, 1988, p. 54)

In British Columbia, the perceived lack of local input to the licensing process provoked enough complaints to trigger a review by the Provincial Ombudsman. While local and regional authorities do have the right to zone their foreshores for specific purposes, aquaculture licences were granted by the provincial government before local authorities had the chance to establish regulations dealing with fish farms. Once in place, the farms were effectively immune from subsequent zoning or community planning (British Columbia, Ombudsman, 1988, p. 52). As a result of public opinion the provincial government slowed the licensing process and local authorities moved to implement plans to deal with aquaculture.

Vague spillover/strong third party

A major, yet unsubstantiated potential spillover involves the effect of salmon aquaculture on the wild stock salmon fishery. It is feared that fugitive pen-reared salmon might either infect the wild stock with disease or interbreed, thereby weakening the genetic stock.

There has been extensive transplantation of salmonid stocks over the years, generally for the purpose of introducing new wild stock species to an area. It is known that these transplanted stocks have carried with them pathogens and parasites which have infected both domestic and cultured stocks (Rohovec *et al.*, 1988, p. 171). To date the direction of infection has been mainly from contaminated, transplanted wild stock to indigenous wild stock and farmed stock, although there is evidence from Norway that *Gyrodactilus* can be spread from farm to wild stocks.

The potential impacts of inbreeding are less clear. Genetic selection is practised extensively in salmonid culture although there are widely divergent views on its efficacy (Hershberger, 1988, p. 177). Wild stock advocates argue that intermingling of the stocks could lead to an adverse reduction of genetic diversity in wild salmon. It is important to note that should pollution of gene material occur, it is irreversible. Once a stock is destroyed, the gene material cannot be recovered. There is also a risk of displacing the wild stock. In Norway, fugitive farmed salmon represent up to 40% of the run in some rivers, although it must be noted that the Norwegian wild stock was extremely weak when farming commenced (Glavin, 1989, p. C1).

Both disease contagion and inbreeding could have deleterious effects on wild salmon, and therefore on the salmon fishing industry. Furthermore, salmon aquaculture can impose a negative pecuniary externality on the wild stock fishery. As farmed salmon production increases, the supply of all salmon will increase and ultimately this will result in a decline in prices for all salmon.

Wild stock salmon fisheries are very important in both Alaska and British Columbia where they account for a substantial proportion of regional employment. In both areas the fishing industry qualifies as a well organized, cohesive, vocal and politically astute interest group. Not surprisingly these interest groups have lobbied for regulation of fish farming on the basis of potential negative externalities. Little is said publicly about negative pecuniary externalities.

Lobbying efforts have been very successful in Alaska. Salmon farming in sea pens is prohibited under state law. In early 1986 in British Columbia, the United Fisheries and Allied Workers Union (UFAWU) argued strongly against the rapid, nearly unconstrained allocation of farm licences which was going on at the time. In response the provincial government imposed a moratorium on the issuance of licences and established an official inquiry in October 1986. Subsequently a report was issued recommending more stringent licensing rules. These were acted upon and the moratorium was lifted. (The inquiry was chaired by David Gillespie and issued the report titled 'An inquiry into finfish aquaculture in British Columbia', 12 December 1986, British Columbia Ministry of Forests and Lands.) Presently there is a growing division between the harvesting and processing sector of the wild salmon industry in British Columbia with respect to the aquaculture issue.

Understandably, in other producing regions without significant wild salmon fisheries, concerns about negative impacts on wild salmon have been fewer. However, there are exceptions. Norway has now taken measures to protect its limited stock of wild salmon. In 1989, farm-free zones were introduced in several river estuaries to avoid the intermingling of fugitive farmed fish with the migrating wild stock. The zones are established for five years during which time research will evaluate the actual threat to the wild stock.

Before leaving the topic of the wild stock fishery, it should be noted, emphatically, that most jurisdictions rigorously regulate the importation of salmon eggs and smolts in the interest of disease control. This is of direct benefit to the salmon aquaculture industry in that it reduces the probability of purchasing infected stocks and it lowers the risk of infecting existing stocks. And, insofar as disease can be transmitted from farmed to wild fish, it also protects the wild stocks.

Vague spillover/diffuse third party

The displacement of recreationalists from coastal areas by fish farms provides a good example of a negative externality suffered by a diffuse third party. The establishment of a fish farm obviously reduces the number of sites for pleasure craft moorage, recreational fishing, diving, and coastal wilderness camping. However, these interest groups (with the possible exception of sports fishermen) are relatively unorganized and enjoy only a very tenuous property right to the use of these sites. Moreover, the spillover is viewed as an inconvenience and not a serious threat to health or economic wellbeing. As a result, in most jurisdictions, these interest groups have not successfully moved governments to protect their claims.

The effect of fish farming on the natural predators of salmon also qualifies as a vague spillover. As yet there is no evidence that predators poaching from the pens suffer any ill-effects, nor is there evidence that predator controls harm these stocks. The interest group in this case, wildlife protectionists, tends to be issue oriented and has not focused on farming practices. The farms are subject, of course, to statutes protecting certain species of wildlife. To date these appear adequate as no wildlife protection regulations specific to salmon aquaculture have been enacted.

Regulation: pursuit of social and political goals

A review of public policies towards salmon aquaculture shows that, without exception, the industry is being supported to achieve broader policy objectives. The common underlying goal is the stimulation of regional economic development. It can be argued, perhaps with some risk of over-generalization, that regions ideally suited for aquaculture are less than ideally suited for other rural economic activities, in particular, agriculture. In Europe, many of the most desirable salmon farm sites are in areas which have depended on the cultivation of marginal agricultural lands, a small-boat fishery or a combination of the two. Communities in these areas have suffered out-migration and unemployment or under-employment and have depended upon sustained public income assistance. Public agencies, both national and supranational (i.e. the EC) view aquaculture as a viable alternative economic activity in these regions and have subsidized it accordingly. Additionally, policy has been implemented in some jurisdictions to achieve more specific social and political goals. Nowhere is this clearer than in Norway.

The Norwegian Ministry of Fisheries severely restricts entry into the industry, allocating licences irregularly and in numbers well below demand. As an example, in 1985–86 there were over 2500 applicants for the 150 available licences (Bjørndal, 1989). Entry is further restricted through regulatory constraints which limit pen size to 12 000 cubic metres per farm. Further regulations restrict ownership in that no person or firm can hold a majority interest in more than one farm. Foreign ownership is indirectly precluded. The stated objective of these policies is to encourage and maintain a locally controlled, single proprietor, fragmented industry structure. Further, licence availability is tied to specific regions. The objective of this policy is to encourage balanced regional growth. Finally, licences are awarded on the basis of the applicant's qualifications (i.e. need, training, background) not on the qualities of a specific farm site. In short, licence allocation in Norway is predicated upon broad political goals such as diffusion of economic power, equitable regional development and the maintenance of population in more remote coastal areas.

The pursuit of these social objectives is in fact costly for the industry. Ceilings on farm size have apparently been set below efficient scale, and restraints on ownership and location of farms have in some instances channelled investment away from the most productive operators and sites (Salvanes, 1988). However, ceilings on farm sizes have been gradually increased (from an original 3000 cubic metres to a current level of 12 000 cubic metres) and there is pressure for further increases.

The Japanese have also imposed ownership regulations in that the licensing of sea-pen operations is restricted to individual fishermen belonging to fisheries co-operatives. In part this policy was enacted to provide employment for those displaced from the high-seas fishery when other coastal nations imposed 200-mile economic zones. Excepting Norway and Japan, the major producing nations do not restrict ownership. Direct foreign investment is welcomed for the most part and there are no industry specific restraints on ownership concentration.

However, other jurisdictions do pursue social and political goals which are equally idiosyncratic, although this is generally done through incentive schemes rather than through direct regulation. In Ireland the Udaras na Gaeltachta, an agency charged with the economic, cultural and linguistic development of the Irish-speaking areas in Ireland, provides incentives to salmon farms established in these regions (An Bord Iascaigh Mhara, 1986, p. 92). In Scotland the Highlands and Islands Development Board provided preferential funding for small crofters entering the industry. In Chile salmon farming benefits from general policies to promote foreign investment and development in the southern regions. There, aquaculture qualifies for a programme whereby foreign investors can purchase Chilean foreign debt at a discount and then convert it at face value into local currency for investment in Chile. Additionally, for economic and geopolitical reasons the Chilean Government seeks development of its southern region. British, Chilean and Argentine claims overlap in Antarctica, and Argentina has been actively maintaining its claim. The Chilean Government believes that by extending its road system to the south and by encouraging economic development in its southern regions its own claims will be strengthened.

In summary, it is difficult to set out a general explanation for socially and politically motivated government intervention in salmon aquaculture. Aside from the pursuit of economic development, motivations range from the pursuit of national sovereignty to the encouragement of a specific culture.

Conclusion

Overall, the regulation of externalities associated with salmon aquaculture is about what would be expected. Where spillovers are significant and/or third parties are politically effective, regualtions have been imposed. There are no glaring errors of omission nor commission.

On the other hand, regulation in the interests of social and political goals is idiosyncratic. There is evidence, particularly from Norway, that this type of regulation has compromised the economic efficiency of the industry.

References

An Bord Iascaigh Mhara (1986) *The Atlantic Salmon Farming Industry*, Dun Laoghaire, An Bord Iascaigh Mhara.

Bjørndal, T. (1990) *The Economics of Salmon Aquaculture*, Blackwell Scientific Publications, Oxford.

British Columbia, Ombudsman (1988) *Aquaculture and the Administration of Coastal Resources in British Columbia, Public Report No. 15*, Victoria, Queen's Printer for British Columbia.

Glavin, T. (1989) Odd salmon escapers jolt netters. *Vancouver Sun*, 14 February, p. C1.

Hershberger, W.K. (1988) Directed and inadvertent genetic selection in salmonid culture. In *Salmon Production, Management, and Allocation* (Ed. by W.J. McNeil), pp. 177–86, Corvallis, Oregon, Oregon State University Press.

Rohovec, J.S., Winton, J.R. and Fryer, J.L. (1988) Potential hazard for spread of infectious disease by transplantation of fish. In *Salmon Production, Management, and Allocation*, (Ed. by W.J. McNeil) pp. 171–76, Corvallis, Oregon, Oregon State University Press.

Salvanes, K. (1988) Salmon aquaculture in Norway: an empirical analysis of cost and production properties. PhD dissertation, Norwegian School of Economics and Business Administration, Bergen.

Chapter 10
Comparative Analysis of the Cost Structure and Profitability in the Salmon Aquaculture Industry

KNUT HEEN

Introduction

Salmon aquaculture for food production is based on a complicated interaction among biological, technological, organizational and economical factors. The farmer must coordinate this interaction process in an efficient way to develop and maintain a competitive position.

The aim of this chapter is to identify and analyse the main cost components in salmon aquaculture and thereby isolate the factors which are critical for efficiency and profitability. Salmon aquaculture can be practised using different species and different levels of technology. Most regions specialize in the species for which they have natural salmon runs. In Europe and the east coast of North America, only Atlantic salmon are grown, while in Japan and the west coast of North America, mainly Pacific salmon are grown. On the west coast of North America, however, farms also grow Atlantic salmon. In areas where there are no natural salmon runs, such as in the southern hemisphere, salmon have been introduced by importing eggs of different species.

The main technology in use today is open sea rearing cages. The cost structure analysis will mainly be based on this technology. The economic implications of land-based farming will also be investigated and discussed. In an analysis of future developments, the perspective of this technology must be included.

In an attempt to discuss the future development of salmon aquaculture by region, the natural and institutional factors as well as transportation costs to the main markets must be included in the analysis.

This chapter proceeds by first introducing the concepts of production functions and cost functions and explaining how these concepts can be used in a fruitful way to reveal the economic characteristics of salmon aquaculture. Thereafter available data on cost structure are presented. Finally, the competitive advantages by region are discussed.

The production process

The analysis of the cost structure will initially focus on the technical side of the production. The production function describes the technological relationship between various levels of factor inputs and the product output. The production function in salmon aquaculture can be written as:

$$x = f(v;e,i) \tag{1}$$

where x is the output, v is the input vector, e is the vector of environmental factors, and i is the vector of institutional factors.

Expression 1 states the maximum output (x) of different input combination (v) given the natural environment factors (e) and the institutional factors (i). In salmon aquaculture, the factor inputs are smolt, feed, labour, capital and others. The output is obviously the salmon itself. The environmental factors are sea temperature and oxygen content, water salinity, water circulation, underwater topography, protections from winds and waves, etc. These factors vary from site to site. The institutional factors include the various government regulations in the industry, for example, limit on size, import restriction on eggs, etc.

As the aim of this chapter is to compare and analyse the competitive advantages by technology, species and regions, it is more useful to think in terms of costs of production rather than in physical terms. The cost function of salmon aquaculture in a particular region can be written as:

$$c_r = f_r(x;q,e,i) \tag{2}$$

where c_r is the total production costs in region r and q is the vector of factor prices.

The emphasis so far has been on the production process and production costs. In an analysis of the profitability, transportation costs and prices of the final product would have to be included.

In the comparison by region, we shall assume that each region has its own set of factor prices. There is also reason to believe that the technology might vary from region to region because of differences in prices of factor inputs.

(For readers with an interest in production and cost theory in economics, the Appendix to this chapter will tie the discussion of the cost of production to the concepts of microeconomics.)

Cost structure by technology, species and region

The aquaculture industry can be divided into production of broodstock, egg, smolt and finally the farming itself. In the comparative analysis of the cost structure of this industry, the focus will be on the farm. It would have been desirable to analyse the cost structure of each stage of the aquaculture process. In this chapter, however, space will not allow us to do so. This will not cause any serious problems in the evaluation of future prospects of the industry because what is output in the first stage is input in the next. Analysing the cost structure in the final stage, the costs of the previous stage are reflected in the price of the inputs. The price of smolt in the farming industry will reflect the cost and efficiency in the broodstock, egg and smolt activities. It should, therefore, be sufficient to study the cost structure of farms. In addition, it is a fact that farming is the dominant activity in terms of value added.

Organizing the chapter as indicated in the title of this section, presents some

Table 10.1 Production cycle.

Year	1	2	3
	a Smolt released in $\frac{1}{3}$ of the pen capacity.	a 1st year smolt half through production b Smolt released in $\frac{1}{3}$ of the pen capacity.	a 1st year smolt slaughtered. b 2nd year smolt half through production process. c Smolt released $\frac{1}{3}$ of pen capacity.

difficulties with respect to finding adequate cost data by technology, species and region. In principle, such an analysis should be performed by changing only one variable at a time, keeping all others constant. For example, the analysis of the cost structure by species should be carried out for a particular region and technology. Changes in the natural environment or technology will alter the production function for the species. This again changes the cost structure. It is difficult to carry out such an analysis because of the limitation of available data. Production data for Atlantic salmon are available only for Norway and Scotland and for Pacific salmon in British Columbia and Chile.

However, analysing the existing data with care should make it possible to perform an economic analysis indicating present status and future prospects for the above-mentioned categories. The main problem in the comparative cost analysis, focusing on any of the following elements, technology, species or region, is the fact that the industry has not yet reached its optimal production level. It is still in an early stage of development, and using empirical cost data from a sample of farms might not reflect the long-run cost curve.

If the farm has a two-year production cycle, it will theoretically take three years from the point in time when the farmer buys the first smolt to the time the farm reaches its optimum production level (see Table 10.1).

In year 1, smolt is put into one-third of the total pens. One-third of the total production capacity is then utilized, and the rest of the capacity will be utilized in future years. In year 2, another one-third of the smolt is put into production. The fish from the first year, have in terms of time, gone through one-half of the production process. In the third year, the first salmon is harvested and is replaced by smolt in the vacant pens. The smolt bought in year 2 has now come half-way through the production cycle. The farm has now reached its two-year cycle that will continue for the next years. For a more comprehensive treatment of optimal harvesting, see Bjørndal (1990).

It would therefore be more satisfactory to study cost data at farm level and adjust for biases that might arise from including farms not yet fully developed. Adjusted data have been published by Bjørndal (1990). However, the only large-scale empirical data set that exists are the annual surveys prepared by the Directorate of Fisheries in Norway. In the analysis of the cost structure in the industry, data from the Fisheries Directorate as well as Bjørndal's adjusted data will be presented and compared.

Before analysing the cost structure by technology, species and region, data will be presented to give the reader an understanding of the main inputs in salmon aquaculture, and how the cost structure could be expected to develop as the industry matures. For these purposes, time series data from the Directorate of Fisheries

Table 10.2 Cost Structure in Norwegian farms, in absolute and relative figures, 1986–90. (US dollars per kg rounded weight)

	1986		1987		1988		1989		1990	
Item	US$	%	US$	%	US$	%	US$	%	US$	%
Smolt	1.50	23	1.35	23	1.20	23	0.85	16	0.80	15
Feed	1.75	27	1.60	27	1.75	34	2.10	40	2.10	40
Labour	0.90	14	0.80	13	0.55	11	0.50	10	0.55	11
Miscellaneous costs	1.00	16	0.95	16	0.70	14	0.80	17	0.90	17
Capital	1.30	20	1.30	21	0.95	18	0.85	17	0.90	17
Total costs/kg	6.45	100	6.00	100	5.15	100	5.10	100	5.25	100

(Fiskeridirektoratet, 1989, 1991) and Central Bureau of Statistics (SSB, 1992) will be presented. The adjusted data of Bjørndal (1990) indicates the expected cost structure in fully developed farms. In this section, only data from Norway are presented because it is in this country the salmon farming is most developed and most data are available. In this section and in the rest of the chapter, costs of production for each region are translated to US dollars by use of average exchange rates for 1990. The metric system is used for measurement.

Table 10.2 presents the cost of production by cost items in absolute and relative figures. The costs can be classified in the following five groups: smolt, feed, labour, miscellaneous cost and capital costs.

Smolt is the term used for the fry adapted to salt water. The smoltification process occurs at different times in the life cycle according to species (see under 'Cost structure by species'). Some farms have their own smolt production facilities, but most salmon farms purchase the smolts. In the case of integrated farms, the opportunity cost of the smolt should be used in the cost analysis.

Dry pellets are the dominant salmon feed in use. They are made from fish meal. Some farms also use wet feed which consists of a mixture of fish and fish meal. The fish used for feed could be either the species that fish meal is made from or waste products from fish processing.

Miscellaneous costs consist of insurance, power, administration, repair and maintenance.

Capital costs consists of imputed interest on working capital and interest and depreciation on investments on long-term assets. Investments can be divided into land facilities and infrastructure like roads, buildings, pier, powerline/production and the sea-pen system itself.

Table 10.2 shows that the total costs per kilogram fell from US$6.45 in 1986, US$6.00 in 1987, US$5.15 in 1988 to US$5.10 in 1989. The total costs per kilogram in 1989, broken down by region in Norway, show that the costs per kilogram in the two southern regions have stabilized at around the US$4.80 level. The reason is that the industry here is more mature and the cost data are closer to the long-run cost curve. This underlines the point made above, new farms entering the industry have higher average costs per kilogram produced. In 1990 the production costs per kilogram were US$5.25. In real terms, these are the same production costs as in 1989.

Data for each of the years show that feed is the single most important cost element. Smolt is the second most important cost element (except in 1989 and 1990). Bjørndal (1990) indicates the following distribution of costs by cost item in mature farms: smolt 10%, feed 40–45%, labour 10–15%, miscellaneous costs 10% and capital costs 17%. Comparing the first three columns (1986–88) in Table 10.2, a striking difference is apparent between the data of the Fisheries Directorate (FD) and Bjørndal's data with respect to smolt and feed costs. In the FD data, smolt cost comprises 23% of total costs while the figure is only 10% in the Bjørndal study. The reason for this discrepancy is the relatively high number of farms that still are being established and developed. The number of smolt purchased, and hence the smolt cost is therefore overstated in relation to a normal production cycle.

The opposite tendency is seen for the feed cost. This cost item accounts for only 27–33% in the FD data (1986–88) and 40–45% in Bjørndal's data. Again, the reason is found in the over-representation of not fully developed farms in the FD data. The lower feed cost is a result of both the lower number of fish, and a disproportionate number of small fish. The 1989 and 1990 data (Table 10.2, column 4) indicates that the farms are maturing as smolt costs are reduced and feed costs increased. If the smolt and feed costs are added, these two cost items comprise between 50–60% of total costs for all six alternatives.

Labour costs comprise between 10–15% of total costs and miscellaneous costs between 10–17%. Capital costs account for only 17–21% of total costs.

Cost structure by technology

The open floating sea-cage system is today the dominating technology in use. Iceland, Scotland, Canada and Norway have some experience in land-based farming of salmon. The author has obtained cost data from one major producer in Iceland, and these data will be presented as a case-study of land-based farming. The costs of production of this case will be compared to cost data of sea-pen farming. Such a comparison is made to indicate some important differences in the cost structure between the two technologies, not to give a definite answer to differences in the cost of production. As the cost data for the latter technology is taken from Norway where the salmon farming industry is most developed, only production of Atlantic salmon is included in this section. In the next section, 'Cost structure by species', available data for Chinook salmon will be presented as well. Introduction of closed technologies such as raceways and land-based facilities implies a more capital-intensive technology which allows a higher degree of control with the natural environmental variables. For the closed technologies, water can be drawn from deep in the ocean or fjord. Surface water pollution or algae can therefore be avoided, and higher temperatures during the winter increase fish growth.

In closed systems, oxygen can be added and temperature adjusted for optimal fish growth and predators like seals, otters and birds can be avoided.

There are obvious advantages to controlling natural environmental variables, but the consequences for fish growth and conversion ratios still remain uncertain. There is little experience with large-scale salmon production in raceways and land-based systems. Evidence from Norway, Scotland, Iceland and British Columbia is inconclusive. Problems with growth rates and diseases have arisen, and many land-based farms have faced economic problems.

Table 10.3 presents the cost structure by technology. The first column presents

Table 10.3 Cost structure by technology, in absolute and relative figures, 1990. (US dollars per kg rounded weight)

	Land-based, Iceland		Fisheries Directorate, Norway	
	US$	%	US$	%
Smolt	0.70[1]	11	0.80	15
Feed	1.60	26	2.10	40
Labour	0.80	13	0.55	11
Pumping costs	0.50	8	–	
Miscellaneous costs	0.55	9	0.90	17
Capital	2.05	33[2]	0.90	17
Total costs/kg	6.20	100	5.25	100

Notes:
[1] The Icelandic farm is an integrated farm producing both smolt and salmon. The internal pricing of smolt does not reflect the opportunity cost because interest and depreciation on the investment is not included. The author has estimated the capital costs.
[2] The capital costs are estimated by the author based on the initial investment and using depreciation over 30 years except for equipment where depreciation over 10 years is used. No data is available for interest on the working capital, and the figures underestimate the total capital cost.

the cost structure for land-based farming in Iceland for 1990. The second column presents the Norwegian data for the same year.

The average cost per kilogram produced is US$5.25 in the FD survey compared to US$6.20 for land-based farm in Iceland. This implies 18% higher production costs for the land-based technology. The cost structure of land-based farming compared to the data for the sea-cage rearing show two striking differences. First of all the pumping cost of US$0.50 per kilogram which occurs only in the land-based farming. Secondly, there is a great difference in the capital costs. The capital costs in land-based farming is US$2.05/kg compared to US$0.90/kg for sea-cage farming. That implies that the former technology has 130% higher capital costs.

As this section is concerned with cost by technology, an evaluation of how the technology can be expected to develop in the future and the economic implication of projected improvements, will be included.

Table 10.4 presents the cost structure by technology improvement. The first two columns indicate the potential of the sea-cage farming technology and the third column, land-based farming. The data in column 1 is taken from a study by Olsen (1986) where future cost reductions due to the use of improved technology, biology and organization in the farming industry were evaluated. These data are quite old, but are included to illustrate the fast development in the salmon aquaculture industry. What were regarded as minimum costs in 1986, are high compared to production costs a few years later.

Column 2 presents the production costs of the 15 best Norwegian farms in 1990. The latter data presents the production costs of disease-free farms with 'optimal' production and can be regarded as 'best-practice' farms. However, in aquaculture

Table 10.4 Cost structure by technology improvement, in absolute and relative figures. (US dollars per kg rounded weight)

	Sea-cage Farming, Norway				Land-based, Iceland	
	Expert evaluation		15 best farms			
	US$	%	US$	%	US$	%
Smolt	0.55	13	0.55	18	0.60	14
Feed	1.90	45	1.30	44	1.20	28
Labour	0.55	13	0.30	10	0.55	13
Pumping costs	–	–	–	–	0.25	6
Miscellaneous costs	0.55	13	0.25	8	0.55	13
Capital	0.70	16	0.60	20	1.10	26
Total costs/kg	4.25	100	3.00	100	4.25	100

with a number of factors beyond the farmer's control, these farms could rather be characterized as the 'lucky' farms. Secondly, the risk and variation is considerably lower on land than in sea, which put restriction on a comparison of land-based farming with the 'best-practice' farms.

Column 3 presents the production costs using adjusted data for land-based farming. The adjusted data are based on the cost data for land-based farming in Iceland in 1990, presented in Table 10.2. The reduced costs are calculated by the author on the basis of the projected goals of a comprehensive research project financed by the Norwegian technological and natural science research council. The aims of this programme are to develop the land-based technology to:

(1) Double the production per unit of volume compared to present sea-pen farming.
(2) Increase the quality of the salmon. The improved quality will lead to a price increase of 10% compared to sea-pen farming.
(3) Reduced mortality from 15–20% today to 2%. The lower mortality is a result of reduced loss due to predators, algae blooms and diseases.
(4) Increase the conversion ratio by reducing the portion of the feed not eaten by the farmed fish from 20–40% to under 5%.
(5) Reduce the wage costs by 30%.

In addition to these objectives, Icelandic farms expect to reduce pumping costs by 50% due to more efficient pumps. If all these goals are reached and applied to the Icelandic data, the cost reduction would be approximately US$1.95/kg salmon, producing at US$4.25/kg. A study by Aaker *et al.* (1990) based on data from large land-based (volume of 12 000 cubic metres) farms in Norway, estimated production costs of US$4.00/kg under optimal conditions. Optimal conditions are comparable to the term technical improvement used in this article. In the Icelandic data, the stocking density was increased to 60 kg per cubic metre compared to 80 in the Norwegian study. In the latter study, it is assumed that oxygen has to be added to the water, while in the Icelandic example the author assumes the volume of water pumped per kilogram fish produced is constant.

Comparing the total production costs in the first two columns in Table 10.4 to the average costs for farms in 1990 (Table 10.3), clearly demonstrates that the costs by improved technology is far below the average costs for 1990. The difference in per kilogram is US$1.00 and US$2.25 respectively. The fast development of salmon aquaculture is clearly demonstrated by the fact that the study by Olsen to evaluate possible future cost-cuts in 1986, gives considerably higher costs than the average for the 15 best farms in 1990. The total average costs for those two studies differ by US$1.25. In relative terms the Olsen scenario gives 25% higher costs. The single cost element which accounted for most of this difference is the feed cost. The '15 best farms' produced 1 kg salmon for US$1.35 in feed, which implies a conversion ratio of 1.1.

The improved technology of land-based farming compared to improved technology of sea-pen farming shows that the total production costs for the Olsen study and those for land-based systems (Columns 1 and 3 in Table 10.4) are the same at US$4.25/kg. Again, the 15 best farms are producing at considerably lower average costs. However, it has to be pointed out that the scenario for the land-based farming in addition to cost reduction, includes a 10% price increase due to improved quality which implies approximately US$0.50/kg. Including this price gain, the land-based farms would be in a competitive position to sea-pen farming.

It is important to point out that the costs of land-based farming in Table 10.4 is similar to the Olsen data for sea-pen farming. We will not draw any further conclusion with regard to the future of the competitive position of these two technologies. The future of the land-based farming rests on its ability to control natural environmental factors. The benefit of this control must be larger than the increased capital and pumping costs that this technology implies. This is true both for raceways and land-based systems.

Cost structure by species

In the comparison of cost by species, Atlantic salmon production in Norway (Fiskeridirektoratet 1989 data) will be compared to chinook production in the Campbell River area in British Columbia (Province of British Columbia, 1989). The Atlantic and chinook species are chosen because of data availability. However, Atlantic salmon is the main species grown on a world basis, while chinook seems to be the Pacific salmon with the best future prospects.

Table 10.5 presents the cost structure of Atlantic and chinook in absolute and relative terms. Transportation and marketing costs are not included. Total production cost is higher for Atlantic salmon than chinook, US$5.10 compared to US$4.35.

In the comparison between Atlantic and chinook, the relatively large difference in smolt and feed costs should be noted. Smolt cost for production of Atlantic salmon comprises 16% of the total production cost, while for chinook it is only 7%. For feed the opposite tendency can be observed. Of the total costs for Atlantic salmon, 40% is feed, while feed for chinook accounts for 55% of the total costs.

There might be several explanations for this difference in the relative importance of smolt and feed costs. The most obvious is the biological difference between the two species. Atlantic salmon have a much longer freshwater cycle than chinook. The Atlantic salmon smolt after being 1–2 years in freshwater while the chinook smolt only 6–8 months after being hatched. As a result, there is considerable difference in

Table 10.5 Cost structure by species, in absolute and relative figures. (US dollars per kg rounded weight[1])

	Atlantic salmon		Chinook salmon[2]	
	US$	%	US$	%
Smolt	0.80	16	0.30	7
Feed	2.05	40	2.40	55
Labour	0.50	10	0.55	13
Miscellaneous costs	0.85	17	0.50	11
Capital	0.90	17	0.60	14
Total costs/kg	5.10	100	4.35	100

Notes:

[1] The Canadian data are given as dressed weight, head on. These data are converted to round weight to make them comparable with the Norwegian figures, using a yield factor of 0.89.

[2] To make the data from the Campbell River Area Study comparable to the Norwegian data, the opportunity costs of the owner/farmer are included in the labour costs. Secondly, the opportunity costs of the working and invested capital are included in the capital costs. As the replacement value is the basis for calculating the annual depreciation in the Campbell River Area Study, the real rate of interest has been used by the author to compute the annual interest.

size when smoltification occurs. The Atlantic salmon has an average weight of 40 grams while the chinook weighs 8–10 grams.

Therefore, the production costs of chinook smolt, and hence the smolt price, are considerably lower than for Atlantic salmon. Starting at this low weight, more feed is needed to grow a chinook to 1 kg than an Atlantic salmon. Therefore, feed for chinook counts for a higher portion of the production costs.

The conversion rates are also important for feed costs. In the chinook data, the conversion rate is 2.05 kg feed/kg round salmon, whereas in the Norwegian study of Atlantic salmon it is 1.7.

The Atlantic salmon grow to 3–5 kg after 2 years (average 2 kg/year), chinook in the British Columbia study grew to 1.5–3.5 kg in 1.5–2 years (average 1.43 kg/year). Early sexual maturation is a particular problem with Pacific salmon including chinook. When the fish mature sexually, it is not marketable and the farmer can be forced to harvest fish at suboptimal weights. Therefore, the chinook generally harvest at lower average weights than Atlantic salmon.

In this section, the subject is cost structure by species. In evaluating future prospects, the price of the fish must be taken into account as well. As salmon prices per kilogram increase by size, the Atlantic salmon disadvantage with regard to production costs might be outweighed by higher prices.

It is surprising to note the differences in capital costs for the two species. For Atlantic salmon the capital cost is US$0.90 and for chinook US$0.60. Due to biological differences, one would expect chinook salmon to have higher capital costs per kilogram because of lower stocking density. For Atlantic salmon, stocking density is 20–28 kg/cubic metre, while for chinook it is 6–8 kg/cubic metre. In other words, the chinook requires more capital investment per kilogram than does the

Atlantic. Secondly, the lower growth rate will also lead to, *ceteris paribus*, higher capital cost for chinook. There are many other factors affecting the capital costs per kilogram, and these are tied to regional factors rather than difference between the species. As the wages are lower in British Columbia than in Norway, the data in Table 10.4 indicates that chinook production has a substantially higher labour/capital ratio than production of Atlantic salmon.

The difference in miscellaneous costs is also notable, 6%. These costs make up a heterogeneous group but such elements as repairs, maintenance, power and administration are partly time-dependent. As was the case for the capital costs, one would expect these cost elements to increase per kg for the species with lower growth rates. Again, differences are probably tied to regional factors. In the next section, the most important cost elements by region will be further investigated.

The analysis of cost structure by species is particularly focused on the implications of biological differences. The cost data in Table 10.4 reflect cost differences due to differences in natural environmental, institutional factors and the prices by region. These factors will be discussed in the next section.

Cost structure by region

In this section the same classification of regions are used as in Chapter 2. The regions are:

(1) Europe
(2) North America
(3) Japan
(4) Southern Hemisphere

Factors not included in the analysis of cost structure by technology and species, must be taken into account in this section. Recalling Expression (2), (p. 221) the cost of production by region is determined by quantities and prices of the inputs, given the environmental and institutional factors of the particular region. This implies that, to the degree that environmental and institutional factors differ, each region has a specific production and cost function.

In addition to the above mentioned factors influencing the cost structure, transportation and distribution costs, tariffs, trade barriers and exchange rates all influence the competitive position of a region. Transportation costs are of importance for two reasons. First, nearly all of the salmon is consumed outside the region of production. Secondly, most salmon is exported fresh. This requires fast and reliable transportation which is costly.

The discussion of cost variations by region will first focus on price and quality differences for the major production costs.

Lack of available and reliable data makes it difficult to compare production cost by regions. In a study by DPA (1988) cost data for five major producing countries were presented. The results are reproduced in Table 10.6.

Table 10.6 shows that Norway, Scotland and Japan are the high cost producers at US\$5–6/kg, while British Columbia and Chile are the low cost producers at costs below US\$3.50/kg.

Before discussing these figures in more detail, one has to be reminded of the

Table 10.6 Cost of production by regions. (US dollars per kg rounded weight)

	Europe				N. America		S. Hemisphere		
	Norway		Scotland		B.C.		Chile		Japan
	US$	%	US$	%	US$	%	US$	%	US$
Smolt	0.67	13	1.22	22	0.20	6	0.19	6	N.A.
Feed	1.71	34	1.87	34	1.13	34	1.21	38	N.A.
Labour	0.68	13	0.52	10	0.29	9	0.60	19	N.A.
Miscellaneous costs	1.19	24	1.06	19	0.89	27	0.86	27	N.A.
Capital	0.80	16	0.79	15	0.81	24	0.34	10	N.A.
Total costs/kg	5.05	100	5.46	100	3.32	100	3.20	100	5.67

problems using unadjusted data for some regions and adjusted data for others. For both Norway and Scotland the costs of production are based on empirical data from the industry, while the costs for Chile and British Columbia are based on costs of production models developed by two consulting firms (DPA, 1988). The latter cost figures could therefore be regarded as future goals for the industry. The British Columbia data do not correspond with those given for chinook in Table 10.4. The substantial problems suffered by the industry in British Columbia in recent years, suggest that the data in Table 10.5 may not be a realistic description of the situation today. More recent data for Chile indicate production costs of US$3.50/kg (Lindbergh, 1991).

Smolt

Smolt prices have varied greatly both over time and by region. The price variations have primarily been a result of disequilibrium in the smolt market. The farming industry in many regions developed faster than the smolt industry. Smolt shortage leads to higher prices. Due to trade restrictions, prices by region did not equalize. Secondly, transportation of smolt over large distances is costly. In the long run one would expect the market to reach an equilibrium situation.

The quality of smolt is perhaps of greater importance than the price. The quality not only varies between regions but also within regions. In Norway, for example, successful farmers have been able to buy smolt or eggs from producers with good broodstock. Good broodstock generally refers to one free of disease which has been selectively bred over several generations. Differences in smolt quality by region may be a result of institutional factors. In some countries there has been a prohibition on smolt import (Iceland) and in other countries import restrictions (New Zealand and Tasmania). Such institutional factors can influence the price of smolt.

Analysing the smolt cost in Table 10.6, substantial differences are found among Norway, Scotland and British Columbia/Chile. The difference between Norway and Scotland is probably due to disequilibrium in the Scottish smolt market at the time the survey data were collected. There are at least two reasons for the difference between Europe and British Columbia/Chile. First, the biological differences between the species which is described in the section 'Cost Structure by Species'. Secondly, the Norwegian and Scottish data are survey data which include developing farms.

The smolt price of Pacific salmon is, however, surprisingly low, only US$0.20 in the British Columbia data in Table 10.6 compared to US$0.67/kg in the Norwegian data. The latter price is close to the price for the 15 best farms (Table 10.4).

Feed costs

Feed is the single most important input in salmon production in terms of share of total costs. Dry pellets made from fish meal are the dominant salmon feed in use. Fish meal is a bulk product in the world market, and feed plants are usually located close to the farms. There are two reasons for this:

(1) The transportation cost of feed is much higher than for fish meal.
(2) Most farmers prefer fresh feed, and therefore the location of the fish feed factories are market oriented.

The advantage that the fish-meal-producing countries were believed to have now seems to be of minor importance. The fish meal industry uses mainly pelagic species (capelin, herring, mackerel, blue whiting, anchovy, etc.). Like the fishery of other wild species, the stock sizes and catches have periodic changes. Norway was a large producer and exporter of fish meal up to the mid-1980s. Since that period, however, it has imported fish meal to the fish feed factories because of the depletion of the capelin stock. This has not significantly affected the competitiveness of the Norwegian aquaculture industries although feed prices have risen slightly due to higher transportation cost. The formula for fish feed might vary from producer to producer, but no region seems to have a competitive advantage related to fish meal quality. The feed costs in Table 10.6 are approximately the same for Norway/Scotland on the one hand and British Columbia/Chile on the other. The reason for the lower feed costs in British Columbia/Chile is improved conversion ratios assumed in the cost of the production model.

Labour costs

Labour costs (wages) are the cost item that varies most by region. The two extremes are Norway and Chile. Even though the labour costs count for only about 10% of total costs, substantial wage differences might be of great importance in competitive markets. With such large differences in wages, one would expect great differences in the quality of labour. According to Lindbergh (1987), however, the Chilean labour is of high standard. The relative difference in the labour cost between Norway and Chile does not reflect differences in labour productivity. It is rather the opportunity cost of labour that explains the great difference.

Capital costs

The capital costs consist of interest and depreciation. Depreciation for each dollar invested differs only by region if the economic life of the asset differs. Weather conditions and other natural environmental factors might lead to differences in life by regions. However, here we will assume no differences in the period of depreciation.

Differences in capital costs would therefore have to be traced back to differences in interest rates or the amount of capital invested.

A number of institutional factors determine differences in the real rate of interest both over time and by region. The most obvious is the fact that the rate of interest at the national level is used as a tool of economic policy. In 1991/92 the real rate of interest in Norway was about 7%. In the period after World War II, it was at an average of 1%. But even if the real rate of interest in Norway is now high, the salmon aquaculture industry enjoys subsidized capital through the Regional Development Fund. The same is the case in other countries.

Norway has a substantial advantage in relation to most other regions with respect to necessary capital investment. The investment can be divided into the sea pens systems themselves and land facilities and infrastructure. Norway has a well-developed infrastructure along the coasts. Many farms are established in connection with previous or existing fish processing plants where there are piers, freezers and buildings and other facilities available. Secondly, these areas have traditionally been linked to fishing and farming and therefore veterinarian services are available in these communities. In areas like British Columbia and Chile, they have to start from scratch in most regions and the investment is higher. In Norway the investment on land will be relatively modest due to the existing facilities.

In sea-pen aquaculture there are certain possibilities of substituting labour for capital, for example by using hand-feeding instead of automatic feeding systems. Closer inspection of these two cost items in Table 10.6, shows that Chile is producing with high labour costs and with the lowest capital costs. At first, this might be surprising because wages are very low in Chile. A probable explanation is, however, that Chile has chosen a sea-pen farming technology intensive in the factor of production that is relatively cheap. That implies a technology which requires much labour and little capital. This is confirmed by Lindbergh (1991). Most farms operate with Chilean-fabricated, inexpensive net-cage facilities. It has been cost-effective to utilize more labour for fish culture activities and facility maintenance than to invest capital funds in the most efficient and long-lasting equipment.

The building materials derived from sources within the country usually cost only a fraction of those which are imported. Wood from local saw mills has been extensively used to fabricate net-cage facilities, but metal and high density polyethylene are increasingly being employed. Construction and installation costs are similarly less expensive.

Of the institutional factors, the limitation on farm size in Norway might be a disadvantage for this particular region. Salvanes (1989) demonstrated that there are economies of scale in the farming industry. In Norway the maximum size of farms was 8000 cubic metres. This was increased to 12 000 cubic metres in 1988 to increase the competitiveness of the industry.

Will the decentralized pattern that can be observed today continue or will there be a tendency towards a more regional concentration? In addition to the costs of production by regions, the transportation and distribution costs, tariffs, trade barriers and the exchange rates will influence the regional distribution of salmon farming. Of particular importance are transportation costs.

Table 10.7 gives the freight and selling costs per kilogram of salmon to major markets. All freight costs in Table 10.7 are by air except from British Columbia to Los

Table 10.7 Freight fares to main markets and landed costs per kg (rounded weight) in major markets. (US dollars)

	Norway US$	Scotland US$	British Columbia US$	Chile US$	Japan US$
Freight and selling costs to:					
Los Angeles	2.75	2.90	0.80	2.10	N.A.
New York	2.00	2.15	1.65	2.10	N.A.
Tokyo	3.75	3.90	2.60	3.70	0.40
Landed cost in:					
Los Angeles	7.80	8.36	4.12	5.30	N.A.
New York	7.05	7.61	4.97	5.30	N.A.
Tokyo	8.80	9.36	5.92	8.70	6.07

Source: DPA (1988)

Angeles and from Japanese producers to Tokyo. Freight and selling costs constitute a substantial part of total landed costs in the markets where air-freight is necessary. The air-freight cost lies between US$2–4/kg except from British Columbia to New York (US$1.65). Japan produces mainly for the domestic market. Because of high transportation cost from other producing regions, only salmon from British Columbia can compete in price with Japanese salmon in the Tokyo market. Table 10.7 demonstrates that the low-cost producer Chile is not the low cost supplier except in the New York market. British Columbia's location gives this region a comparative advantage in major markets in the USA and Japan.

These transportation costs point in the direction of regional decentralization. Even though the salmon is mainly consumed outside the region of production, all major producing regions, except Chile, have a market located close to the producing region where road/boat transportation is possible. For the European producers, the markets are France, Germany, Spain, Denmark and England. For the Canadian producers the markets are the large cities of the United States like New York and Los Angeles. The producers in the States of Washington and Maine and Japan are relying entirely on the home market. As the majority of the farmed salmon is sold fresh, the air transportation costs limit the competition in these markets as long as the fish can be trucked, the transportation costs are modest. For example, the transportation costs from northern Norway to southern Europe is US$0.50/kg. It is, therefore, reasonable to believe that there will not be a geographical concentration in one of the aforementioned regions. There are some exceptions to this. Even though the European market is dominated by the fresh fish from the European producers, there are certain periods of the year (winter time) when the harvesting is low in the north and at a peak in the Southern Hemisphere. In that situation it might be profitable for the Chilean industry to export to the European market. The future competition will probably take place in the market where air freight is necessary from the main producer regions.

The above analysis relates to fresh salmon. Some regions accept frozen salmon as well, for example, Japan. In 1991, 60% of the Chilean harvest was frozen and shipped by sea to Japan at a cost of about US$0.50/kg (Lindbergh, 1991). If the preference changes in the direction of accepting frozen salmon, the competitive advantages among regions will also change.

In evaluating the development of salmon aquaculture by region, the reader should draw on the information in Chapter 2 concerning environmental factors, constraints and growth potential for each producing region.

Summary and conclusions

The aim of this analysis has been to give the reader an understanding of the cost structure in salmon aquaculture and to evaluate the profitability of this industry by technology, species and regions. The data presented must be used critically with appropriate explanations.

The limitation in the interpretation and use of the data in the comparative studies are related the problems concerning:

(1) The number of farms included in each sample;
(2) The industry has reached different levels of development in the regions analysed;
(3) The figures in the tables include both primary data collected from farms and cost estimated based on cost of production (COP) models.

The cost of production in sea-pen farming today is substantially lower than land-based production. The future of land-based farming rests on its ability to increase stocking density with or without adding oxygen to the water and to control natural environmental factors. The benefit of this increased stocking density and control must be larger than the increased capital, pumping costs and possible other costs that this technology implies.

The cost of production by species demonstrates differences in biology between Atlantic and Pacific salmon. From the data presented, we cannot draw any conclusion to whether one species is more suitable for farming than others due to the influence of regional factors in the cost of production.

The costs of production differ substantially by region mainly due to the data problems described above.

Most salmon is marketed fresh, and the high transportation costs point in the direction of regional decentralization.

Appendix

In this appendix the concepts of production functions and cost functions will be used in the analysis of economics of production in salmon aquaculture.

The production function in salmon aquaculture can be written in a general form as:

$$x = f(v) \tag{A1}$$

where x is the output vector and v is the input vector.

Expression (A1) gives the maximum output (x) of different input combinations (v).

Including environmental factors in the production function, expression (A1) can be rewritten as:

$$x = f(v;e) \tag{A2}$$

where e is a vector of environmental factors.

Expression (A2) gives the maximum output (x) of different input combinations (v) given the natural environmental factors (e).

Including institutional factors, the production function can be rewritten as:

$$x = f(v;e,i) \tag{A3}$$

where i is a vector of institutional factors.

A product can usually be produced by different technologies. When a particular technology has been selected, the substitution among inputs is limited.

In production theory, two polar cases exist, the fixed proportion case with no possibility of substitution and the case where the substitution is continuous.

By assuming only two factor inputs, labour (L) and capital (K), Fig. 10.1

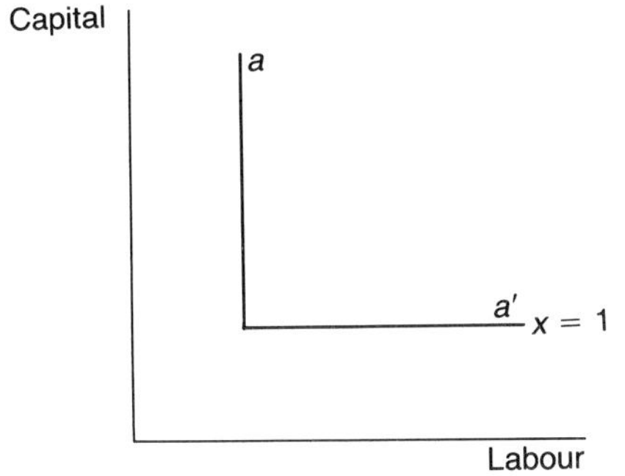

Fig. 10.1 Production curve of 1 unit of salmon (isoquantcurve) without substitution possibility.

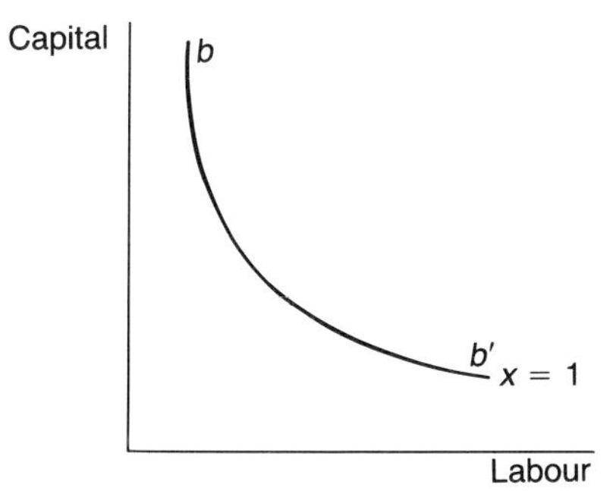

Fig. 10.2 Production curve of 1 unit of salmon (isoquantcurve) with continuous substitution possibility.

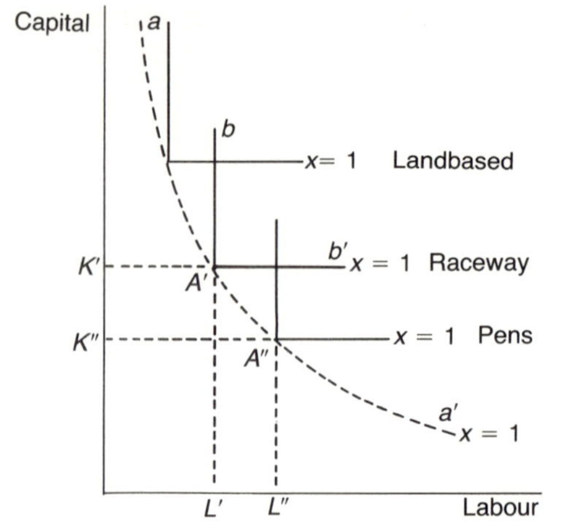

Fig. 10.3 Production curve of 1 unit of salmon before investment (ex-ante) and after investment (ex-post).

illustrates an isoquant (given output) in the fixed proportion case. The isoquant a–a' illustrates the production of one unit of salmon ($x = 1$).

Figure 10.2 illustrates the case where substitution is continuous. The isoquant b–b' illustrates the production of one unit of salmon ($x = 1$).

By using the concepts of fixed proportion and continuous substitution, the production decision for a farmer can be illustrated before investment (ex-ante) and after investment (ex-post) in Fig. 10.3.

The curve a–a' in Fig. 10.3 illustrates the possible capital/labour ratios or the technological alternatives *before* investment is made. If a raceway is chosen, the isoquant b–b' gives the possible production technology (a capital/labour ratio which produces one unit of salmon. A' is the most efficient point. Any other point on the isoquant using the raceway technology, would either use more capital or more labour for producing one unit of salmon and is therefore not technically optimal. At point A', K' amount of capital and L' amount of labour is used.

The open rearing pen technology gives the optimal solution in point A'', using K'' units of capital and L'' units of labour. As $K' > K''$, the raceway technology is more capital intensive than the open rearing pen technology. On the contrary, the open rearing pen technology is more labour-intensive than the raceway technology ($L'' > L'$).

The description of the choice of production technique and optimal adjustment is based on some assumptions. First, only two inputs in production were assumed. In the real world with n inputs, there will generally be some substitution possibilities after the investment is made (ex-post). Substitution between feed and labour is an example in aquaculture of substitution possibilities after the investment is made. Supposing hand feeding technology is used, management has the choice between feeding a few times a day with larger rations or many times using smaller feed rations. By feeding a few times one would expect that more feed would pass through the pens to the bottom and thereby reducing the conversion ratio. On the other, hand feeding a few times would save labour. The management would therefore have a substitution possibility between labour and feed. However, the substitution possibilities in the short run will normally be very limited.

What technology should then be chosen? This depends, obviously, on the price of factor inputs. In a simplified example with only labour and capital as inputs, the prices of these inputs will determine the production technology. In areas with relative low wage, and relatively high interest rates, one would expect to find more labour-intensive technologies than in areas with a lower wage/interest rate ratio.

As the aim of this chapter is to compare and analyse the competitive advantages by species, technology and regions, it is more useful to think in terms of costs of production rather than in physical terms.

The cost function of salmon aquaculture in a particular region can be written as:

$$c_r = f_r(x;q,e,i) \qquad \text{(A4)}$$

where c_r is the total production costs in region r.

Expression (A4) says that the cost of production in region r (c_r) is a function of the regional production (x), given the factor prices (q), environmental (e) and institutions factors (i) of the region.

The emphasis so far has been concerned with the production process and production costs. As pointed out in Chapter 11, the farmer will adjust in a way that maximizes the profit. In addition to cost data, the price of output or the price/quantity relationship would have to be known.

Production and cost theory assumes that the producers have full information about technology, costs and prices. As pointed out under the presentation of the production process, the production functions describe the technical relationship between the various level of factor inputs and the product output. Assuming full information, all producers will choose the optimal technology. Introducing prices for the inputs in the production function, leads to cost functions. The producer will choose a production technology that minimizes the cost for a given level of output.

References and selected bibliography

Aaker, H. Edvardsen, Ø., Wold, A. and Meland, H. (1990) Økonomisk analyse av land-og sjøbasert lakseproduksjon. NF-report no. 20.07/90 Nordland Research Institute, Bodø (in Norwegian).

Bjørndal, T. (1990) *The Economics of Salmon Aquaculture*, Blackwell Scientific Publications, Oxford.

Chaston, I. (1984) *Business Management in Fisheries and Aquaculture*, Fishing News Books, Oxford.

DPA Group (1988) Cost of Production Model for Pen-Rearing of Salmon in Alaska and Currently in Producing Regions, Report prepared for State of Alaska, Department of Commerce Economic Development.

Fiskeridirektoratet (1989) Lønnsomhetsundersøkelse for matfiskanlegg 1988. Report No. 7/89, Fiskeridirektoratet, Bergen (in Norwegian).

Fiskeridirektoratet (1991) Lønnsomhetsundersøkelse for matfiskanlegg 1989. Report No. 1/91, Fiskeridirektoratet, Bergen (in Norwegian).

Lindbergh, J. (1987) The economic potential for the commercial production of

Atlantic and Pacific salmon in Chile. Paper presented at the conference organized by the Fundacion Chile on salmon aquaculture, Santiago, 17–19 March 1987.

Lindbergh, J. (1991) Salmon farming in Chile, Unpublished working paper.

Olsen, S.O. (1986) Lønnsomhet i oppdrettsnæringen. Working paper, Fiskeriteknologisk forskningsinstitutt, Tromsø (in Norwegian).

Province of British Columbia (1989) *Estimated Costs and Returns for Chinook Salmon Production in the Campbell River Area*. Ministry of Agriculture and Fisheries, Victoria, British Columbia.

Salvanes, K. (1989) The structure of the Norwegian fish farming industry: an empirical analysis of economies of scale and substitution possibilities. *Marine Resource Economics*, 6(4), 349–73.

Shang, Y.C. (1981) *Aquaculture Economics: Basic Concepts and Methods of Analysis*, Boulder, Colorado, Westview Press.

Shaw, S.A. and Muir, J.F. (1987) *Salmon: Economics and Marketing*, London, Croom Helm.

SSB (Statistisk Sentralbyrå) (1992) *Fiskeoppdrett 1990* (Fish Aquaculture 1990), Weekly Bulletin of Statistics No. 6, Kongsvinger, Norway (in Norwegian).

Chapter 11
Salmon Sea Farm Management: Basic Economic Concepts and Applications

TROND BJØRNDAL and RUSSELL S. UHLER

Introduction

The major objective of the managers of existing salmon farms with an established sea-pen capacity is to maximize the farm's overall profitability. To do this, the salmon farmer must make operating decisions involving the choice of many factors. Particularly important factors that are directly controlled by management are: (1) the feed ration and type of feed to use, (2) the pattern of feeding, (3) the number of smolts to purchase and the stocking density of the fish, and (4) the time to harvest the fish. Factors that are not controllable, or that are less subject to the control of management but which also affect profitability are fish prices, feed and other input prices, site environment, mortality from disease and algae blooms, etc.

The factors that are under the direct control of managers might be generally classified as *control variables* whereas this latter group might be classified as *economic and natural environmental variables*. Thus, one distinction between fish farms that use sea pens and those that use land-based tank systems is the degree of control that is exerted over the natural environmental variables. Land-based tank systems, in principle, permit the regulation of water temperature, salinity, and oxygen content. They also have water filtration that can eliminate algae and other harmful material. However, this additional control is achieved at significant extra capital equipment costs. It remains true that neither type of farm has any control over the economic variables since salmon farmers typically act as price-takers in the markets for their product, and in the markets where labour, materials, etc. are purchased.

The purpose of this chapter is to accomplish two things: (1) to set out a structure, or model, for examining how a salmon farmer might ideally go about choosing the level or amount of the variables over which he has control and (2) indicate more or less how these choices are actually made in practice and how closely they conform to the ideal. It should be noted that ideal choices usually require extensive information, the nature of which will be made clear in the following sections, whereas actual practice typically involves making decisions with incomplete information. Thus the closeness of actual practice to the ideal is likely to be related to the amount of information that is available to the farmer.

(In the pursuit of the first of these objectives, our intention is to outline in a non-mathematical way, the basic economic concepts that are needed to manage salmon sea farms efficiently. The reader who is familiar with the methods of dynamic optimization will be able to see how the concepts discussed here can be formulated into a concise mathematical optimization model.)

Short-run decisions in salmon farm management

Before discussing choices made with complete information, it is important to distinguish between the short run and the long run. The reason for this distinction is that it affects the kinds of choices that need to be included in the management decision process. In the context of salmon farming, the short run involves the choices of all the variables for a given capacity of the sea-pen system, where capacity is measured in terms of water volume. In other words, the short run is viewed in the traditional way in that the capital equipment that provided a given farm capacity is fixed and ideal choices involve making decisions about the control variables mentioned above in order to maximize farm profits subject to the capacity constraint provided by the sea-pen system. In the long run, all factors are variable including the farm capacity (possibly subject to some upper limit as determined by site or regulatory constraints). This long-run choice is discussed later in the chapter.

Fish growth and management decisions

It is well known that individual fish in a population grow at different rates, primarily because of genetic differences. Thus, at any point in time in the grow-out cycle of the salmon stock, there will be a distribution of fish across a range of weights. Ways of dealing with this fact in modelling the choices made by farmers will be discussed later in this chapter. For the time being, we will simplify the situation by using the growth relationship of the average fish so that the fish population biomass is simply the product of the weight of the average fish and the number of fish in the stock.

The growth rate of the average fish depends on: (1) its weight, (2) its ration of feed, (3) the stocking density of the fish population and, (4) a variety of environmental variables such as water temperature, water current, salinity, etc. The farmer normally has little control over environmental variables once a farm site has been chosen. There has been considerable research into examining how these factors affect fish growth, most of which has been done in a hatchery or laboratory setting. The relationship between growth and fish weight, including the effect of water temperature, has been examined by Iwama and Tautz (1981). Brett, Shelbourne and Shoop (1969) examined the effect of ration on growth for fingerling sockeye salmon. Austreng, Storebakken and Asgard (1987), and Storebakken and Austreng (1987), reported on a series of experiments carried out in tanks and sea pens relating fish growth and feed ration for Atlantic salmon and rainbow trout. One of their experiments is notable because it involved rainbow trout as large as 5 kg.

We have indicated that fish growth is affected by stocking density, but since in the short run, the capacity of the farm is fixed and the weight of the average fish is already specified in the growth relationship, dependency on the stocking density can be converted into dependency on the remaining number of fish in the population. Since the remaining number of fish in the population is the initial number of fish less the cumulative mortality, this reduces to two variables in the growth relationship over which the farmer has control; the feed ration and the initial number of fish. First we will discuss the choice of the feed ration and in the section, 'Stocking density', of this chapter, we will discuss the choice of the initial number of fish and the resulting stocking density.

Feed ration

We now ask the question: ideally, how should the farmer feed his fish? If the farmer wants to maximize his profits, then the answer is: in each period of time the ration should be such that the value of the additional growth increment due to extra ration just equals the cost of the extra ration. The additional growth increment is valued at the price received at the harvest time and both this price and the feed price are discounted to a common time, say, for example, the beginning of the grow-out cycle. It is clear that in order to feed in this way, the farmer must have a substantial amount of information. In particular, he must know how the growth of the average fish in each short period of time, say a month, responds to additional amounts of feed. He must also know how this relationship shifts with respect to fish weight, with respect to stocking density, and with respect to the environmental variables. In other words, he must know the complete fish growth function.

Current feeding practice is usually based upon feeding tables that specify daily feed requirements per unit of weight which depend upon the weight of the fish and the water temperature. These unit feed requirements rise with water temperature and decline with fish weight. Alternatively, conversion ratios which are the ratio of feed ration to weight gain are used to determine feed requirements. For example, if the expected conversion ratio is 1.7 and the weight gain of the average fish in a particular month is expected to be 0.5 kg, then the feed requirement for the month would be 0.85 kg, or about (0.85/30) kg/day. The kinds of feeding rules that come from feed tables and the conversion ratios appear to rely upon only a minimum amount of information and do not appear to be closely related to the ideal determination of feed requirements discussed above. It is therefore of interest to examine these rules in more detail and attempt to establish their relationship to ideal feeding rules. First the connection between the feed table rule and the conversion ratio rule will be examined. The discussion will continue to be in terms of the average fish in the population.

A typical manufacturers' feed table might indicate that at 14°C water temperature, a 1 kg salmon should receive 0.012 kg/day or 0.36 kg/month. Although there is no indication of how much monthly weight gain is expected from this feeding programme, let us suppose experience shows that fish at this weight and at this water temperature can be expected to gain 20% in the month, so that the weight gain can be expected to be 0.2 kg which produces a conversion ratio of 1.8. Taking this calculation backwards produces the conversion ratio rule. Suppose that experience shows the conversion ratio should be about 1.8 and the weight gain should be about 0.2 kg so the feed requirement is calculated to be about 0.36 kg for the month. Although their equivalence is established, what is the basis of these rules? In other words, what, if anything, is being optimized? Of course it is not possible to answer this question based upon the information given here, but hints from the literature are of interest.

A survey of the biological fish growth literature indicates two suggestions for the optimal feed ration. One suggestion is the feed ration that maximizes fish growth by providing a ration that is based on the physical capability of a fish to take in and evacuate food from its stomach. Vahl (1979) has presented a model of this variety and there has been considerable experimentation using this feed ration reported in

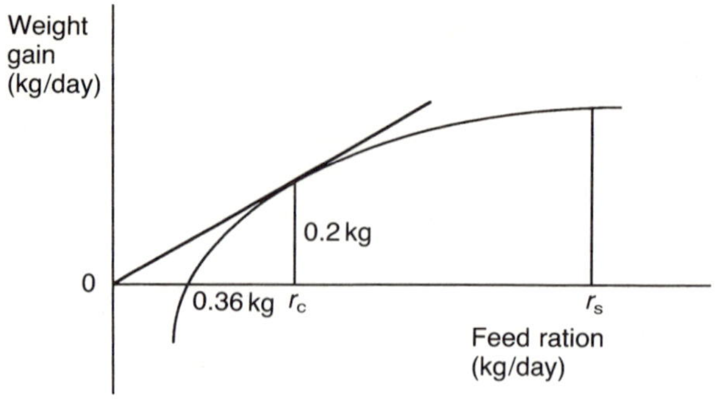

Fig. 11.1 Weight gain – ration contour of the fish growth relationship.

the literature. We refer to this as the *satiation* ration and it is clearly optimal in the sense that it maximizes growth.

The other suggestion for the optimal ration is the ration that maximizes the *conversion efficiency* which is defined as the ratio of weight gain to the feed ration. This is the reciprocal of the conversion ratio discussed above. These two suggestions are illustrated in Fig. 11.1. The weight gain curve presented in Fig. 11.1 assumes that weight gain is an increasing function of the ration but that there are diminishing returns to feeding in that the increment of weight gain declines at the margin with increases in ration. This curve is a weight gain–ration contour of the growth function and is consistent with the model of Brett *et al.* (1969) and the results recently obtained experimentally by Austreng *et al.* (1987). The feed ration, r_s, is the satiation ration since weight gain is maximized at this feed amount and r_c is the ration that maximizes the conversion efficiency. How do these rules compare with the ideal rule discussed above that is derived from maximizing discounted profit and equates the marginal value of weight gain due to extra ration with its marginal cost? In order to make a comparison among these rules it is convenient to restate the ideal rule to equate the ratio of the additional weight gain to the additional ration with the ratio of the appropriately discounted feed price to the appropriately discounted price of the fish prevailing at the optimal harvest time. Thus, with a growth function like the one in Fig. 11.1, unless the feed price is zero, the satiation feed level will never conform to the ideal, and only by chance will the feed ration that maximizes the conversion efficiency conform to this ideal.

To illustrate these points further, return to the example used above and suppose that Fig. 11.1 represents a single weight gain–ration contour of the growth function for a 1 kg fish. As the fish increases in weight, this function will shift to the right and upward so that the contours intersect. Let us also suppose that the feed table discussed in our example was actually designed to maximize the conversion efficiency so that r_c is 0.36. We are now in a position to compare r_c with the ideal feed ration. Neglecting discounting, it is easily seen that the maximized conversion efficiency ration, r_c, and ideal ration will coincide only if the ratio of the feed price to the fish price is equal to $0.2/0.36 = 0.56$. Since the actual price ratio for salmon is much lower than this, the ideal ration will exceed r_c, but of course, be less than r_s. Although the inclusion of discounting will move the ideal feed ration closer to r_c, under current ratios of feed prices to fish prices, the ideal ration will still be greater than r_c. On the other hand, if the feed table had been designed to maximize fish growth instead of the conversion efficiency, the ideal ration would have been less than that ration recommended by the feed table. It is again emphasized that this feed ration is ideal only in the sense that it is the ration that maximizes the discounted profit of the salmon farm.

Figure 11.1 illustrates that the difference between the ideal, the maximum conversion efficiency, and the satiation rations depends on the curvature of the weight gain–ration contour. For example, if this contour is linear up to the amount of the satiation ration, as in Corey *et al.* (1983), then the three rations will all coincide.

It is clear that the computation of the ideal feed ration requires information about the relationship depicted in Fig. 11.1 and since this is a contour of the growth function, it means that information about the growth function itself is required. How can such information be obtained? As already indicated above, some information is available from experimental evidence but more is needed from experiments conducted under actual farm conditions. Fish farmers tend to feed and stock their fish in accordance with standard practices so that there is not enough variation in the data to properly determine the growth relationship from farm data alone. Experimental scientists need access to facilities that are similar to commercial operations both in terms of site location and numbers of fish so that fish of different weights, stocked at different densities, receiving different rations can generate the data that is required to estimate the growth relationship. Complete knowledge of the growth relationship would permit the establishment of ideal feeding rules.

The discussion of feeding rules has focused on the amount of feed. Besides the nutritional content of the feed, another aspect of the feeding programme involves the time pattern of feeding a particular ration. The typical commercial feeding pattern is to feed a given daily ration evenly during the daylight hours. Of course, there are many alternatives to feeding in this way. Not feeding for a few days followed by heavy feeding, a process called pulse-feeding, is one such alternative that may hold promise for improved feeding efficiency. But there are many other alternatives and learning the time pattern that produces the best results is a matter of experimentation both in laboratory and commercial settings.

Stocking density

It was noted earlier that fish growth is expected to depend upon stocking density. Although the impact of stocking density on fish farming operations usually focuses on how stocking density affects mortality, there is good reason to believe that it also affects fish growth (see, for example, Papoutsoglou, 1987). Given that the salmon farmer has decided on a farm capacity, he must choose the number of smolts to place on the site. This choice will then help to determine the stocking density that evolves over the grow-out cycle of the fish population. The benefit of increased numbers of smolts is of course the higher biomass that might be expected at the harvest time. However, this biomass presumably experiences lower growth rates and higher mortality because of higher stocking densities. Figure 11.2 is a contour of the expected relationship between the initial number of smolts and weight gain, given the values of the other variables in the growth relationship. The graph indicates that weight gain will not be affected by larger numbers of smolts until some critical number, N_0, after which weight gain will decline with increases in numbers of smolts

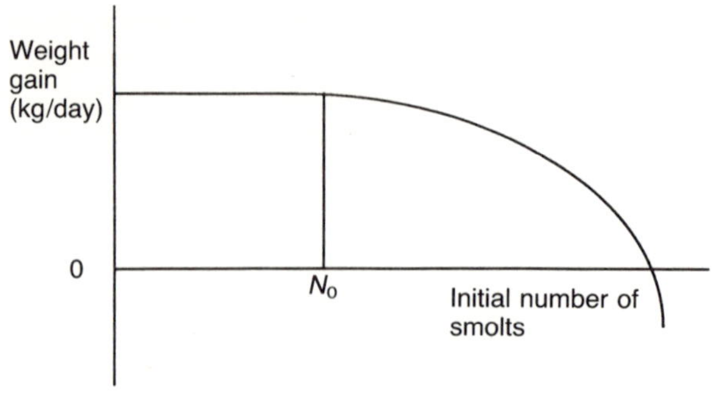

Fig. 11.2 Weight gain – initial smolt contour of the fish growth relationship.

because of increased stocking density. Again it should be noted that this is a contour of the growth relationship and hence is constructed for a given fish weight and ration.

In the discussion of feeding it was possible to choose an ideal ration provided enough information was available on the fish growth relationship. The same is of course true for choosing the initial number of smolts and hence the time path of the stocking density over the grow-out cycle. The nature of the tradeoff between benefits and costs has already been indicated so that the ideal number of initial smolts involves the number that maximizes the net discounted value of the fish population at the time of harvest. As indicated, more smolts released means more net revenue at harvest time, but this may be offset by slower growth and higher mortality. Thus the farmer must choose the number of smolts that strikes a balance among these factors and maximizes profitability. To know exactly how to do this requires a large amount of information that is currently not available.

It is likely that the relationship between weight gain and initial numbers of smolts depends upon the weight of the fish. For example, when the fish are small, the indication is that they are less tolerant of high stocking densities. Assuming that the ultimate capacity of the farm is put in place before any smolts are purchased, this would not normally be a problem for the first year class of fish in a new farm but it might well be a consideration for future age classes. Thus the effect of higher fish weight on the contour in Fig. 11.2 is to shift it to the right. But relatively little is known about the effect of stocking density on weight gain and without weight gain data generated with large enough variations in stocking densities across different fish weights and feeding programmes, adequate information to permit the choice of ideal stocking densities will remain unavailable.

Fish mortality and management decisions

Fish mortality arises from diseases, predators and algae blooms. To some extent, disease can be controlled by use of medicines but since stocking density may affect the outbreak of disease, it may also be partially controlled by management decisions on stocking density. For example, an important cause of salmon mortality is bacterial kidney disease (BKD) and there is some evidence that high stocking densities lead to increases in the outbreak of this disease. Algae bloom, the other major cause of mortality, results in suffocation when fish gills become clogged with algae. Low

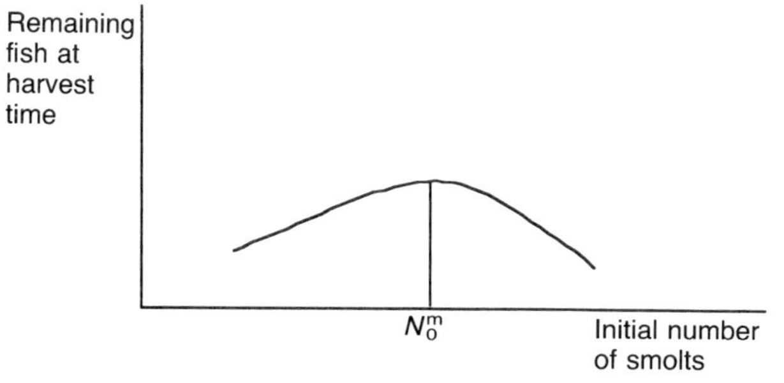

Fig. 11.3 Relationship of remaining fish at harvest time to initial smolt, given density dependent mortality.

stocking densities may help reduce mortality from this source in that lower densities permit fish to have greater dissolved oxygen at depths where the concentration of algae may be lower.

Information on the relationship between stocking density and mortality comes both from commerical farm operations and from laboratory and hatchery experiments. (See Fagerlund *et al.* (1981) for some evidence of the effect of density on mortality of hatchery coho salmon.) Although this relationship is obviously complex and more information is needed over a full range of fish weights, the establishment of such a relationship would make it possible to determine an ideal level of stocking density. In terms of our previous discussion, for a given farm capacity this would involve the choice of the initial number of smolts which then determines the initial stocking density. As growth occurs and stocking density rises, if densities become high enough, this may lead to a rise in mortality. Thus the choice of the initial number of smolts to place into a given volume of water (farm capacity) involves a tradeoff of higher mortality over the grow-out cycle against the numbers of fish surviving until harvest time. Figure 11.3 illustrates this idea by plotting a hypothetical relationship between the remaining number of fish at harvest time and initial number of smolts. For low values of the initial number of smolts, increases in this number results in a rise in the remaining number of harvestable fish but eventually a point is reached where the stocking densities implied by still higher initial numbers of smolts leads to mortality over the grow-out cycle that is high enough to reduce the remaining fish at harvest time. Thus, there is some initial number of fish that maximizes the remaining number at the time of harvest. This maximum is probably close to the ideal number of smolts, but since the conclusion does not take into account the effect of density on fish growth, one cannot be certain. For example, if the choice of a number of smolts that maximizes the remaining fish at harvest time resulted in densities that materially reduced fish growth, then the ideal number of smolts, which would come from maximizing the discounted net value of the biomass at harvest time, would be less than N_0^m in Fig. 11.3.

Harvesting and management decisions

The time to harvest the fish is one of the most important management decisions in salmon farming operations. Smolts are usually put in seawater in the spring of the year with their weight depending upon the species. For example, Atlantic salmon smolts would average about 40 grams whereas Pacific chinook smolts would average only about 8 grams. The time of harvest is dependent upon a number of factors but most farmers tend to harvest the majority of the fish between the autumn of the second and the spring of the third calendar year after the fish were put in seawater. By this time, the fish are in weight ranges that usually command a good price. Even though the fish might be held for another year at which time they would command even higher prices, the risks of mortality due to disease and algae bloom and reduced growth rates are a major concern to farmers. Since feed costs during this additional year are also quite high, cash flow concerns are often important in the decision of when to harvest.

Is there a harvesting rule that can be derived on the basis of a specific management objective such as the maximization of profit and, if so, then what information is required to implement the rule? Setting aside the problem of the risk of loss of fish due to major disease outbreak or algae bloom and recognizing only normally expected fish mortality, it is quite easy to devise such a rule (see Bjørndal, 1988). If the unit of time for decision-making purposes is a month, then the rule is that the population of fish should be held so long as the gain in the present value of the population exceeds the present value of the costs of holding the fish for another month. These costs should only include the variable costs specifically associated with holding the fish such as feed, insurance, and medical costs. The present value calculation is made to bring all value and cost increments to a common time. It should be noted, however, that since growth can vary over the seasons of the year because of changes in water temperature, if applied myopically, this rule may lead to an incorrect decision. The more general rule is to hold the fish to the month in which the net present value of the fish stock is maximized.

It was noted earlier that the growth rate of the individual fish in any given year class is not homogeneous so that at any point in the grow-out cycle, the fish in the population will be of different weights. As the stock grows and stocking density requirements dictate the movement of some portion of the population to different pens, it is common practice for salmon farmers to separate fish into weight groups. Thus the ideal harvesting rule above applies to each of the different weight groups so that the weight group should be harvested when the net present value of the fish stock in the group is maximized.

Some salmon in a given year class will reach sexual maturity and cease to gain weight at a time when their cohorts are still growing and consuming feed at a rate that indicates that they should not be harvested. The increment to the value of this group from holding them for another unit of time is zero (or negative) while the cost of holding them remains positive so that to the extent that these fish can be identified and separated from the remainder of the stock, they should be immediately harvested. Actually they may have passed the optimal harvest time, but the practical problems of identifying these fish may not permit the strict application of the harvesting rule.

It is common industry practice to harvest fish at a time other than that suggested by the profit-maximizing rule. For example, we have indicated that harvest typically takes place 18 to 22 months after smoltification whereas, the ideal harvest time might well be 30 to 32 months. There are many reasons for this, not least among them is the risk of holding a large biomass through another summer season when the risks of mortality are high. Another reason for harvesting at this time is the need for cash flow, especially in newer farms. As is commonly the case, let us suppose that the fish have been separated into groups by weight and that the harvest is to occur over a period 18 to 22 months after smoltification and that this is several months prior to the ideal harvest time. Thus, in all weight groups, the discounted value of the growth increment in succeeding months would exceed the extra cost of holding the fish for another month. But given that a decision has been made to harvest, in what order should the groups be harvested? The general rule would be to first harvest the group with the smallest difference between the value of the growth increment and the extra cost of holding the fish for another month. In other words, harvest the group that contributes the smallest amount to additional profits first. If marginal weight gain is dominant in this calculation, then this would normally mean that the fish with the lowest growth rate should be harvested first. However, not harvesting them may permit them to grow into a weight range that commands a higher price so that even if their weight gain is lower, their gain in value may be higher than fish with a higher growth rate. Thus, a general group harvesting rule based on weight is not possible (Bjørndal, 1988).

Long-run decisions in salmon farm management

Natural environmental factors and the choice of site

Natural environmental factors such as water temperature and current, salinity, etc. affect the economic return to a site through their effect on fish growth and mortality. Once a farm site is chosen, these variables are beyond the control of managers. However, as site returns are currently captured by the farm owners it is obviously in the long-run economic interest to understand how these factors affect growth and mortality. There has been considerable research into the effects of environmental factors on growth. For example, Iwama & Tautz (1981), Brett *et al.* (1969), Elliott (1982) and others have studied the effect of temperature on growth and Leon (1986) has studied the effect of water currents on growth of salmonids and trout. However, most of these studies have been done in a laboratory or hatchery environment.

It is interesting to note that there may be a reasonable scope for research into the effects of environmental factors on growth and mortality using the data from actual farm operations. Given that salmon farmers tend to follow similar practice with respect to feed ration and stocking density, the variation in growth and mortality among farms will be mainly due to the differences in environmental factors and fish genetics. If genetics can be controlled in the analysis, then data from actual farm operations could be used to estimate the effect of environmental factors on growth and mortality.

Capacity and scale of operation

Once a site has been chosen, management must decide on the capacity of the salmon farm operation. Farm capacity is regulated in Norway to a maximum of 12 000 cubic metres so the decision must fall within these regulations. In those countries where capacity is not regulated directly, the physical size and other features of the site are of consequence in the capacity decision. Supposing that the site is large enough not to provide constraints on the size of the pen system, what capacity should be installed at a site? This is a separate question from that which asks, what is the best size of the salmon farming firm which might involve several sites? This last question will not be addressed in this chapter.

The capacity of the pen system is established by choosing the number and size of pens in the system and by choosing the depth of the pen nets. Since pens are secured to each other to form the pen system which involves open water areas for the nets and walkways for personnel and equipment, it is possible to extend the area of the system by simply attaching additional pens. For any given pen size, the depth of the nets is constrained by water depth or the weight of the nets or both. In any case, the main point is that the technology of the system is such that it is usually easily expanded or contracted by simply adding or subtracting pens to the existing structure.

Site capacity should be chosen to maximize the long-run profitability of the site. We have already seen that profitability is dependent on a number of decisions affecting growth and mortality, so it is natural to ask if there is an expected or observed relationship between site capacity and fish growth or mortality. If there does not appear to be such a relationship, increasing the capacity of the system would be expected to generate proportional increases in revenues. But would such an expansion of capacity be expected to generate proportional increases in total costs? It has already been noted that the capital equipment costs (interest opportunity costs plus depreciation) for sea pens would be expected to rise more or less proportionately with increases in capacity. Feed costs, which are of course the major cost component, may rise less than proportionately with capacity as quantity discounts can be realized up to a point but thereafter these costs would also be expected to rise proportionately with capacity. The major cost savings from increasing capacity are probably derived from some capital costs (building(s) and other fixed installations) and labour costs. However, these costs only make up a relatively small proportion of the total so the cost savings would not be expected to be very large.

Moreover, if hand feeding is practised, then labour costs should also rise more or less proportionately with capacity. All of these unit cost reduction features of the increasing capacity of salmon farm sites combine to suggest that significant economies of scale would not be expected. Analyses of cost of production based on Norwegian data have found economies of scale in salmon production, although they appear to be limited (Salvanes, 1989; Bjørndal, 1990, ch. 5).

Although significant economies of scale at the site level are not to be expected, what then about diseconomies of scale? Feed waste and excretion from fish farms cause pollution at the farm site. The seriousness of this pollution is influenced by environmental variables, in particular the flushing action. This means that site productivity is of importance in determining total site capacity. Besides the factors already mentioned that are related to site size, risk of disease or algae bloom are also important in limiting site capacity. Since disease can spread from pen to pen on a site, a large biomass concentrated on a single site is a high risk situation and if an algae bloom occurs, even if the bloom is widespread, it is likely to be less serious at some sites than at others. If it is possible for a firm to obtain licences for other sites, the risk of mortality from major disease outbreak and algae can therefore be reduced. The benefits of site diversification are very much like the risk reduction benefits that occur through the diversification of financial portfolios.

A microcomputer model for salmon farm management and planning

In this section, a description of a microcomputer model that is designed to assist in the management and planning of salmon farms will be discussed. In describing the model's structure, emphasis will be given to pointing out the relationship between this model and a hypothetical ideal model as it could be, particularly if more information on the growth and mortality relationship as discussed above were available. The name of the model is the Information Model for Seafarm Planning and Management (IMSPM), and it currently runs under the spreadsheet programs Lotus 123 on IBM PC or compatible computers and Microsoft EXCEL on Macintosh computers. It is a useful management and planning tool but it suffers, as do all management models of which we are aware, from some serious shortcomings that are mainly a result of incomplete information regarding the growth and mortality relationships.

The user of the model specifies a number of input variables that describe the environmental, financial, and economic conditions of a farm and the model produces for a period of five years a variety of information ranging from feed management to optimal harvesting schedules as well as cash flow and profit and loss reports.

In the sections above, we discussed the operational decisions that farm managers must make and how these decisions can be made in order to maximize farm profits. In IMSPM these decisions are referred to as input variable choices since they supply user input to the model. For example, the user must supply the number of smolts that are planned for each of five year classes and the month that they will be placed in seawater. This is identical to one of the decisions discussed earlier where the choice of the number of smolts to release to sea pens is a management control variable that will, for a given farm capacity, determine the pattern of stocking density over the grow out cycle. The input variables are:

Biological and environmental variables

(1) Number of smolts released in:
- Year class 1
- Year class 2
- Year classes 3, 4, & 5

(2) Average size of smolts released in:
- Year class 1
- Year class 2
- Year classes 3, 4, & 5

(3) Month of smolt release

(4) Growth, mortality, and maximum stocking density

(5) Water temperature

(6) Feed factors

Economic and financial variables

(1) Capital outlay

(2) Insurance premium:
- Fish (%)
- Upfront fish insurance payment (fraction)

(3) Price of feed (US$/kg)

(4) Harvesting expenses (US$/kg)
(5) Price of fish
(6) Seasonal price factor
(7) Loan interest
(8) Labour expenses
(9) Other expenses
(10) Fish harvest time
(11) Discount rate (%/month)
(12) Interest rate on line of credit (%/month)
(13) Interest rate on bank balance (%/month)

Growth, mortality, and maximum stocking density

The unit of time for the model is month and the user is asked to input the expected monthly growth of the average fish in the population over a 36-month period. Thus, given the initial weight of the average smolt and its monthly growth, the weight of the average fish is projected for each month.

In our previous discussion, fish growth was not a decision variable but depended upon decision variables such as feed ration and stocking density, plus a number of environmental factors. Herein lies the first departure of IMSPM from an ideal model. The growth information supplied by the user is based upon knowledge of how fish have grown in the past under the feed rations, stocking densities, etc. that have been used by farmers in the past. Thus, it is apparent the IMSPM cannot optimize with respect to the control variables affecting fish growth because insufficient information about the growth relationship currently forces such a modelling requirement.

The user must also input the expected monthly mortality for the fish population in each year class. But again our previous discussion suggested that mortality was at least partly a result of the management decision on the stocking density resulting from the choice of the number of smolts. As in the case of the fish growth, the mortality input data in IMSPM is based on the industry experience that has resulted from the stocking density and other decisions that have been made in the past. There is not functional relationship between stocking density and either mortality or growth.

The user of IMSPM also inputs the maximum stocking density that will be permitted in each month. This input is based on industry experience and varies among the months because of changes in fish weight and water temperature. Its purpose is to permit the model to alert the user to those months where stocking density will exceed the maximum permissible given the capital equipment acquisition program (to be discussed below).

Water temperature and feed factors

Water temperatures vary between sites and are required as input information by IMSPM. The purpose of this information is not to act as an input to the fish growth relationship since the fish growth data supplied to ISMPM by the user are not directly functionally related to the specification of water temperature. However, water temperature is required by IMSPM to assist in determining the required feed ration since this ration will depend on both water temperature and the average weight of the fish as specified by the feed factors. The feed factors are nothing more than user-supplied information relating daily feed requirements in kilograms of feed per day per kilogram of fish biomass to fish weight and water temperature. In other words, this is the standard information that comes from the feed tables discussed earlier.

It is important to understand exactly how IMSPM determines daily and monthly feed requirements. From the previous discussion it is clear that the model does not determine feed requirements to meet a particular objective like the maximization of profits. What it does is to use the information on average fish weight that comes from the user supplied initial smolt size and subsequent growth information, in combination with user supplied information on water temperature and remaining fish derived from the user supplied mortality information and then derives from this input information the feed requirement. As mentioned earlier, this feed requirement is usually based on the feeding recommendations of manufacturers, but how this recommendation is related to the ideal or optimal feed ration is unknown. Some observers have speculated that it leans towards a ration designed to maximize fish growth; that is, a satiation ration.

The capital acquisition plan

The IMSPM provides the user with the opportunity to specify a six-year plan of capital equipment and smolt purchases on a monthly basis. Although much of the equipment would be in place prior to the release of the first year class of smolts to seawater, it is common to add pen capacity as required and to spread the acquisition of certain other types of equipment over time. In any case, ISMPM gives the user the option of specifying a variety of plans and determining the implications of these plans for cash flow and yearly profitability. The user specifies the monthly farm capacity that is associated with the capital acquisition plan so that it can compare this with the monthly required capacity that is based on the size of the biomass and the maximum stocking density for the month. This comparison therefore alerts the user to any incompatibility between planned capacity and the required capacity that is determined by the choice of the number of smolts, their subsequent growth and mortality, and from the choice of the harvesting programme.

Fish insurance

For those farms carrying fish insurance, it is a major expense. The IMSPM permits the user to plan for fish insurance payments by specifying the insurance rate and the percentage of the yearly payment that must be made at the beginning of the period. These insurance payments are included in the projected cash flow and profit and loss statements. Fish insurance is also a variable cost associated with holding the fish population for another month and is therefore important in determining the ideal time to harvest the fish which will be discussed later in the chapter.

Prices and expenses

Certain prices and expenses are also specified as input information to IMSPM by the user. Although it is recognized that feed prices vary slightly depending on pellet size and manufacturer, the model requires the specification of a single average price that is expected to apply over the grow-out cycle. Harvesting, loan interest, labour, and other expenses are standard expenses that are specified on a monthly basis by the user.

Fish prices vary depending upon the weight range. Currently, salmon price differentials exist for five different weight ranges so in order to properly value the biomass in each year class of fish, in each month the model estimates the value of the fish population by determining and then totalling the value of the biomass in each weight range. These values are also adjusted by a seasonal price factor in order to recognize that prices may vary at different times of the year.

The harvesting plan

Earlier in this chapter we discussed various harvesting plans including one that maximizes farm profits. This is the plan that sets the harvest month at that month when the net present value of the fish population is maximized where it is recognized that the population may actually consist of a number of sub-populations as a result of the fish having been separated by weight over the course of the grow-out cycle. In this case, the sub-populations are harvested when their net present value is maximized.

In contrast with its inability to determine profit maximizing feed rations and stocking density, IMSPM does have the capability to determine the optimal harvest time. It does this by keeping track of the net discounted value of the population and then finding the month when this quantity is maximized. However, although this is a useful feature it still must be interpreted cautiously because it is a maximum that is chosen on the basis of the choices of feed ration and the stocking density that are not necessarily the optimal choices. To get an appreciation of the possible difficulties here let us suppose that the feed ration used is the satiation ration; the one that maximizes growth. Relative to our definition of the optimal ration this means overfeeding the fish. This means that in each month, both feed costs and fish growth are higher than they should be, but the value of the growth increment (zero at satiation) is less than the cost of achieving that growth increment so that the longer this goes on, the worse is the profit picture for the farm. Even though the profit increment from letting the fish grow from month to month is too small with this ration, it is not clear whether this distortion from the optimal causes the month of maximum profit indicated by the model to occur sooner or later than it would have, had the other variables, particularly feed ration, been chosen optimally. The answer to this question depends upon how the contour in Fig. 11.1 shifts through time as fish weight increases and only complete knowledge of the growth relationship can provide this information.

The IMSPM calculates cash flow and profit and loss projections for any harvest time. Thus it permits the user to compare the financial implications of the optimal harvest time with any other harvest time.

Output reports

The IMSPM provides three printed output reports (1) the feed information summary, (2) the cash flow report, and (3) the profit and loss report. Feed must be ordered and stored so that farms need to be able to plan for feed requirements. The feed information summary provides monthly feed requirements for each year class and the total monthly feed costs involved as well as the implied conversion ratios (kg of feed/kg of weight gain) indicated by the growth projections and the derived feed requirements.

The cash flow report presents the cash flows associated with fish sales and various expenditure categories including feed, smolts, capital equipment, labour, harvest, fish insurance, etc. The cash flow permits the user to determine the timing and amounts of various outlays. Particularly useful is that for feed because of the large expenditures involved and for fish insurance because of the complex way in which fish insurance premiums are currently set. The farm must estimate the value of the fish stock 12 months hence and pay some percentage of the premium as an upfront payment. Monthly estimates of biomass value are then submitted to the insurance company and at the end of the year an insurance calculation is made based on the average value of the fish over the year in which case the farm either pays an additional premium amount or receives a rebate. The IMSPM calculates these insurance amounts on the assumption that the actual growth of the value of the fish equals that which is projected in the model on the basis of the growth, mortality, and price information provided.

The profit and loss report gives a five-year projection of farm profitability based on the projected sales and expenses associated with the user specified harvest plan. This report is of standard, yet simplified, form and specifies seven expense categories.

References

Austreng, E., Storegbakken, T. and Asgard, T. (1987) Growth rate estimates for cultured Atlantic salmon and rainbow trout. *Aquaculture*, 60, 157–60.

Bjørndal, T. (1988) Optimal harvesting of farmed fish, *Marine Resource Economics*, 5, 139–59.

Bjørndal, T. (1990) *The Economics of Salmon Aquaculture,*, Blackwell Scientific Publications, Oxford.

Brett, J.R., Shelbourne, J.E. and Shoop, C.T. (1969) Growth rate and body composition of fingerling sockeye salmon, (*Oncorhynchus nerka*) in relation to temperature and ration size. *Fish Res. Bed. Can.*, 26, 2363–94.

Corey, P.D., Leith, D.A. and English, M.J. (1983) A growth model for coho salmon including effects of varying ration allotments and temperature. *Aquaculture*, 30, 125–43.

Elliott, J. (1982) The effects of temperature and ration size on the growth and energetics of salmonids in captivity. *Comp. Biochem. Physiol.*, 73B(1), 81–91.

Fagerlund, U.H.M., McBride, J.R. and Stone, E.T. (1981) Stress-related effects of hatchery rearing density on coho salmon. *Trans Am Fish Soc.*, 110, 644–9.

Iwama, G.K. and Tautz, A.F. (1981) A simple growth model for salmonids in hatcheries. *Can. J. Fish. Aquat. Sci.*, 38(6), 649–56.

Leon, K.A. (1986) Effect of exercise on feed consumption, growth, food conversion, and stamina of brood trout. *Prog. Fish Cult.*, 48, 43–6.

Papoutsoglou, S.E., Parparaskeve-Papoutsoglou, E. and Alexis, M.N. (1987) Effect of density on growth rate and production of rainbow trout (*Salmo gairdneri* Rich), over a full rearing period. *Aquaculture*, 66, 9–17.

Salvanes, K. (1989) The structure of the Norwegian fish farming industry: an empirical analysis of economies of scale and substitution possibilities. *Marine Resources Economics*, 6(4), 349–73.

Storebakken, T. and Austreng, E. (1987) Ration level for salmonids. II, Growth, feed intake, protein digestibility, body composition, and feed conversion in rainbow trout weighing 0.5–1.0 kg. *Aquaculture*, 60, 207–21.

Storebakken, T., Asgard, T. and Austreng, E. (1987) Ration level for salmonids. I. Growth, survival, body composition, and feed conversion in Atlantic salmon fry and fingerlings. *Aquaculture*, 60, 207–21.

Vahl, O. (1979) An hypothesis on the control of food intake in fish. *Aquaculture*, 17, 221–29.

Chapter 12
Marketing

SUSAN A. SHAW

Introduction

'Marketing' is often defined by practitioners in two different but related ways. The first is a definition of a philosophy of doing business and centres on the notion of successful exchanges based on willing consumers and producers (Baker, 1985). The second defines marketing as a management process which 'identifies, anticipates and supplies customer requirements efficiently and profitably' (Chartered British Institute of Marketing). Both definitions help to explain the role of marketing in salmon aquaculture. The first emphasizes consumer sovereignty: that salmon aquaculture can only develop if consumers want its products. The second emphasizes the need for researching and understanding the consumer and for core management programmes at the correct times to meet those needs. These are a prerequisite for the success of the business.

This chapter has three objectives. The first is to identify the main markets for salmon, the second is to examine the ways in which the marketing of salmon is undertaken at present and the third is to consider the way in which markets and marketing are likely to evolve in the future.

Salmon products

Salmon markets are complex, with a major factor being the variety of forms in which salmon are consumed. Since, as will be discussed subsequently, this variety adds greatly to the strategic strength of the industry, it is necessary to describe this variety. Salmon is marketed to consumers in the following forms.

Fresh salmon

Most salmon is bought by final consumers or caterers fresh, whole, gutted on ice. Occasionally the salmon is steaked or filleted prior to sale, usually by regional wholesalers.

Chilled, packaged salmon

Fresh salmon can be gas-flushed and packed in sealed containers (controlled atmosphere packs) or packed in vacuum packs. This prolongs shelf life and enables salmon to be sold in retail chill cabinets alongside non-fish products. Normally this treatment is applied to portion-sized fillets or steaks but ranges of ready-meals using salmon are now available using this process. Such packaging increases the range of retail outlets in which salmon can be sold and offers salmon in a very convenient form (if more highly priced) to consumers. Sales are mainly in Europe, particularly the UK, and are a small proportion of the total, but are growing rapidly.

Frozen salmon

Sales of frozen salmon vary from sales of frozen whole fish to steaks, joints and prepared dishes. This is not a major end-use, but sales in this form have been growing.

Smoked salmon

Salmon smoking is a major outlet for farmed salmon, particularly in European markets where it can account for over 60% of total salmon usage. Smoking processes vary between smokers and regions, but in general salted fillets of salmon are smoked over smoldering wood, with spices and flavours often added at the salting stage. In Europe and on the eastern seaboard of North America, salmon are cold smoked (at less than 30°C). The West Coast of North America produces a hot smoked 'kipper' type product. Products are normally sides or slices of salmon in vacuum or other forms of packaging.

Other processing

Salmon can be processed in other ways such as by marinading, the production of pâtés, mousses, terrines, etc. The volumes involved are small, but growing.

Wild chum, pink and sockeye salmon are canned in substantial quantities but this is not currently an attractive outlet for farmed salmon because of the much higher prices they command in other forms.

Growth potentials vary since demand varies for different products and between countries. The demand for farmed salmon does not derive from a single market but is the aggregation of the demand for salmon in each of these different markets.

The markets for farmed salmon

Overview

The markets for farmed salmon are international ones. Fresh, smoked, frozen and other forms of salmon are consumed in many different countries throughout the world. These markets are supplied with salmon from sources which in some cases are very distant from the markets themselves. Fresh farmed salmon are transported by air from Norway and Scotland to Japan and from Chile to North America and France. Farmed supplies compete to varying extents with wild salmon (Table 12.1) which are also widely traded internationally. Frozen wild salmon are shipped from Alaska to the major European, Japanese and North American markets.

There are some tariffs on trade between countries, particularly on more highly

Table 12.1 World supplies of salmon to 1995 (000 tonnes).

	1987	1989	1990	1991	1993 (Estimate)	1995 (Estimate)
Farmed salmon						
Norway	48	116	158	155	115	115
Scotland	13	28	32	36	35	35
Canada	2	18	25	30	33	36
Chile	2	9	13	23	23	23
Other European	7	15	20	29	14	14
Japan	21	21	22	N.A.	21	21
Other	4	5	5	6	5	5
Total	97	212	275	279	246	249
Wild salmon						
Total	600	780	800	800	800	800
World total	697	992	1075	1079	1046	1049

Sources: Federation Europêene de la Salmoniculture
Department of Fisheries & Oceans, Canada
US Department of Commerce
BC Fisheries Council
Industry discussions

processed salmon products and some non-tariff barriers based on different regulations and different trading histories. In 1990 and 1991, there has been increased activity in this area as a result of dramatic increases in supplies which have depressed international prices (q.v.). Successful actions have been brought against Norwegian exports to the United States because of low export prices which, it has been claimed, have been below production costs (*Fish Farming International*, 1991). A similar threat to introduce a minimum price system for Norwegian salmon within the European Community (*Fish Trader*, 1990) has now receded as a result of a Norwegian undertaking to control exports (*Fish Farming International*, 1991). In general, however, markets are interrelated with changes in trading conditions in one market having knock-on effects on others. Thus, while individual farmers may target their products to a limited number of end uses and geographical areas, those markets will be affected by the wider world demand and supply situation of which they are a part.

The major salmon markets and their sources of supply

The major trading flows for all types of salmon, farmed and wild, are considered first while the next section examines the distinctive markets for farmed salmon.

The largest markets for salmon are in the United States, Europe and Japan (Table 12.2).

There is little statistical data from which estimates can be made of the end uses for salmon in the United States, but it is estimated that consumption of fresh or frozen

Table 12.2 Major salmon markets 1989: tonnes live product weight equivalent, including wild and farmed salmon, but excluding supplies going to canned salmon.

Country	Domestic supplies	Imports	Exports	Balance for domestic consumption
Canada	49 000[3]	–	18 500	30 500
France	100	60 000	–	60 100
Germany	–	12 000	–	12 000
Italy	–	5 000	–	5 000[1]
Japan[2]	223 000	149 000	–	372 000
Spain	200	13 000	–	13 200
United Kingdom	30 000	8 000	11 000	27 000
United States	190 000[3]	35 000	135 000	90 000

Notes:
[1] Estimated
[2] Ignoring inventory adjustments
[3] Estimated for non-canned supplies

These figures include consumption in further processing in country of destination.

salmon dominates, with smoked salmon accounting for less than 20% of total consumption. Consumption has traditionally been concentrated in the west and north-east of the United States but markets are now being developed in Florida, Texas and the Midwest. The United States market is mainly supplied with Pacific salmon from the West Coast and Alaskan salmon fisheries. It is in fact a hierarchy of different markets, reflecting the heterogeneity of the Pacific salmons. Chinook, coho and sockeye salmon are the most highly regarded and command the highest prices while pink and chum salmon are usually (depending on quality) available at lower prices. However, increasing quantities of Atlantic farmed salmon are also now consumed (see the section, 'The distinctive position of farmed salmon') from Norway, Scotland and Eastern USA and farmed Pacific salmon from British Columbia and Chile is also imported. This farmed salmon occupies a premium market position and sells at relatively high prices. Canada and the United States are substantial exporters of salmon, mainly frozen, to Europe and to Japan.

The European market, despite the European Community customs union, remains a collection of different markets with different consumption levels and species preferences. As a generalization, Atlantic salmon is the most highly regarded in these markets, supplied from salmon farming in Norway, Scotland, Ireland, the Faroes and Iceland and from limited wild supplies. Substantial quantities of frozen Pacific salmon are also consumed, mainly imported from the United States and Canada (Table 12.3), but the recent growth in these European markets is primarily associated with Atlantic salmon.

Substantial proportions of both Atlantic and Pacific salmon are smoked. This smoked salmon is mainly consumed in domestic markets, but quantities of smoked salmon are exported. Figure 12.1 shows an estimate of flows of salmon in France as an indication of this.

Table 12.3 Usage of fresh and frozen salmon in selected European markets 1987 and 1989: tonnes rounded weight equivalent, including consumption in further processing but excluding imports of smoked salmon.

	Pacific tonnes	Atlantic tonnes	Total tonnes
1987			
Germany	2 200	8 800	11 000
France	26 000	20 000	46 000
Spain	100	4 500	4 600
United Kingdom	4 400	11 600	16 000
1989			
Germany	1 000	11 000	12 000
France	10 600	49 400	60 000
Spain	1 000	12 000	13 000
United Kingdom	2 300	24 700	27 000

Source: Trade statistics, UK Government

Note: Pacific salmon is almost entirely wild salmon. Atlantic salmon is almost entirely farmed salmon.

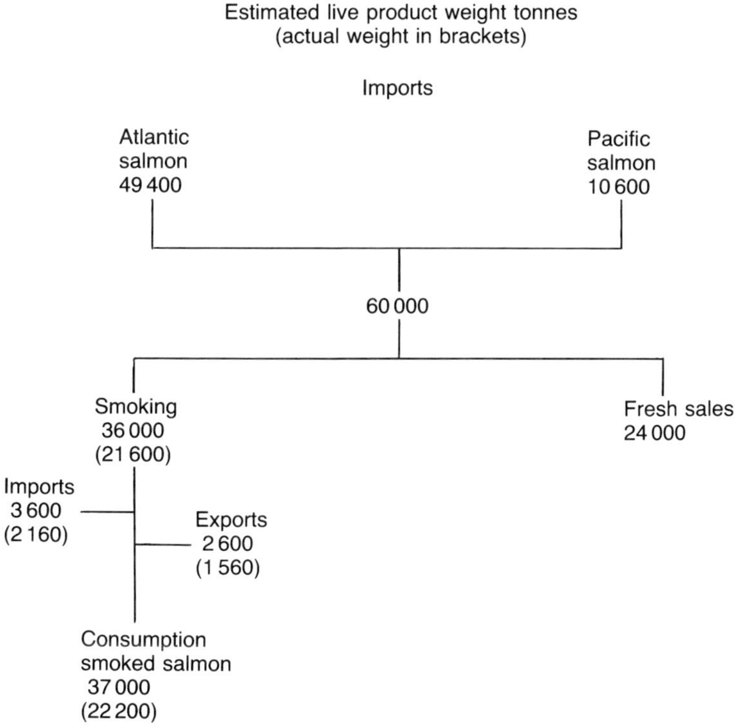

Fig. 12.1 Flows of salmon to the French market.

In Japan, per capita levels of salmon consumption are higher than in the other markets. The Japanese mainly consume salmon which has been salted but non-salted fresh and frozen salmon are becoming increasingly popular. Japanese consumers are supplied with chum salmon from the substantial domestic salmon ranching industry and farmed supplies of coho salmon are growing. Japan is also the largest importer of salmon. Imports of wild fresh and frozen salmon come mainly from the United States and Canada, with sockeye the preferred species for which they are prepared to pay the highest prices. In recent years, imports of farmed salmon, particularly from Norway have been increasing, but it remains a specialist product, often consumed raw.

There are smaller, but growing markets for salmon in many other countries, notably Hong Kong, Singapore, Australia, New Zealand and the countries of the Middle East. As an example fresh salmon is exported from the United Kingdom to 20 different countries and smoked salmon is exported to 35 different countries. With the growth in supplies of farmed salmon, these markets are expected to grow in importance.

The distinctive position of farmed salmon

There are strong competitive relationships between farmed salmon and the higher value wild salmons (Atlantic salmon, chinook, coho and to a lesser extent, sockeye) but farmed salmons have also created new and distinctive markets for themselves because of unique product attributes. These attributes include the following.

Year-round supplies

Unlike wild salmon, farmed salmon can be available to customers year-round from most production locations. This has created an 'out-of-season' market for salmon but is also attractive to final consumers and trade customers who want to buy products in steady, continuous supply.

Quality and consistency

The precise production control possible through aquaculture systems creates a high quality product which is also consistent in quality and size. This meets the needs of modern consumers who demand high standards in foodstuffs.

These advantages have been helped by rising consumer incomes and static supplies of high quality wild fish, particularly in Europe. They have led to the rapid development of markets for farmed salmon. Farmed salmon was initially supplied primarily to fresh markets but, as the availability of farmed salmon has increased, so also have the quantities which are used by salmon smokers (Table 12.4) and other forms of processing.

Markets for farmed salmon were first developed by the Norwegians who were early exporters of fresh and frozen salmon to the major European markets because of the small size of their own domestic market. In conjunction with the major airlines, new technologies were developed for air-freighting salmon which led to the development of the United States market for Atlantic farmed salmon from Norway.

Table 12.4 End uses for farmed salmon in the United Kingdom: estimated live product weight equivalent.

	1984		1985		1986		1988		1989	
	tonnes	%	tonnes	%	tonnes	%	tonnes	%	tonnes	%
Exports fresh, frozen	950	24	3 420	33	4 000	31	5 000	27	9 000	28
Salmon smoking	1 000	26	4 100	40	4 700	37	7 000	38	10 000	31
Other processing	500	13	720	7	1 400	11	1 800	10	3 000	9
Fresh fish retail	800	21	1 240	12	1 600	13	3 000	16	7 000	23
Fresh fish catering	400	10	670	7	1 021	8	1 700	9	3 000	9
Local direct sales	250	6	150	1	–	–	–	–	–	–
Total	3 900	100	10 300	100	12 721	100	18 500	100	32 000	100

Source: Estimates of author based on industry discussions

Note: – = Very small

Table 12.5 Norwegian farmed salmon exports 1987, 1989, 1990 and 1991 – fresh and frozen.

	1987	1989	1990	1991
	tonnes	tonnes	tonnes	tonnes
France	11 424	27 420	33 679	30 933
United States	8 424	12 960	8 990	1 295
Denmark	7 903	16 316	22 427	36 495*
Germany	4 715	10 046	10 204	13 668*
Sweden	2 052	4 255	3 686	9 091*
Spain	1 993	7 187	10 333	13 381
United Kingdom	1 642	2 936	3 227	2 769
Japan	1 553	5 700	4 657	5 250
Belgium	1 451	2 781	3 557	4 301
Switzerland	1 137	1 567	1 677	2 221
Netherlands	868	3 972	2 215	2 349
Italy	496	2 989	5 062	6 596
Others	402	849	1 585	3 318
Total	44 060	98 978	111 299	131 667

Source: Norwegian Fish Farmers Sales Organization
1991 Figures – Norwegian Fish Farmers Association

Notes:
* Includes net increase in stored frozen salmon. Large quantities of salmon exported to Denmark and Sweden are re-exported, final destination is not known.

Initial market targets were high quality restaurants, outside the wild salmon season. These have been broadened to include year-round sales to a wider range of outlets which have in common the desire to purchase salmon of high and consistent quality. Norway now exports substantial quantities to all the major markets (Table 12.5).

The Scottish industry has developed in a similar manner, except that the greater size of the domestic market has led to a lower percentage of exports. Nevertheless,

Table 12.6 Per capita rates of consumption of salmon: some comparisons 1985–89.

	Kg per head per annum			
	Salmon (all uses except canned)	All fish	Poultry	Red meats
France	1.07	17.9	17.9	81.5
Germany	0.20	6.4	9.7	89.1
Spain	0.34	25.4	21.9	49.3
United Kingdom	0.46	15.0	16.0	58.2
Japan	3.05	35.9	12.1	23.0
USA	0.37	7.1	31.7	80.7

Note: Salmon consumption figures are for 1989 and based on estimates of production, landings and trading flows. All other figures are taken from OECD Food Consumption Statistics for 1985, the latest date for which consistent cross country data are available.

Table 12.7 Relative prices of fish and meat products in the United Kingdom.

	US$ per kg				
	1985 March	1987 March	1988 May	1988 November	1990 July
Fish farmed salmon (steaked)	10.1	10.2	17.1	16.4	15.9
Whole chicken	3.2	3.6	5.5	3.6	3.3
Chicken fillets	7.5	8.2	13.9	13.4	14.0
Beef topside steak	3.4	6.4	8.2	10.7	8.9
Pork tenderloin	5.7	6.5	10.0	10.2	12.2
Fish: sole	6.2	10.0	14.6	16.6	16.1
Fish: trout fillets	N.A.	8.7	13.2	13.2	12.8
Fish: haddock	4.3	5.0	10.3	9.2	10.1

Source: Stirling Retail Price Survey

Scotland exported around 9000 tonnes in 1989 to a variety of European markets and the United States and Japan. British Columbia salmon farming is in a much earlier state of development with sales mainly to the domestic market and to the United States. Significant exports to Japan also seem a likely scenario for the future.

Other farmed salmons are joining in the development of international markets. Chile now exports to the United States and to France. Ireland exports to France and the United States, etc. As an example, consumers in the US market now have three or four different types of farmed salmon from which to choose.

While market development has been rapid, it is still on a relatively limited scale. Farmed salmon is still consumed in small quantities compared with other comparable foodstuffs. Per capita rates of consumption of salmon in comparison with other meat and fish products are low in all the markets with the exception of Japan (Table 12.6). Often consumption is limited to narrow groups in the population.

For instance, a recent survey of consumption of farmed salmon in the UK (Gentles

& Shapley, 1990) found than older and more affluent people living in the richer parts of the country were more likely to eat salmon than other groups. It is still an expensive product (Table 12.7) in the United Kingdom, for example, and while consumption has been increasing, it remains a minor item overall in the diet.

The marketing of salmon

Business structures and marketing organization in salmon farming

This discussion starts with an analysis of business structures since they influence marketing organization and marketing strategies. As three examples, contrasts

Table 12.8 Industry characteristics and marketing structures of salmon farmers.

	Norway	Scotland	British Columbia
1. *Industry Structure*			
Size Distribution	Most farms restricted to less than 12 000 cubic metres.	Variety of sizes with businesses ranging from 10 000 tonnes to 50 tonnes output per annum. Industry dominated by three large producers but many small producers.	Variety of sizes with both large and small producers.
Ownership	Most farms privately owned.	Tendency for larger businesses to be owned by multinational companies, some of them Norwegian.	Tendency for larger businesses to be owned by multinational companies, some of them Norweigan.
2. *Vertical Integration*	Farmers carrying out primary processing only. Export companies mainly marketing fresh and frozen fish, with limited smoked output.	Most farmers carrying out primary processing only, although some smaller farmers smoke salmon and market leader sells fresh, smoked and other processed forms of salmon.	Farmers carrying out primary processing only.
3. *Marketing Organisation*	Dominant role played by export companies, five accounting for bulk of sales. Industry trade association backed by statutory powers for generic promotion, quality control, regulation of distribution and minimum prices.	Largest companies do own but groups of smaller and medium sized becoming increasingly important. Industry trade association with voluntary membership undertakes generic promotion, quality control and education.	Most carried out by individual companies but some contractual relationships with seafood processing companies. Industry trade association with voluntary membership carries out generic promotion quality control and education.

between the three major producers, British Columbia, Scotland and Norway can be seen in Table 12.8. Salmon farming businesses are very heterogeneous in British Columbia & Scotland while there is a greater degree of homogeneity in Norway (Bjørndal and Schwindt, 1988).

In both British Columbia and Scotland, businesses range in size from those with production of more than 10 000 tonnes of salmon, often subsidiaries of large multinational business organizations, to family operated farms producing less than 100 tonnes, with much more limited long-term ambitions. In Scotland, for instance, there are around 100 companies ranging from the market leader, Marine Harvest, which currently produces around 9000 tonnes with a 25% share of Scottish production to a fringe of over 50 businesses producing less than 50 tonnes (Shaw, 1990). 60% of the output in Scotland comes from the three largest companies. In 1986 in British Columbia it was estimated that over 60% of the production capacity was concentrated among less than 15% of the farms (DPA, 1986). In Norway, by contrast, most salmon farming businesses are small, since legal restrictions limit the size of farms to 12 000 cubic metres.

In all three countries, most farms sell fresh whole or gilled and gutted salmon to exporters, importers or other trade customers. There is very little vertical integration forward into salmon processing. Norwegian salmon farmers sell to licensed exporters who in turn market salmon internationally. The licensed exporters on the whole, however, do little processing other than primary processing and packing. Substantial quantities of Norwegian salmon are smoked, but mainly in the export markets to which fresh and frozen salmon is destined (see Fig. 12.2).

In the Scottish industry, a substantial quantity of salmon is gutted and packed by farmers or by marketing groups acting on their behalf. The market leader, Marine

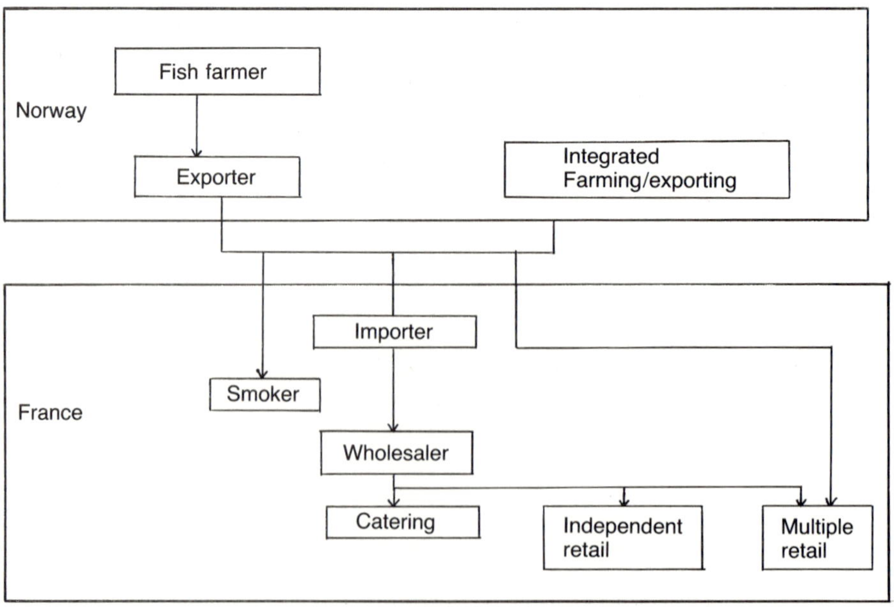

Fig. 12.2 Distribution channels for Norwegian farmed salmon to French consumers.

Harvest, sells salmon smoked for (but not by) them on contract and processes fresh salmon into controlled atmosphere packs, but is somewhat unique in this degree of forward integration. Some farmers smoke salmon but these tend to be very small producers (Shaw, 1989). In British Columbia, farmed salmon in sold for fresh consumption but often primary processing (gutting) is carried out not by the farmers but by seafood processing companies.

There are a number of explanations for these patterns. In the Norwegian case, the small size of most farmers and the remoteness of markets dictates that export companies, of whom five dominate, buy salmon from growers and carry out downstream marketing. The 20 largest companies in Norway have 74% of sales, the ten largest 51%, and the five largest 32% (Askiidsen & Hersoug, 1988). Higher tariffs on processed than on fresh products in export markets have inhibited the development of processing in Norway. In the United Kingdom, the main processing issues are associated with the very different natures of processing and farming activities. There are some economies of scale in salmon processing including smoking (Schwindt, 1982) so that the processor of 2000 tonnes of product has lower costs than the 500-tonne operator. Thus there is a mismatch between the supply needs of larger processors in terms of the volumes, sizes and timings and the optimal output patterns of growers which has kept these activities largely separate from growing for all but the largest farmers and marketing groups. Farmers in any case have been typically more concerned to develop the farming activities with which they are most familiar and where their competitive strengths lie.

As far as marketing organization for the selling of fresh salmon is concerned, in both Scotland and British Columbia, considerable variety exists, but there is a general trend towards increasing concentration of selling. This reflects the existence of substantial economies of scale in marketing. Large size also gives access to larger individual accounts and the resources to develop export markets (Shaw, 1990). This has been achieved through the large size of the main companies and through the growing importance of marketing groups of small producers. It is estimated that eight or nine buying points now account for 90% of sales of Scottish farmed salmon.

In all three countries, trade associations have played important marketing roles. The Norwegian Fish Farmers Sales Association (Fiskeoppdretternes Salgslag) had up to 1992, statutory powers to set minimum prices, rights to approve exporters and operates quality control, education and promotional programmes (Hempel, 1988) and by law all salmon farmers had to belong to it. After the bankruptcy of this organization in late 1991, the future organization of sales and marketing has been uncertain (see Chapter 2). The Scottish Salmon Board and the BC Salmon Farmers Organization do not have such statutory powers and membership by farmers is voluntary. However, they are also engaged in quality control, education and promotional programmes. Their activities are discussed further in the next section.

Marketing strategies

The main marketing strategies followed by salmon farmers in the three countries are detailed in Table 12.9. There are many similarities since they are all engaged in commodity marketing, that is, the role played by product differentiation is small. The fresh, unprocessed and typically non-packaged nature of the product gives limited opportunities for the establishment of brand loyalty to individual farmers. This means

Table 12.9 Marketing strategies of salmon growers.

	Norway	Scotland	British Columbia
1. *Target Markets*	Global with main emphasis on Europe and United States.	Global with main emphasis on Europe and United States.	United States and Canada; limited markets in early development stage; Japan likely future target.
2. *Product Strategy*			
Timing	Year round supplies by export companies.	Year round supplies but objective not achieved by smaller growers.	Supplies mainly during out of season for wild salmon but eventual target year round supplies.
Sizes	Full range by export companies.	Full range but not achieved by smaller growers.	Full range but not achieved by smaller growers.
Quality Control	Industry quality control scheme.	Industry quality control scheme.	Industry quality control scheme.
3. *Pricing Strategy*	Sales through limited numbers of sales points and minimum prices imposed by law by Norwegian Fish Farmers Sales Association.*	Individual pricing but some attempt to impose industry pricing discipline via price information service.	Large numbers uncoordinated selling points.
4. *Channel Strategy*	Sales via limited number of export companies; export companies sell direct to major processors in international markets; other sales mainly to importers and wholesalers; little use of brokers.	Very varied; direct sales to major UK and international processors and direct sales by large farmers to major UK retailers; other sales via wholesalers.	Sales via local (BC) brokers and wholesalers to wholesalers in US and Canada; direct sales to US wholesalers; some contract sales to BC seafood companies.
5. *Promotional Strategy*			
Company	Most marketing to trade customers; Brand awareness weak and only at trade level.	Most marketing to trade customers; brand awareness weak and only at trade level; market leader now attempting to build consumer awareness of its brand.	Most marketing to trade customers; brand awareness weak and only at trade level; unsophisticated.
Industry	Active generic marketing by industry trade association; compulsory farmer participation.	Active generic marketing by industry trade association; voluntary but majority participation by farmers.	Active generic marketing by industry trade association in early stages of development; voluntary participation by farmers.

Note:
*Up to 1992 – see Chapter 2 for recent changes in Norway.

Table 12.10 Salmon prices at Billingsgate Market (wholesale), London.

	Price range per kg (£)
1981	3.22–4.63
1982	2.76–3.99
1983	3.09–4.12
1984	3.02–4.41
1985	4.06–4.92
1986	4.34–5.47
1987	3.48–5.42
1988	4.12–6.13
1989	3.99–5.40

Source: Fishmongers Company

that farmers operate in turbulent trading environments. Changing demand conditions lead to rapid price adjustment and individual farmers do not have scope for producing sufficiently different products from those of their competitors to protect themselves from adverse general fluctuations in trading conditions. This turbulence is exacerbated by the international nature of trading which causes further uncertainties through exchange rate movements.

Farmers themselves create fluctuations through erratic supplies to markets. In principle, as indicated earlier, any farming industry should be able to plan supplies in line with demand forecasts but in practice this has not always been achieved. For example, Scottish supplies have peaked in the summer, depressing prices while there are still shortages at other times of the year. British Columbia salmon farmers have in many cases not yet developed year-round supplies to customers and must also try to adjust supplies in the light of forecasts of the often unpredictable wild salmon catch with which they compete. Long-term planning has also shown a lack of foresight. High prices in the early and mid-1980s led to great optimism about global potential and little understanding of the need to develop markets properly in order to achieve this successfully. Large additional supplies placed on markets with relatively little marketing support in 1988–90 led to rapid reduction in salmon prices (Table 12.10). Low prices have persisted, leading to bankruptcies, especially in Norway and British Columbia and an anticipated reduction in supplies by the mid-1990s (Table 12.11).

The situation may now have changed. Relevant factors are the high capital costs of new farms, the lack of profitability in the industry in 1988–91 together with increasing professionalism through the emergence of larger selling organizations. This may lead to better future planning and more stability in prices.

The Norwegian minimum selling price system, which is mandatory on growers in the only formal system in operation for stabilizing prices. In practice, buoyant demand conditions prior to 1988 meant that price fluctuations were usually above minimum price levels. After 1988 the minimum price system came under considerable pressure. Minimum prices had to be reduced, but it was not always possible for the trade association to ensure that deals were struck at or above these levels. In practice, the problem with the scheme was the excess supply situation which had developed. It is very difficult to establish prices above market equilibrium levels without supply control which the Norwegian Fish Farmers' Sales Organization did not have, since control over licences to produce rested not with themselves but with the government. The Norwegian government has suspended any issue of new licences. All trade associations in the different countries have urged restraint in production increases. However, the main mechanism for restoring prices is likely to be the voluntary restraint adopted by producers world-wide in the light of low or negative profitability.

It is because of limited individual product differentiation that industry associations play a strong role in promotion, quality control and education. In order to stimulate higher consumption of salmon and to establish strong generic identities for farmed salmon (Norwegian, Scottish, etc.) consumers and trade customers have to be made aware of the product, have favourable attitudes and remember to keep buying. All three trade associations have been active in such promotion. Initially the Norwegian and Scottish associations concentrated their efforts on trade promotions and public relations work. Budgets have grown in line with production increases, so that these associations can now spend money on advertising directly to final consumers, including through the use of television. The Norwegians spent US$5.3 million and the Scottish association US$2.7 million on promotion in 1988. These budgets are small in comparison with sums spent on other food advertising (the British biscuit industry spent US$36 million on advertising biscuits in Britain in the same period) but important if markets are to be developed.

Generic advertising to be successful requires that the actual quality of the products match the images created. The creation of effective quality control schemes has therefore been an important objective of all trade associations. These are based on random, independent inspections of both plant and product and systems of labelling and numbering to ensure that problems with consignments at any point in the distribution channel can be tracked back to the farmer responsible (Scottish Salmon Quality News, 1990).

Farmers with little experience of marketing can underestimate the importance of quality to the consumer. It is for this reason that educational programmes for members are incorporated into the activities of trade associations. These cover quality control issues but can also include wider marketing programmes.

This is particularly necessary when, as in the Scottish and British Columbia cases, individual growers are responsible for their own marketing. Smaller businesses in particular may lack marketing expertise and experience. In a situation where demand has been buoyant and sales achieved easily, this did not cause major difficulties but this issue has become more critical as markets and competition have evolved.

Future markets and the evolution of marketing systems

Future markets

The marketing success of salmon farming, together with the rapid diffusion of farming technologies led to ambitious plans for the development of supplies to world markets (Table 12.1) which led to the current crisis. Over the long term, however, there can be some optimism about the potential for steady if not spectacular growth.

Table 11.6 showed low per capita consumption in some countries. At the same time, salmon in both fresh and processed (non-canned) form is highly regarded by consumers (Shaw, 1989; Gentles & Shapley, 1990). There is a noticeable trend towards the consumption of lighter meals and white meats and away from red meats in a number of countries including the United States and the United Kingdom (OECD, 1989) from which salmon should benefit. The variety of forms in which salmon can be processed will also help in demand development because it increases the range of interesting salmon products which can be made available to consumers. On the supply side, world landings of fish are static and supplies of wild salmon may have reached a historic high from which they are now declining (Shaw & Currie, 1989).

Nevertheless, it is likely that competition will increase substantially and that there will be continuous pressure on prices. For farmers to survive and continue to make profits, they will require to operate at high levels of efficiency and with low operating costs. There will also be a need to develop new marketing strategies and extend the scope of existing strategies. To extend consumption to wider segments in the population, there will be a need for higher levels of promotion at both individual and group level. New product developments to provide more variety for consumers are necessary. While there are many possibilities, sales currently are still very dependent on markets for fresh, unprocessed fish. Finally, if fluctuations and uncertainties in prices and excess capacity are to be avoided, there will be a need for planned, coordinated development of supplies by the world industry.

The evolution of marketing systems

Marketing systems are changing as a consequence of the need for new and more active marketing strategies. Three changes are most apparent: the development of shorter marketing channels, the development of integrated relationships and the increasing marketing disadvantages of smaller producers.

Taking the United Kingdom as an example, growing quantities of Scottish farmed salmon have been reaching the consumer by shorter marketing channels. Sales direct to the multiple food retail sector in the UK have been increasing as the latter have internalized their distribution systems buying directly from producers without the use of intermediaries. These are bulk sales and, in line with modern retailing requirements, based on planned ordering and scheduling of consistent qualities on a year-round basis. Smoked salmon is also purchased by retail outlets in the same way with planning integrating the activities of the retailer, smoker and farmer. As volumes ordered have increased, so the direct links between Scottish salmon farmers and salmon smokers in other major European markets have strengthened. Finally, a tendency is noted for growing volumes to by-pass major wholesale auction markets and go directly to regional wholesalers or to major catering and retail accounts (including non-multiple retailers). This tendency has probably gone furthest in the marketing of Scottish farmed salmon because of its large domestic market, the extensive network of third party chilled transport systems available and because of high retail concentration in the UK. It is however a general tendency which is noted in all distribution systems: as concentration in retailing and catering increases, the case for using several wholesale intermediaries is much weakened.

The trend towards more integrated relationships has developed from this. Again the example of multiple food retailing in the UK is relevant. British multiple retailers are interested in developing sales of farmed fish because of buoyant demand (O'Hara, 1991). They do, however, have certain requirements of their suppliers to develop this interest, whether for fresh salmon, smoked salmon, or chilled packed salmon:

- Supplies of standardized product;
- Regular year-round deadlines;
- Bulk deliveries
- High levels of consistency of product and packaging;
- Stable and competitive prices;
- Prompt responses to changing demand conditions.

If these advantages are to be achieved this means the development of very close working relationships with suppliers. Trade customers are becoming more demanding: they are less willing to tolerate lack of availability of products when they want them. They are less willing to tolerate disruptive and frequent price changes. The costs of changing suppliers and the costs of dealing with many suppliers are rising as trade demands become more stringent and delivery logistics become more complex. Trade customers will react unfavourably to salmon growers showing short-term opportunism in a search for short-term price advantages. The implication for growers is of a need to work closely with trade customers in the development of markets, both to develop demand through trade willingness to handle the product and to protect their own market shares.

Similar trends are beginning to be observed in Norway with the development of closer coordinated and planned relationships between exporters and salmon farmers so that exporters in turn are able to take a more planned approach towards the development of relationships with their customers. This will facilitate the development of more direct relationships between Norwegian exporters and customers further down the distribution channel, similar to those which are developing in the United Kingdom. Similar moves can be expected in the development of marketing of British Columbia farmed salmon into year-round sales into the supermarket sector (Archibald, 1988). Such relationships confer benefits on salmon farmers themselves. The product is more attractive to final consumers and closer relationships can have important spin-offs in innovation and new product development for growers. For example, the main pressure towards the development of controlled atmosphere packaging and the retailing of small consumer packs of smoked salmon came originally not from farmers or indeed processors but from British retailers. The latter were willing to work closely with suppliers to develop products they wanted. This included advice from retailers on product development, production methods and development of high standards of hygiene and quality control, right back to the farm itself.

These changes leave the small producers at an increasing competitive disadvantage. While production economies of scale in salmon farming are relatively modest above 1000 tonnes output (Shaw & Bailly, 1991), marketing economies are much more significant. There are some economies of scale at site level in salmon farming (Salvanes, 1989) but beyond 500–1000 tonnes, they are probably not substantial (Shaw & Bailly, 1991). With large systems, higher output per site creates pressures on the marine environment and can lead to reduced growth rates and efficiencies. Large growers can offer large retail customers continuity of supply year round at no cost penalty while smaller growers are effectively excluded from these markets. There are substantial set-up costs associated with running an effective marketing operation, staffed with qualified personnel and able to undertake necessary market research, sales visits, advertising and promotion. Having established such an operation, the incremental costs of handling additional volumes are low. There are substantial scale economies in primary processing. Our analyses (industry sources) suggest that cost penalties of processing less than 2000 tonnes per annum are substantial and even beyond this point, the larger the plant, the lower are the operating costs.

Should the operation wish to extend into further processing such as smoking or the production of branded convenience goods the economies of scale are equally substantial (Traill, 1989). There is considerable interest in this issue at present with some growers examining the potential for entry into sectors with higher added value and the possibility of developing stronger consumer brand identities through the sale of packaged products. This option appears more difficult for a smaller grower.

There is greater bargaining power associated with a large market share. Larger volumes make it possible to supply a greater number of market outlets on a cost-effective basis. This allows the larger grower to spread risks.

Research findings support this conclusion (Shaw, 1989). Larger growers in Scotland were more likely to export, to supply multiple retailers and salmon smokers than were smaller growers in a 1987 survey of the industry. They were less dependent on primary wholesalers than smaller growers and were more likely to supply a spread of different types of outlets.

The advantages of size are likely to increase in importance. The growth in multiple retailing with large unit orders and direct dealing with growers has already been mentioned, but parallel changes are also beginning to take place in the sectors supplied by wholesalers. Regional wholesalers are becoming more conscious of the need to take more positive approaches to seafood marketing, rather than being passive order-takers. As a consequence they are likely in the future to pay greater attention to planning sales and obtaining consistent qualities and year-round supplies. As export markets become more competitive, there will be a greater premium on marketing expertise, again favouring the larger growers.

It is certainly the case that as individual growers scale up, they will find it easier to support the expenses of proper marketing operations and cover a wider spread of markets. The position of the grower producing 2000 tonnes per annum, while different from the 9000 tonnes producer, is also very different from the 500-tonne producer. In general, however, the trend is toward large sizes. What is happening in the salmon business is part of a general phenomenon. Companies in food retailing, wholesaling and processing are becoming larger and trade markets are becoming more concentrated.

The advice usually given to small firms with larger competitors in the same sector is to seek specialist market niches where they can cater for the special needs of small market segments, avoiding head-on competition with larger companies. Unfortunately, this option is difficult for many smaller growers to pursue. Large growers can supply every segment as effectively as smaller salmon growers because the product is relatively undifferentiated (Shaw, 1990). Indeed, their greater sales volume and year round supply enables them to be more flexible in supplying all accounts than smaller companies. Another favoured option for small farmers is direct farm-gate sales or direct sales to local outlets, but this is precluded by the remoteness of most salmon farms from major population centres.

Thus it seems that smaller growers will in time have to restructure to replicate the marketing advantages of larger growers. There are a number of ways in which restructuring is likely to occur. Farms may be bought out or merged. Small farmers may produce for larger growers or for processors under contractual arrangements where they grow to specified requirements and marketing is carried out by larger organizations with appropriate resources. The latter is a likely solution in Norway, given the restrictions on horizontal mergers among growers.

Alternatively, small growers who wish to retain their independence can group together to market their product collectively by setting up joint marketing companies or marketing cooperatives. This trend, as previously indicated, is already seen in Scotland. Solutions such as these are likely to benefit farmers individually and collectively. At an individual level, they remain competitive. At a collective level by reducing the number of selling points and concentrating resources in larger organizations with greater strength, the dangers of weak selling are avoided. Larger producers controlling larger volumes are more likely individually and collectively to be able to control markets rather than being controlled by them.

Summary and conclusions

The 1980s saw the rapid development of salmon farming as a consequence of strong demand for a high quality fish product, a demand which has been developed in the major markets of the world. Buoyant demand created conditions where most salmon farmers found readily available and profitable markets for their products. However, conditions changed as industries developed and supplies increased, and markets became more competitive. This has created pressures for greater efficiency in production and marketing. As a consequence, more sophisticated marketing practices and structures are emerging. Higher levels of promotion, new product development, orderly marketing and pricing will be the foundations of marketing strategies. These activities are increasingly likely to be carried out by larger organizations with the resources to plan and to influence markets, rather than reacting passively to developments as they occur.

References

An Bord Iascaigh Mhara (1987) *The Atlantic Salmon Farming Industry, Past Performance and Future Potential.* Irish Sea Fisheries Board, Dublin.

Archibald, A. (1988) Negotiating the distribution dilemma, *Canadian Aquaculture*, Jan/Feb.

Askildsen, J.E. and Hersoug, B. (1988) Naeringsstruktur i norsk fiskeoppdrett, Centre for Applied Research, Report No. 15, 1988, Norwegian School of Economics and Business Administration, Bergen.

Baker, M.J. (1985) *Marketing Strategy and Management*, Macmillan, London.

Bjørndal, T. and Schwindt, R. (1988) An international analysis of the industrial economics of salmon aquaculture. *Working Paper No. 191 1988*, Centre for Applied Research, Norwegian School of Economics and Business Administration, Bergen.

DPA Group (1986) *Industrial Organization of the BC Salmon Aquaculture Industry*, report prepared for Fisheries and Oceans, Canada.

DPA Group (1987) *Market Access and Penetration Strategy BC Farmed Salmon*, report prepared for BC Salmon Farmers Association.

Fish Farming International (1991) Norway fresh salmon faces US duties. vol. 18, no. 4, April.

Fish Trader (1990) EC proposes levy on Norwegian salmon, October 20–November 2, issue no. 5, 526.

Gentles, P. and Shapley, G. (1990) Salmon and the consumer, paper presented at the Salmon '90 Conference, Trondheim, Norway.

Hempel, E. (1988) *The Marketing of Aquaculture Products*, OECD Committee for Fisheries, Extended Meeting on Aquaculture, Paris.

OECD (1989) *Food Consumption Statistics 1975–85*, Organization for Economic Co-operation and Development, Paris.

O'Hara, F. (1991) *Retail multiples and the demand for farmed fish*. Unpublished paper, Scottish Fish Farming Conference, Glasgow, February 1991.

Salvanes, K.G. (1989) The structure of the Norwegian fish farming industry: an empirical analysis of economies of scale and substitution possibilities. *Marine Resource Economics*, 6 (4),349–73.

Schwindt, R. (1982) Industrial organization of the Pacific fisheries, Report prepared by McDaniels Research Ltd for the Commission of Pacific Fisheries Policy, Canada.

Scottish Salmon Quality News (1990) *Ten Reasons to Buy Only From Quality Approved SSB Members*. Scottish Salmon Board, Perth.

Shaw, S.A. (1989) Markets and marketing systems of Scottish farmed salmon. In *Aquaculture: A Review of Recent Experience*, Organization for Economic Co-operation and Development, Paris.

Shaw, S.A. (1990) *Marketing: A Practical Guide for Fish Farmers*, Fishing News Books, Oxford.

Shaw, S.A. and Bailly, D. (1991) Evaluation of the effectiveness of Commission aquaculture support policies with special reference to Regulation 4028/86. Unpublished study for European Commission DGXIV Structures.

Shaw, S.A. and Currie, A. (1989) *Markets in Europe for Selected Aquaculture Species*, Globefish Research Programme Volume 1, Food and Agriculture Organization of the United Nations 1989, R.

Shaw, S.A. and Gabbott, M. (1990) An International study of the marketing of farmed salmon, Centre for Applied Research, Report No. 590, Norwegian School of Economics and Business Administration, Bergen.

Traill, B. (Ed.) (1989) *Prospects for the European Food Chain*, Elsevier, Amsterdam.

Index